AF594931

Ship Dynamics for Performance Based Design and Risk Averse Operations

Ship Dynamics for Performance Based Design and Risk Averse Operations

Editors

Spyros Hirdaris
Tommi Kristian Mikkola

MDPI • Basel • Beijing • Wuhan • Barcelona • Belgrade • Manchester • Tokyo • Cluj • Tianjin

Editors
Spyros Hirdaris
Aalto University
Finland

Tommi Kristian Mikkola
Aalto University
Finland

Editorial Office
MDPI
St. Alban-Anlage 66
4052 Basel, Switzerland

This is a reprint of articles from the Special Issue published online in the open access journal *Journal of Marine Science and Engineering* (ISSN 2077-1312) (available at: https://www.mdpi.com/journal/jmse/special_issues/ship_dynamics).

For citation purposes, cite each article independently as indicated on the article page online and as indicated below:

LastName, A.A.; LastName, B.B.; LastName, C.C. Article Title. *Journal Name* **Year**, *Volume Number*, Page Range.

ISBN 978-3-0365-0616-6 (Hbk)
ISBN 978-3-0365-0617-3 (PDF)

Cover image courtesy of Mikko Suominen.

Contents

About the Editors

Spyros Hirdaris (Aalto University, Finland) is an Associate Professor of Maritime Technology (Ship Safety) with teaching duties on Principles of Naval Architecture and Ship Dynamics. In his research, he combines knowledge from advanced ship and safety science, marine hydrodynamics, and structures for the prediction of sea loads, safety, and performance of ships and offshore structures operating in extreme conditions. He completed his PhD in 2002 on Ship Science (Hydroelasticity of Ships) at the University of Southampton. He is a Chartered Engineer, Fellow of the Royal Institution of Naval Architects (UK), and a Member of the Society of Naval architects and Marine Engineers (USA). He served the International Ship and Offshore Structures Congress as a member and chairman of various committees with a focus on sea loads and responses since 2008. From 2004–2018 he worked for Lloyd's Register Classification Society internationally (UK, Poland, and South Korea) and before this spent short spells with UK-based engineering consultancy firms. This work involved research and product development, planning and strategy for R&D, consultancy, and marine new construction activities. He has been the recipient of the Lloyd's List (2013) and the UK IMechE/Ross Brown F1(2012) Innovation Awards for his contributions to EU FP7 project Lynceus and the Royal National Lifeboat Institution (RNLI – UK) charitable activities respectively.

Tommi Kristian Mikkola (Aalto University, Finland) is a University Lecturer of Fluid Mechanics teaching fundamental and advanced fluid mechanics as well as computational marine hydrodynamics. His research interests are focused on the application of computational methods and high-performance computing to solve problems related to ship hydrodynamics effectively and reliably. He is particularly interested in interactions between waves and ships. He completed his DSc (Tech) in 2009 on Naval Architecture (Computer algorithms and code verification) with distinction at Helsinki University of Technology (now Aalto University). He has served as a technical committee member of the International Towing Tank Conference (ITTC) for three terms between 2006 and 2014. In 2007, he was awarded the Landrini award for his contributions to the field of marine hydrodynamics. He has also been awarded the Aalto University School of Engineering Award for Achievements in Teaching twice (2012 and 2018). In his spare time, he puts the theory into practice by racing onboard a 41ft sailing yacht in the northern Baltic Sea.

Journal of
Marine Science and Engineering

Editorial

Ship Dynamics

Spyros Hirdaris * and Tommi Mikkola

School of Engineering, Maritime Technology Group, Aalto University, 02150 Espoo, Finland; tommi.mikkola@aalto.fi
* Correspondence: spyros.hirdaris@aalto.fi

Academic Editor: Tony Clare
Received: 18 January 2021; Accepted: 19 January 2021; Published: 20 January 2021

More than a century-and-half ago, William Froude and his son Robert [1,2] conducted the first scientifically designed towing tank experiments using scaled ship models travelling in calm water and waves. Since then, advances in mathematics and technology led to the development of various methods for the assessment of the dynamic behavior of ships. Today it is recognized that continuous improvement of direct assessment methods validated by model experiments or full-scale measurements are the doctrine of naval architecture. Yet, as we enter the 3rd decade of the 21st Century the advent of goal-based regulations and the emergence of safe and sustainable shipping standards still confront our ability to understand the fundamentals and assure absolute ship safety in design and operations.

To instigate renewed interest on the well-rehearsed subject of ship dynamics, this Special Issue presents a collection of 12 high-quality research contributions with a focus on the prediction and analysis of the dynamic behavior of ships in a stochastic environment. The papers presented are co-authored by leading academics and practitioners from Europe, the Far East and USA. The contributions discuss recent developments on:

- Ship wave loads in confused seas, including nonlinear effects [3], steady state hydroelastic responses [4], sloshing [5] and slamming [6].
- Dynamic coupling and resonant phenomena associated with seakeeping performance of ships in wind and waves [5,7].
- Ship maneuvering in level ice, waves and open water conditions [8–11].
- Dynamic stability in waves [12,13].
- Ship energy efficiency in abrupt wave conditions [14].

The methods presented use combined knowledge from theoretical hydrodynamics, computational aero/hydrodynamics, fluid/structural dynamics and their interactions, as well as results from model tests and full-scale measurements. Strong emphasis is attributed on understanding nonlinearities and flow dissipation associated with stochastic responses in confused seas.

Based on these independent contributions it may be concluded that contemporary developments will be influenced by new science at the interface of multifield problems and the implementation of improved design criteria in advanced decision support systems. We therefore hope that this Special Issue will renew the interest of academics, practitioners and regulators in their pursuit to push forward ship science, technical services and safety standards of relevance.

Author Contributions: S.H. and T.M. wrote and reviewed this editorial. Both authors read and agreed to the published version of the manuscript.

Funding: S.H. acknowledges the support of the Academy of Finland under university competitive funding award (SA Profi 2-T20404).

Conflicts of Interest: The authors declare no conflict of interest.

References

1. Froude, W. *The Papers of William Froude*; Institution of Naval Architects: London, UK, 1955.
2. The Royal Institution of Naval Architects. In Proceedings of the William Froude Conference: Advances in Theoretical and Applied Hydrodynamics–Past and Future, Organised by RINA, Lloyd's Register, Qinetiq and CD Adapco, Portsmouth, UK, 24–25 November 2010; ISBN 978-1-905040-77-3.
3. Ley, J.; el Moctar, O. A comparative study of computational methods for wave-induced motions and loads. *J. Mar. Sci. Eng.* **2021**, *9*, 83. [CrossRef]
4. Tilander, J.; Patey, M.; Hirdaris, S. Springing analysis of a passenger ship in waves. *J. Mar. Sci. Eng.* **2020**, *8*, 492. [CrossRef]
5. Igbadumhe, J.-F.; Sallam, O.; Fürth, M.; Feng, R. Experimental determination of non-linear roll damping of an FPSO pure roll coupled with liquid sloshing in two-row tanks. *J. Mar. Sci. Eng.* **2020**, *8*, 582. [CrossRef]
6. Chen, L.; Wang, Y.; Wang, X.; Cao, X. 3D Numerical simulations of green water impact on forward-speed wigley hull using open source codes. *J. Mar. Sci. Eng.* **2020**, *8*, 327. [CrossRef]
7. Bigi, N.; Roncin, K.; Leroux, J.-B.; Parlier, Y. Ship towed by kite: Investigation of the dynamic coupling. *J. Mar. Sci. Eng.* **2020**, *8*, 486. [CrossRef]
8. Suominen, M.; Li, F.; Lu, L.; Kujala, P.; Bekker, A.; Lehtiranta, J. Effect of maneuvering on ice-induced loading on ship hull: Dedicated full-scale tests in the Baltic Sea. *J. Mar. Sci. Eng.* **2020**, *8*, 759. [CrossRef]
9. Ren, Z.; Wang, J.; Wan, D. Investigation of the flow field of a ship in planar motion mechanism tests by the vortex identification method. *J. Mar. Sci. Eng.* **2020**, *8*, 649. [CrossRef]
10. Liu, S.; Papanikolaou, A. Prediction of the side drift force of full ships advancing in waves at low speeds. *J. Mar. Sci. Eng.* **2020**, *8*, 377. [CrossRef]
11. Nam, B.W. Numerical investigation on nonlinear dynamic responses of a towed vessel in calm water. *J. Mar. Sci. Eng.* **2020**, *8*, 219. [CrossRef]
12. Kapsenberg, G.; Wandji, C.; Duz, B.; Kim, S. A comparison of numerical simulations and model experiments on parametric roll in Irregular Seas. *J. Mar. Sci. Eng.* **2020**, *8*, 474. [CrossRef]
13. Acanfora, M.; Balsamo, F. The smart detection of ship severe roll motions and decision-making for evasive actions. *J. Mar. Sci. Eng.* **2020**, *8*, 415. [CrossRef]
14. Sun, C.; Wang, H.; Liu, C.; Zhao, Y. Dynamic prediction and optimization of energy efficiency operational index (EEOI) for an operating ship in varying environments. *J. Mar. Sci. Eng.* **2019**, *7*, 402. [CrossRef]

Journal of
Marine Science and Engineering

Article

A Comparative Study of Computational Methods for Wave-Induced Motions and Loads

Jens Ley [1,2,*] and Ould el Moctar [2]

1 Development Centre for Ship Technology and Transport Systems, 47057 Duisburg, Germany
2 Institute of Ship Technology, Ocean Engineering and Transport Systems, University of Duisburg-Essen, 47057 Duisburg, Germany; ould.el-moctar@uni-due.de
* Correspondence: ley@dst-org.de

Received: 15 December 2020; Accepted: 9 January 2021; Published: 14 January 2021

Abstract: Ship hull structural damages are often caused by extreme wave-induced loads. Reliable load predictions are required to minimize the risk of structural failures. One conceivable approach relies on direct computations of extreme events with appropriate numerical methods. In this perspective, we present a systematic study comparing results obtained with different computational methods for wave-induced loads and motions of different ship types in regular and random irregular long-crested extremes waves. Significant wave heights between 10.5 and 12.5 m were analyzed. The numerical methods differ in complexity and are based on strip theory, boundary element methods (BEM) and unsteady Reynolds-Averaged Navier–Stokes (URANS) equations. In advance to the comparative study, the codes applied have been enhanced by different researchers to account for relevant nonlinearities related to wave excitations and corresponding ship responses in extreme waves. The sea states investigated were identified based on the Coefficient of Contribution (CoC) method. Computed time histories, response amplitude operators and short-term statistics of ship responses and wave elevation were systematically compared against experimental data. While the results of the numerical methods, based on potential theory, in small and moderate waves agreed favorably with the experiments, they deviated considerably from the measurements in higher waves. The URANS-based predictions compared fairly well to experimental measurements with the drawback of significantly higher computation times.

Keywords: ship motions; sectional loads; hydroelasticity; boundary element methods; CFD; validation; regular waves; steep irregular waves

---

1. Introduction

Recent designs and operational profiles of ships require that their safety is evaluated under adverse conditions. In this regard, assessing ship safety in terms of the integrity of a ship's hull structure and motions is of primary importance. Ships encountering extreme seas are exposed to considerable risk. The International Union of Marine Insurance [1] documents an increase of seaway-induced losses over a five-year period between 2011 and 2015 compared with losses for previous years between 2001 and 2010. Hence, severe weather conditions are most likely responsible for part of ship losses.

Generally, it is advisable to explicitly estimate seaway-induced loads, especially for modern newbuilds that may differ substantially from those for which Classification Society design rules were prepared. For large modern ships with pronounced bow flare and a large, flat overhanging stern, effects of hull flexibility and the associated structural vibratory responses have become important because the associated wave-induced hull girder loads significantly contribute to the life-cycle hull girder load spectra. Already in the 1970s, Bishop and Price [2] developed a hydroelastic theory using a beam model to idealize the ship's hull. Over the last years, numerous studies were performed to assess the influence of wave-induced hydroelastic effects on ships, e.g., [3–15]. An overview of achieved progress to evaluate hydroelastic effects of ships is presented in [16], another comprehensive overview was published by the International Ship and Offshore Structures Congress, [17]. Available methods for wave-induced global and impact loads on ships and offshore structures are discussed in [18].

Numerical methods to assess wave-induced ship motions and loads may be categorized as strip theory methods (e.g., [6,19–23]), boundary element methods (e.g., [24–30]) and viscous field methods based on the solution of the Navier–Stokes equations (e.g., [13–15,31–43]). Several international benchmark studies were carried out to validate numerical methods. However, most of these benchmark studies addressed linear or weakly nonlinear problems in regular waves and the simulated time frames covered only a few wave periods (e.g., [44]). This paper addresses these gaps and aims to verify the suitability of state-of-the-art numerical methods of increasing complexity to assess ship motions, associated loads and hydroelastic effects in moderate and extreme seas of long durations for different ship types. To our knowledge such studies have not been undertaken till now. Here, we see the novelty of our paper. The presented experimental and numerical results for different ship types may be used by other authors for benchmarking.

Extreme sea conditions were identified based on the Coefficient of Contribution method [45,46]. The simulation results obtained from different numerical methods were compared systematically against experimental data. For the study, the codes applied were enhanced to allow for the prediction of nonlinear ship responses in extreme seas. The nonlinear strip theory method and the Rankine source BEM were developed and applied by the university Instituto Superior Técnico (IST) [47–50] and the class society DNV GL [51–53], respectively. The results of the Green function boundary element method as well as of the unsteady Reynolds-Averaged Navier–Stokes (URANS) solver were obtained by the authors.

The investigations cover four different ships, namely a typical medium sized cruise ship, a containership, a liquid natural gas (LNG) carrier and a chemical tanker. For the cruise ship and the containership, model tests were carried out at Canal de Experencias Hidrodinámicas Del Pardo (CEHIPAR); the LNG carrier and chemical tanker were experimentally investigated at the Technical University of Berlin (TUB) [54]. All ship models were segmented to measure sectional loads at segment cuts. The cruise ship and containership were equipped with backbones. Ship motions, pressures at bow and stern, as well as water column heights above the weather deck were monitored. Except for the containership, ship models were stiff to replicate rigid body responses.

The above mentioned measured and computed quantities were compared to assess the quality of the numerical predictions. For the investigated ship models, we considered regular waves to obtain response amplitude operators (RAOs) and extreme irregular sea states to determine short-term statistics of ship responses. The statistical analyses of ship responses in irregular sea states enabled assessing the feasibility of predicting extreme ship responses in a short-term statistical sense, as this is the basis for the estimation of long-term extreme responses.

2. Numerical Methods

The following numerical methods have already been described in detail in other publications, which are referred to here. Therefore, we will limit ourselves here to the major features of these methods.

Strip theory and the boundary element methods (BEM) are based on potential theory, while field methods solve the Reynolds-Averaged Navier–Stokes equations. A classical frequency-domain formulation of the potential flow problem yields results for linear ship responses in low and moderate sea states, and potential flow methods are still the method of choice to estimate RAOs due to their high computational efficiency and robustness. During the last two decades, however, more advanced time-domain simulations based on potential theory have emerged. Nonlinear boundary conditions, impact pressure loads (slamming) and green water loads are examples of nonlinear additions that require time-domain computations.

Potential flow methods based on two-dimensional strip theory are widely used for seakeeping computations. The development of such methods started about 60 years ago, e.g., by Gerritsma and Beukelman [55]. However, these methods have limitations in principle, for instance, for certain ranges of forward speed, wave-ship length ratios and wave modeling. Three-dimensional Rankine source methods are not restricted to low Froude numbers; nevertheless, when dealing with ship responses in extreme seas, the free-surface conditions need a special treatment to account for high and steep waves.

Increased computational power in recent years made it possible to employ advanced field methods based on the solution of the RANS equations. Nonlinear effects, such as wave propagation, wave breaking, green water loads, etc. are inherently included in these time-domain methods. Field methods are computationally inefficient and, therefore, scarcely applied, especially when analyzing time-domain simulations requiring long runs to reliably estimate extreme ship responses. However, recent work made it feasible to predict short-term statistics with field methods [15].

2.1. Strip Theory Method

The strip method applied calculates the instantaneous pitch, heave and surge motions as well as corresponding vertical bending moments. The existing code was enhanced to account for second-order drift forces in longitudinal direction to improve the surge motions. It was assumed that surge motions significantly influence the vertical bending moment in extreme waves.

Comparisons of numerical results of ships in extreme seas with experimental data revealed that sagging moments were remarkable overestimated, while hogging moments agreed quite well. To overcome this problem, a simplified method to compute the nonlinear radiation and diffraction forces was implemented. Relevant hydrodynamic properties (added masses, radiation restoring forces, etc.) of the wetted surface were pre-calculated and updated for each time step. More details may be found in [47–49].

Nonlinear effects related to slamming, water on deck or hydroelasticity were not taken into account. Shallow water effects were also neglected.

2.2. Rankine Source Boundary Element Method

The Rankine source BEM computes ship responses, taking into account the ship's forward speed. Nonlinear Froude–Krylov forces are solved together with radiation and diffraction forces. Surface panels on the hull and the free surface discretize the computational domain. Typically, in low and moderate waves, a linear wave model is applied for incident waves, resulting in reliable predictions of ship motions and loads. Aiming to improve the code for the computation of extreme wave scenarios, the boundary element method was extended to take into account nonlinear terms in the free-surface conditions. The free surface elevation was computed using a High-Order Spectral Method (HOSM). More details may be found in [51–53].

2.3. Field Method

An unsteady computational fluid dynamics approach simulates free surface flows by coupling Reynolds-averaged Navier–Stokes equations solver with a Volume of Fluid (VOF) method. A finite volume method (FVM) discretizes the solution domain, using a finite number of arbitrarily shaped control volumes. A Semi-Implicit Pressure Linked Equations (SIMPLE) algorithm was used to couple the velocity and the pressure field. The High Resolution Interface Capturing (HRIC) differencing scheme served the discretization of the transport equation for the volume fraction. Boundary conditions that provide the appropriate time-dependent velocity field and free surface elevation at the inlet and a non-reflective boundary at the outlet were defined. An active wave absorption method based on additional source terms was implemented and used in the far field to prevent wave reflections. At the inlet boundary, uniform or non-uniform velocity profiles are specified to define regular, focused, or irregular waves. Second-order Stokes waves were used. A linear superposition of wave harmonics according to this theory calculated the initial surface elevation and velocity field. In [14] it was shown that wave skewness, i.e., the wave-crest asymmetry, appears immediately behind the inlet boundary. The RANS equations are coupled with the nonlinear equations of motions and the linear equations of elastic deformations in an implicit way. For cases with flexible hull girder, a Timoshenko-beam model was used to represent the governing structural properties, namely the bending and shear stiffness. A grid morphing method was employed to allow for rigid body motions and elastic deformations. An extensive description of the numerical method can be found in [15,35].

2.4. Green Function Boundary Element Method

The linear frequency-domain panel code uses zero-speed Green functions and a forward speed correction based on the so-called encounter frequency approach. A velocity potential is found by distributing singularities (sources and sinks) of constant strength over the mean wetted surface of the hull. The velocity potential is separated into a time-independent steady contribution caused by the ship's forward speed and a time-dependent part associated with the incident wave system and the oscillating ship motions. Computed sectional loads are corrected to account for the nonlinear effect originating from the non-wallsided hull shape of the ships' fore and aft body in finite amplitude waves [56]. More details about the numerical method may be found in [30].

3. Investigated Ships and Model Tests

The test cases comprise different ship sizes and hull forms (bulky and slender bodies). Table 1 lists principal particulars of the four investigated ships. The LNG carrier and the chemical tanker were comparatively small ships, while the cruise ship was of medium size and the containership was a large post-Panamax design. Small ships are prone to experience high translational and rotational accelerations in waves because sea states with waves in the relevant length range are relatively steeper. Large containerships operate at relatively higher speeds in more severe sea states and longer waves with potentially higher energy.

Table 1. Main particulars of the investigated ships.

	Cruise Ship	Containership	LNG Carrier	Chemical Tanker
Length overall [m]	238.00	349.00	197.10	170.00
Length bet. perpendiculars [m]	216.80	333.44	186.90	161.00
Moulded breadth [m]	32.20	42.80	30.38	28.00
Design draft [m]	7.20	13.1	8.40	9.00
Block coefficient [-]	0.65	0.62	0.73	0.75
Displacement [t]	34,087	125,604	35,355	30,707
Mass moment of inertia (Ixx) [kgm^2]	5.62×10^9	3.65×10^{10}	4.90×10^9	2.73×10^9
Mass moment of inertia (Iyy) [kgm^2]	1.00×10^{11}	8.59×10^{11}	5.95×10^{10}	3.30×10^{10}
Longitudinal Center of Gravity [m]	99.60	161.94	94.88	82.51
Vertical Center of Gravity [m]	15.30	19.20	8.24	6.20

The cruise ship was tested at one full-scale speed of 6.0 kn; the containership, at two full-scale speeds of 15.0 and 22.0 kn; and the LNG carrier and the chemical carrier at zero speed. Model tests of the cruise ship and the containership at a scale of 1:50 and 1:80, respectively, were performed at the model basin CEHIPAR [54]; model tests of the LNG carrier and the chemical tanker at a scale of 1:70, at Technische Universität Berlin (TUB) [54]. Except for the containership, computations and model tests treated the ships' hull girder as rigid. All four ship models were constructed as segmented models to experimentally measure sectional hull girder loads. The containership model, comprising six segments, was equipped with an aluminum backbone that reflected vibration modes and natural frequencies of the full-scale ship, see [57]. The cruise ship model, consisting of four segments, was equipped with an aluminum backbone as well. However, this backbone was characterized by high stiffness to represent a rigid hull. Models of the LNG carrier and the chemical tanker consisted of two segments joined amidships.

These four ships with their different bow and stern flares covered a broad range of hull forms. The ship lines for aft and forward sections are illustrated in Figure 1. Water depths of the model basins differed significantly. At CEHIPAR basin, water depth was 5.0 m; at TUB basin 1.0 m. Furthermore, models investigated at CEHIPAR were self-propelled; models tested at TUB were kept on position with soft springs.

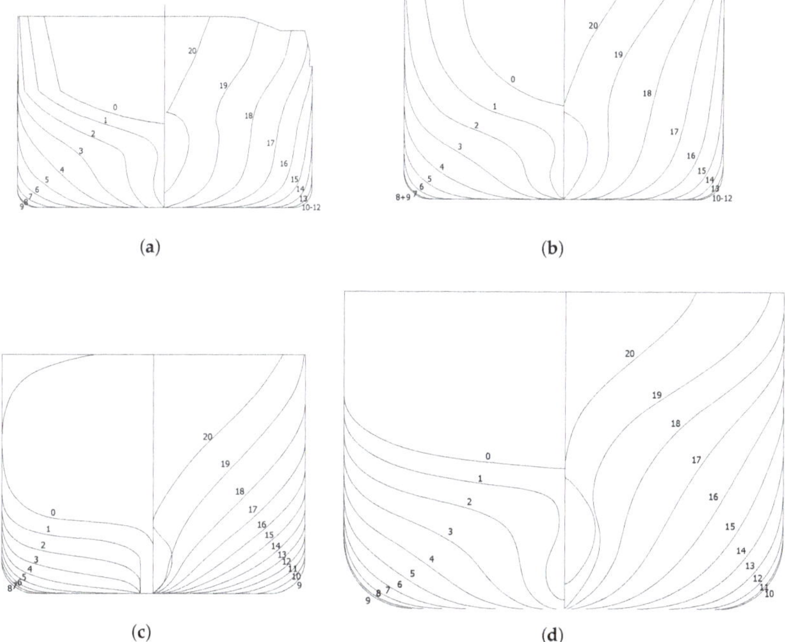

Figure 1. Body plans: (**a**) Chemical tanker, (**b**) LNG carrier, (**c**) Cruise ship, (**d**) Containership.

4. Computational Setup

4.1. Strip Theory Method

Time-domain strip method computations were performed by IST for the cruise ship, the LNG carrier and the chemical tanker. Transverse strips distributed along the overall ship lengths idealized their hulls. Two-dimensional computed added mass and damping coefficients for each strip are integrated over the ship's length to yield approximate three-dimensional added mass and damping coefficients for each ship hull. The hull of the cruise ship, for example, was represented by 38 strips, and each strip was approximated by 46 straight line boundary elements extending from keel to second deck. For these time-domain strip theory computations, a time step size of 0.1 s was selected to ensure convergence of ship responses in irregular seaways of three hours duration. Each seaway was composed of at least 1000 harmonic wave components. Regular waves and irregular sea states were computed for the comparative study. The simulated irregular sea states were reconstructions of the experimental sea state realizations. Heave, pitch and surge motions were computed for regular and irregular waves computations.

4.2. Rankine Source Boundary Element Method

Computations with the Rankine source boundary element method were performed by DNV GL for three ships, namely the cruise ship, the chemical tanker and LNG carrier. Structured surface meshes discretized the hulls and the free surface. The overall number of panels varied between 4000 and 6000. Panel sizes depended on wave lengths (regular waves) and significant wave lengths (irregular waves) tested. Typically, at least six panel lengths equaled the shortest expected wave length. This yielded a panel length near the hulls of about 2.4 m. Finite water depth, if needed, was accounted for by additional Rankine functions. The potential distribution on the boundaries was described by B-spline patches. The time step size applied varied between 0.01 and 0.10 s. Between 100 and 200 harmonic wave components were superposed to establish an irregular seaway. Heave, pitch and surge motions were computed for regular and irregular waves computations.

The simulations with irregular seas were conducted with random realizations and do not match the exact sea state realization applied in the model tests. As a consequence, time series of wave elevation and ship responses can not be compared.

4.3. Field Method

Field method computations were performed for all four ships. We used a Cartesian coordinate system $S(x, y, z)$ fixed to the ship's body. Its x-axis is directed forward, its y-axis to backboard, and its z-axis normal to the plane decks upward. Its origin is at the centre of gravity G.

A large number of numerical simulations under various wave conditions were carried out. Different significant wave heights H_s and zero-uprossing periods T_z necessitated different grids with adapted grid topologies. The influence of the spatial and temporal discretization on nonlinear wave propagation and on ship responses in regular and irregular waves was extensively studied, discussed and published by the authors in [14,15]. This discretization study was the basis for the selected temporal and spatial discretization in this paper. For this reason, we refrained from performing a similar study here. The papers address the free surface elevation and ship responses in regular waves as well as the relative energy loss for different sea states and discretization levels at different distances from the inlet boundary. While ship motions and loads in regular waves are less sensitive to discretization errors, the free surface elevation depend significantly on the order of approximation as well on the discretization. Further on, it was found that discretization errors increase the energy loss (down stream) of irregular waves with high steepness defined as

$$s = \frac{2\pi\, H_s}{g\, T_z^2}, \tag{1}$$

with gravitational acceleration g. The difficulty to distinguish between energy loss related to wave-breaking and numerical diffusions was discussed.

Common for all cases was the refinement area near the free surface ahead of the ships. A refinement box at and underneath the free surface resolved wave velocities, and pressure fields needed to be sufficiently fine to avoid loss of wave energy. The vertical extension of this refinement box depended on a sea state's significant wave height and mean wave period. To capture the interaction between waves and hull, the refinement box around the hull had to be extended in the positive z-direction. The ships' center plane defined a vertical symmetry boundary. Towards the far side of the domain (outlet), stretched cells and additional source terms dampened the waves. Figure 2 shows sample numerical grids for each ship, including simulated free surfaces for irregular long-crested waves and typical grid extensions as multiples of overall ship length, L_{oa}.

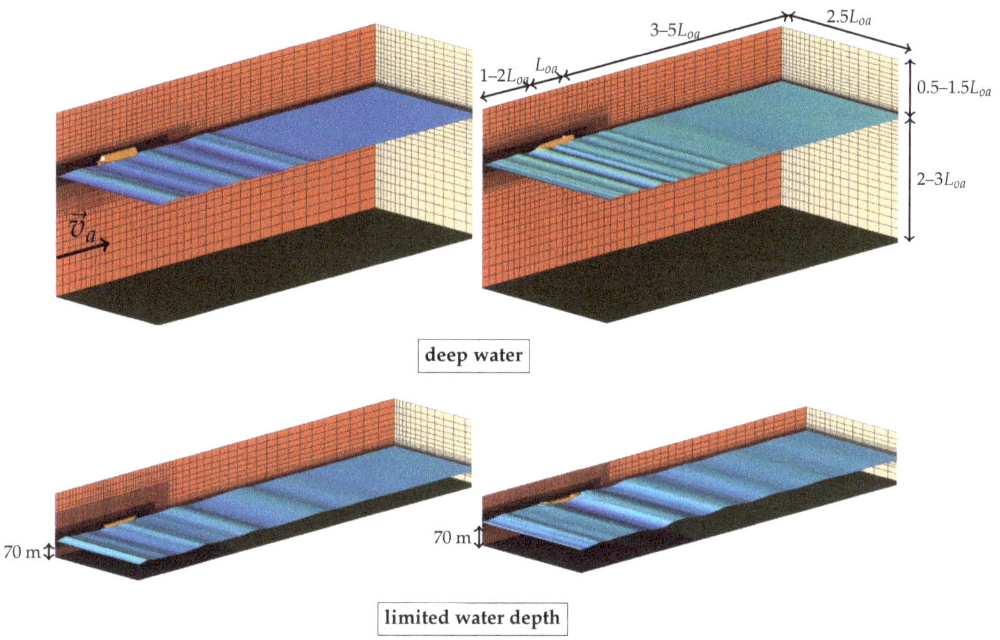

Figure 2. Overview about numerical grids for cruise ship (**top left**), containership (**top right**), liquid natural gas (LNG) carrier (**bottom left**), chemical tanker (**bottom right**).

The effect of wave damping in vicinity to the outlet boundary is clearly visible. It starts about 1.5–2 times the ship length behind the model. Interfering effects of active wave damping on the simulation results can be ruled out. Model tests of the LNG carrier and the chemical tanker had to account for the limited water depth of the basin at TUB. Wave probes were mounted numerically to monitor the wave elevation at different locations.

Table 2 summarizes discretization parameters used for mesh generation. Cell lengths, Δx, and cell heights, Δz, indicate characteristic spatial resolutions and were related to the sea state parameters significant wave height H_s (or regular wave height H), wave length λ, and peak period T_p (or regular wave period T). Between 250 and 800 harmonic wave components were superposed to model an irregular seaway.

Surge motions were suppressed in regular wave simulations. For the cruise ship and containership sailing with forward speed in severe irregular seas, the surge motion was prescribed using the measured time signal of the physical self-propelled model[1].

Table 2. Discretization parameters.

Grid	$H_s/\Delta z$	$\lambda/\Delta x$	$T_p/\Delta t$	Number of Cells
Rigid hulls	10 to 20	70 to 160	800 to 950	600,000–1,800,000
Flexible hulls	15 to 25	100 to 200	950 to 1260	800,000–2,000,000

The simulations with irregular seas were conducted with equivalent sea state realizations as used in the model tests. Thus, time series and short-term statistics for wave elevation and ship motions can be compared with model test results, see Sections 5.2.2 and 5.2.3.

4.4. Green Function Boundary Element Method

The Green function boundary element method was used to compute the RAOs for the four test case ships. We discretized the wetted hull surface using about 4500 surface panels and accounted for the limited water depth of the towing tank for the chemical tanker and LNG carrier. To investigate the water depth effects on ship responses, we performed additional computations under deep water conditions.

5. Results

The numerical methods were validated based on Froude-scaled model test results. The flow around ships in waves (and related vertical motions and loads) is pressure dominated. We can assume that viscous effects are negligible.

5.1. Regular Waves

For all four ships, systematic tests were conducted to obtain the ship response RAOs. As ship responses in regular waves with small and moderate heights were expected to be almost linear, deviations between numerical and experimental RAOs helped to quantify uncertainties in predictions of linear or weakly nonlinear responses before starting to address strongly nonlinear responses obtained in irregular seas. Not only biases and uncertainties that may have affected numerical predictions have to be taken into account, but also uncertainties in measurements and post processing of data.

Simulations and model tests were conducted in regular waves of constant steepness and varying wave period and in transient wave trains comprising harmonic wave components with amplitudes and frequencies in accordance with the spectral energy of a specified seaway. The latter approach significantly saves computational time; instead of simulating each wave period, only one simulation with a transient wave train is required.

Experiments and computations were obtained only in head waves ($\mu = 180\,\text{deg}$) of varying frequencies, with the cruise ship and the containership advancing at a forward speed of 6.0 and 15.0 kn, respectively. The LNG carrier and the chemical tanker were investigated at zero forward speed. Evaluated responses comprised midships vertical bending moment, M_y, pitch motion, ϑ, heave motion, ζ and (partly) surge motion, ξ, see Table 3.

---

1 The propeller's rate of revolution of the physical models at CEHIPAR was PID-controlled to maintain the mean forward speed. It was aimed to bypass the uncertainty of this condition influenced by the specific control mechanism.

Table 3. Parameters for the determination of response amplitude operators (RAOs) and applied methods.

Vessel	μ [deg]	v [kn]	Response Quantity	Field Method	Rankine Source Method	STRIP Method	Green Function Method	Experiment
Cruise Ship	180	6.0	M_y, ϑ, ζ	✓	✓	✓	✓	✓
			ξ		✓	✓	✓	✓
Containership	180	15.0	M_y, ϑ, ζ	✓			✓	✓
			ξ				✓	
LNG carrier	180	0	M_y, ϑ, ζ	✓	✓	✓	✓	✓
			ξ		✓	✓	✓	
Chemical tanker	180	0	M_y, ϑ, ζ	✓	✓	✓	✓	✓
			ξ		✓	✓	✓	

5.1.1. Cruise Ship

For the cruise ship, Figure 3 shows the resulting RAOs of heave and pitch motions; Figure 4 the resulting RAOs of midships vertical bending moment. Heave RAOs were normalized against wave amplitude, ζ_a; pitch RAOs against wave slope, $k\zeta_a$; vertical bending moment RAOs, against the product $\rho g B L_{pp}^2 \zeta_a$, where k is the wave number. The numerically predicted heave and pitch RAOs agree well with each other and with experiments. However, the strip method yields lower pitch motion amplitudes in longer waves. Comparative experimental data were unavailable for these wave lengths. The midship vertical bending moment RAOs were difficult to appraise. Although maximum values from field method, Rankine source and Green functions boundary element methods agree well, they are about 20% below measurements. Only results from strip method compare well to measured maximum bending moments. The measurements are characterized by a pronounced peak, whereas all numerical predictions (apart from strip method) have smoother slopes. Mismatches of ship model mass distributions might have caused the relatively large deviations of the maximum midship vertical bending moment RAOs. Since all heave and pitch motion RAOs generally compare favorably, poor performance of numerical methods was ruled out.

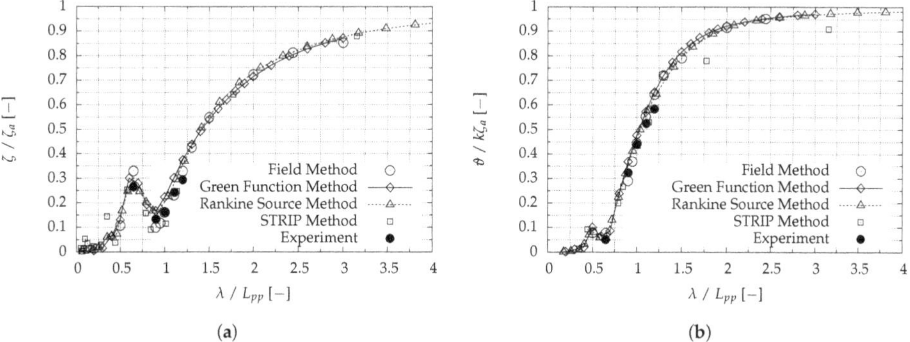

Figure 3. Cruise ship: computed and measured RAOs. (**a**) Heave motion ζ and (**b**) pitch motions ϑ. ζ_a is the wave amplitude, k the wave number, λ the wave length and L_{pp} the ship length.

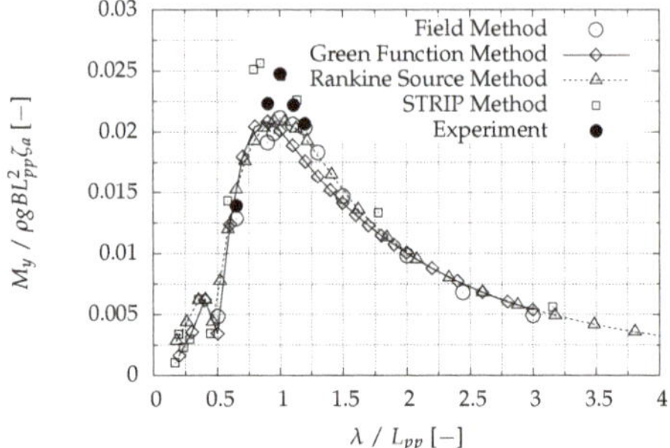

Figure 4. Cruise ship: computed and measured RAOs for midship vertical bending moment M_y. ρ is the water density, g the gravity constant, B the ship's breadth and L_{pp} the ship length.

5.1.2. Containership

The containership model was equipped with a flexible backbone to measure the fundamental elastic behavior of the full-scale ship as described in Section 3. For the midship vertical bending moment, the low-frequency ship responses (without hydroelastic effects) were used for comparison of RAOs. Experimentally obtained time series of the ship advancing at a full-scale speed of 15.0 kn were analyzed. Comparative field method computations generally compare well to measurements, as seen by the heave and pitch motion RAOs shown in Figure 5 and the midship vertical bending moment RAOs shown in Figure 6. While the Green functions boundary element method computations predicted a local heave maximum at wave length to ship length ratio, λ/L_{pp}, of 0.58, experimental and field method results indicate that this peak is shifted towards longer waves of λ/L_{pp} close to unity. The discrepancy between the field method computed maximum RAO for the vertical bending moment and model test results was less than 2%, for the Green function method 7.5%.

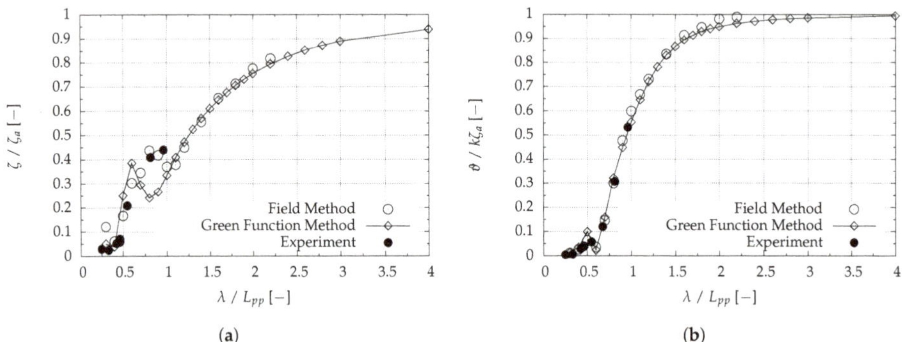

Figure 5. Containership: computed and measured RAOs: (**a**) heave motion ζ and (**b**) pitch motion ϑ. ζ_a is the wave amplitude, k the wave number.

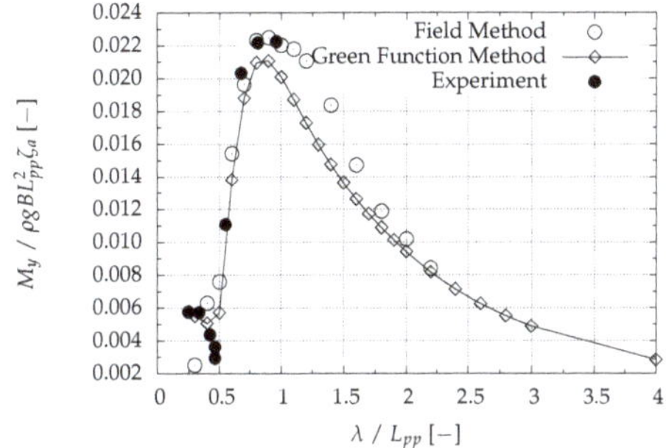

Figure 6. Containership: midship vertical bending moment M_y. ρ is the water density, g the gravity constant, B the ship's breadth and L_{pp} the ship length.

5.1.3. LNG Carrier

Figures 7 and 8 show computed and measured RAOs of heave, pitch and surge motions as well as midship vertical bending moments. Numerically predicted maximum RAO values of heave, pitch and vertical bending moment from field method, from Green functions boundary element method and from Rankine source boundary element method agree fairly with experimental measurements. Comparable pitch and surge motions from strip method are underpredicted, while heave motions in long waves are overpredicted. This is caused by the neglection of shallow water effects. Computed maximum midship vertical bending moments using field and Green functions boundary element methods compare favorably with experimental data. The differences obtained are less than 3%. In shorter waves ($\lambda / L_{pp} <$ 0.8), agreement between results obtained from all numerical methods is satisfactory, in longer waves the discrepancies increase. At a scale of 1:70, the model tests at TUB represent a water depth of about 70 m, which is less than $0.5L_{pp}$ for the LNG carrier.

Additional comparative calculations for finite and infinite water depths were performed. Shallow water effects begin occurring at different wave lengths, depending on the ship response considered, see Figures 9 and 10. Here, red lines mark the lower wave length limits where responses in finite water depth started to significantly deviate from responses in infinite water depth. While shallow water effects decreased the heave amplitudes in longer waves, pitch motions and midship vertical bending moment amplitudes increased. For wave lengths larger than $\lambda / L_{pp} > 0.95$, heave is already affected by the limited water depth, whereas pitch and surge start to visibly deviate from $\lambda / L_{pp} > 1.5$. When heave was influenced by shallow water, vertical bending moments are affected as well. This correlation appeared to be reasonable. Consistent with classical theory of gravity water waves and ship dynamics, water depth h starts to affect ship responses in waves for $h < 0.5\lambda$.

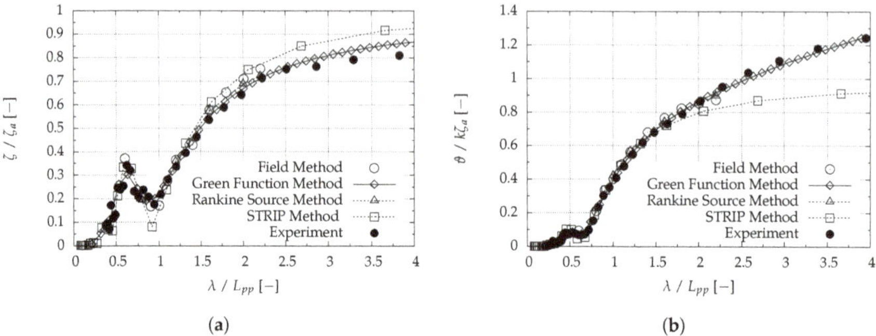

Figure 7. LNG carrier: computed and measured RAOs. (**a**) Heave motion ζ and (**b**) pitch motion ϑ. ζ_a is the wave amplitude, k the wave number, λ the wave length and L_{pp} the ship length.

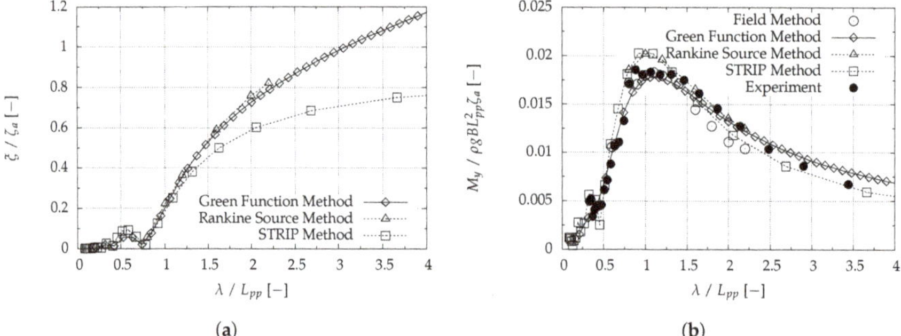

Figure 8. LNG carrier: computed and measured RAOs. (**a**) Surge motions ξ and (**b**) midship vertical bending moment M_y (right). ρ is the water density, g the gravity constant, B the ship's breadth and L_{pp} the ship length.

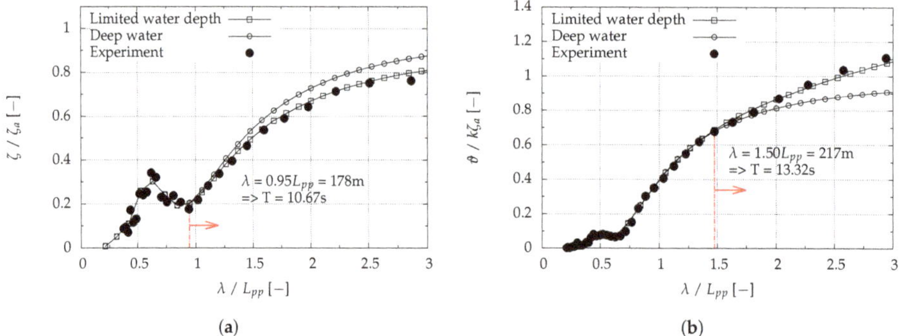

Figure 9. LNG carrier: with Green functions boundary element method computed effects of water depth on (**a**) heave and (**b**) pitch motions.

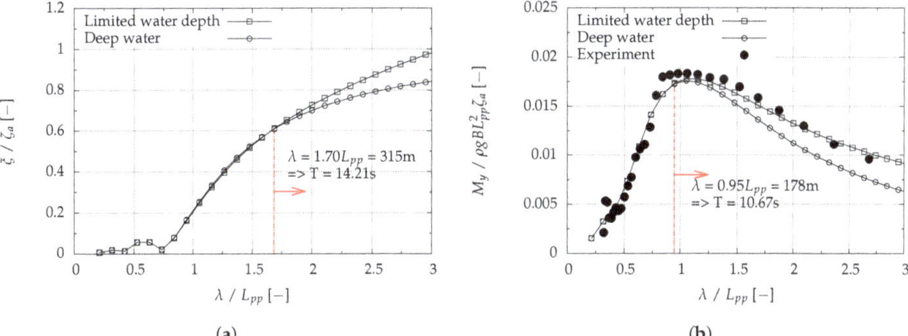

Figure 10. LNG carrier: with Green functions boundary element method computed effects of water depth on (**a**) surge motion and (**b**) midship vertical bending moment.

5.1.4. Chemical Tanker

Figures 11 and 12 show computed and measured RAOs of the chemical tanker responses. Except for strip method predictions of ship motions and midship vertical bending moment in long waves, numerical and experimental results agree well. This is valid for both concepts, single regular waves and transient wave trains. We carried out field method simulations for this ship in the same transient wave train as in the model tests performed at the TUB basin [54]. Figure 13 shows the measured and computed midship vertical bending moment. Agreement between computed and measured vertical bending moments is impressive, indicating that using a field method in wave trains to determine RAOs bears a great potential to reduce computation times because only one single run is required instead of a set of runs in regular waves. For ship responses dominated by restoring forces and small memory effects (such as heave, pitch and midship vertical bending moment), this example demonstrated the usefulness of this approach. Field method simulations of the chemical tanker accounted for shallow water effects.

The exemplary vertical longitudinal cut through the fluid domain in Figure 14 shows pressure and velocity distributions while the ship encounters an extreme wave of about 15 m height and 200 m length. Pressure disturbances at the basin's bottom caused by incident waves are visible, illustrating the influence of limited water depth on wave kinematics. The pressure field was obtained by subtracting hydrostatic (calm water conditions) pressures from total pressures. Red colored areas indicate positive pressures; blue colored areas, negative pressures. Black vectors represent velocity directions and magnitudes. Water depth corresponded to the basin depth of the experiments.

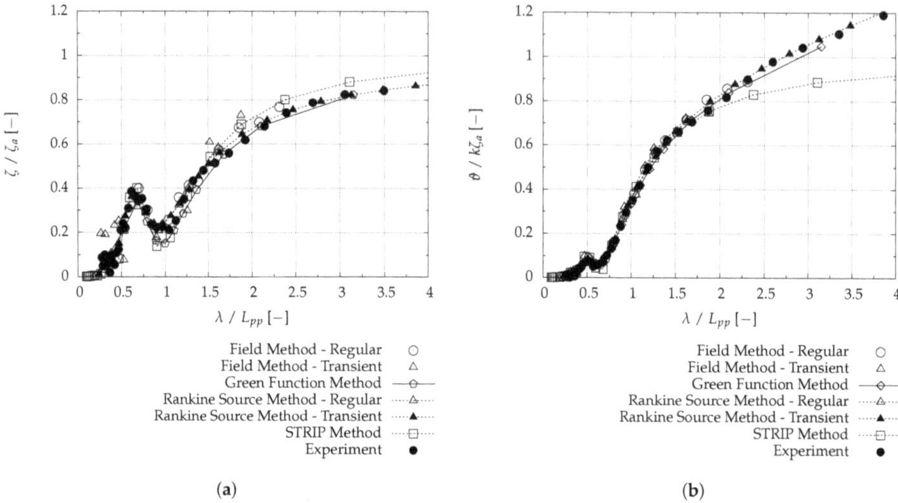

Figure 11. Chemical tanker: computed and measured RAOs, (**a**) heave motion ζ and (**b**) pitch motion ϑ. ζ_a is the wave amplitude, k the wave number, λ the wave length and L_{pp} the ship length.

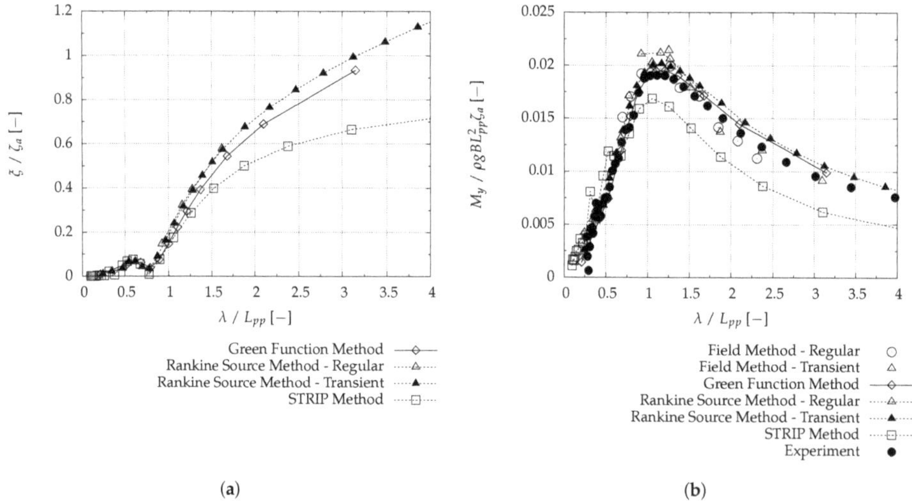

Figure 12. Chemical tanker: computed and measured RAOs, (**a**) surge motion ξ and (**b**) midship vertical bending moment M_y. ρ is the water density, g the gravity constant, B the ship's breadth and L_{pp} the ship length.

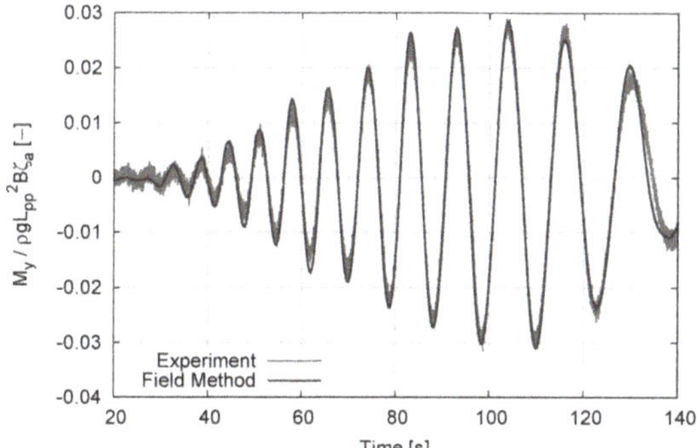

Figure 13. Chemical tanker: time histories of field method computed and measured (normalized) midship vertical bending moment M_y. ρ is the water density, g the gravity constant, B the ship's breadth, L_{pp} the ship length and ζ_a the free surface elevation.

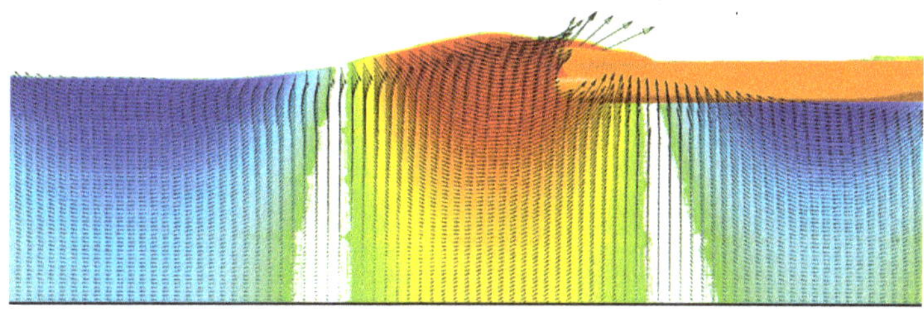

Figure 14. Exemplary field method computed pressure and velocity distribution in the fluid domain surrounding the chemical tanker at the symmetry plane (y = 0). An extreme wave ($H \approx 15$ m and $\lambda \approx 200$ m) impinges the vessel's bow. The orbital velocity field of the wave crest and wave troughs are cut off from the sea-bed (indicated by solid bottom-line).

5.2. Irregular Extreme Waves

To assess the ability of the developed methods to predict ship responses in extreme seas, reliable predictions of short-term response statistics of the four investigated ships in irregular seaways were thought to be the best measure.

Usually statistical measures, such as the spectral energy density distribution, $S(\omega)$, describe an irregular sea state. Most common are the Pierson–Moskowitz spectrum [58] that only depends on wind speed, the modified Pierson–Moskowitz spectrum that depends on significant wave height and zero up-crossing wave frequency and the JONSWAP spectrum for limited fetch and wind duration [59]. The spectral energy density of the JONSWAP spectrum in is given as

$$S(\omega) = \frac{\alpha g^2}{\omega^5} \cdot \exp\left[-\frac{5}{4} \cdot \left(\frac{\omega_p}{\omega}\right)^4\right] \cdot \gamma^{\exp\left(-\frac{(\omega-\omega_p)^2}{2\sigma^2\omega_p^2}\right)} \tag{2}$$

with peak frequency ω_p, energy parameter α, peak enhancement factor γ and shape factor β. The parameter σ depends on the peak frequency ω_p:

$$\sigma = \begin{cases} 0.07: & \omega \leq \omega_p \\ 0.09: & \omega > \omega_p \end{cases} \tag{3}$$

For wave load predictions of ships, the International Association of Classification Societies (IACS) recommends the modified Pierson–Moskowitz spectrum, which corresponds to a JONSWAP spectrum with a peak enhancement factor of unity. By concentrating the wave energy in a smaller frequency band, an enhancement factor larger than 3.3 increases the modulation instability of sea states, promoting wave group formation with extraordinary high waves. When this frequency band coincides with frequencies relevant for ship responses, response amplitudes tend to be large compared with those in a sea state of the same energy but smaller enhancement factor.

Sea states are characterized by their peak period, T_p, while their steepness is related to the zero-up-crossing period, T_z, see Equation (1). The ratio of T_p and T_z depends on the peak enhancement factor γ and reads

$$\frac{T_z}{T_p} = 0.6673 + 0.05037\gamma - 0.006230\gamma^2 + 0.0003341\gamma^3. \tag{4}$$

Ocean waves are often not unidirectional; their energy is directionally spread about the principal wind direction. Although a cosine square distribution of wave energy over wave encounter angle is often assumed, the actual spreading strongly depends on wind conditions. Here, for the sake of simplicity, only head seas ($\mu = 180$ deg) were investigated. Based on the JONSWAP spectrum formulation, these random sea states were generated. The maximum peak enhancement factor γ used was 5.0.

5.2.1. Investigated Sea States

We used the Coefficient of Contribution method to identify each sea state's relevance for long-term extremes of ship responses, thereby focusing the numerical and experimental investigations on sea states with large potential for extreme events. This screening relied on the linear three-dimensional Green functions boundary element method and wave statistics according to the IACS North Atlantic scatter diagram [60].

Table 4 lists the parameters identifying the irregular sea states investigated for comparative purposes. The run duration T_d, i.e., the effective time the ship encounters the waves, is listed.

Table 4. Parameters of investigated irregular sea states and applied methods.

Ship	H_s [m]	T_p [s]	γ [-]	s [-]	v [kn]	T_d [s]	Field Method	Rankine Source Method	STRIP Method	Experiment
Cruise ship	10.5	12.22	3.3	0.075	6.0	1600	✓	✓	✓	✓
Containership	12.5	11.80	5.0	0.089	15.0	1400	✓			✓
LNG carrier	10.5	12.22	3.3	0.075	0	2700	✓	✓		✓
Chemical tanker	10.5	12.22	3.3	0.075	0	2700	✓	✓		✓

5.2.2. Time Histories

Time histories are the basis for statistical analyses. As described in Section 4, the strip method and the field method relied on reconstructions of experimental sea state realization. These reconstructions were based on wave probe measurements during the tests, safeguarding the best possible identity of wave processes in model tests and computations, eliminating uncertainties in numerical wave modeling and improving correlation of numerically and experimentally predicted ship response processes.

The boundary element method did not allow using replicas of the experimental wave processes. Instead, random sea state simulations based on the sea state parameters of the experiments had to be used. This made direct comparisons of time histories impossible.

In addition to ship responses, the incident wave elevation, $\zeta(t)$, at the ships' midship position was monitored to relate ship responses to the wave excitations and to assess the numerical wave models. This was of interest in sea conditions in which significant nonlinearities of the wave process were expected. Furthermore, surface elevation monitoring enabled assessing the numerical damping of waves likely to occur in field method simulations. This section shows exemplary results obtained for the cruise ship and the LNG carrier.

Cruise Ship

Figure 15 shows the computed and measured free surface elevation and the vertical bending moment amidships. As the free surface elevation obtained in the experiments is predefined in the STRIP method, the corresponding time history is not shown. The field method solves the nonlinear wave propagation inside the computational domain. Hence, the resulting wave field is of interest and compared with measurements.

Except for a few events, a satisfying agreement between measured and computed free surface elevation as well as the corresponding wave-induced motions and bending moment was obtained, see Figures 15 and 16. There is a good agreement in terms of phasing and amplitudes between measurements and field method computed results.

The comparison between strip method results and model tests reveals a favorable agreement in phasing and small ship response amplitudes as well. For moderate and large ship motions and loads, the deviations increase. In principle, large motion and load amplitudes are overestimated. Sagging moments (positive values) agree better than hogging moments (negative values) with model test results.

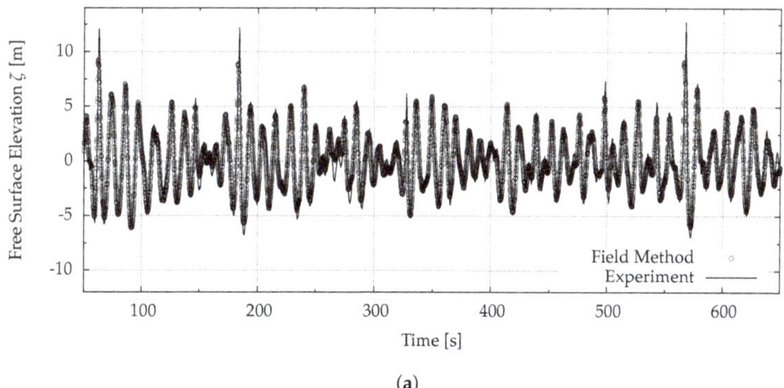

(a)

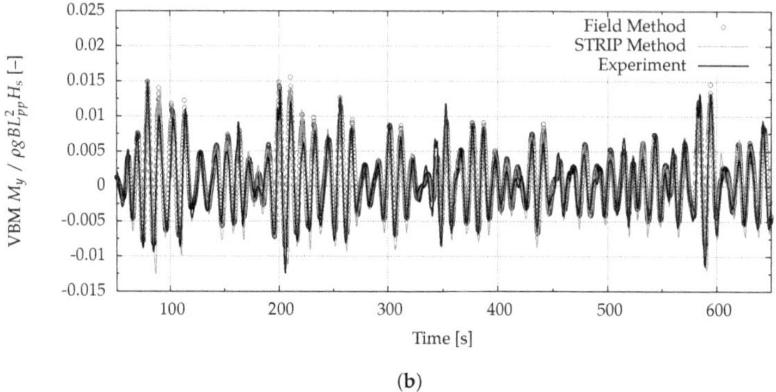

(b)

Figure 15. Cruise ship: comparison of time histories obtained with field method and experiments for (**a**) the free surface elevation. The lower figure (**b**) shows the field method and strip method computed vertical bending moment amidships in comparison with model test results.

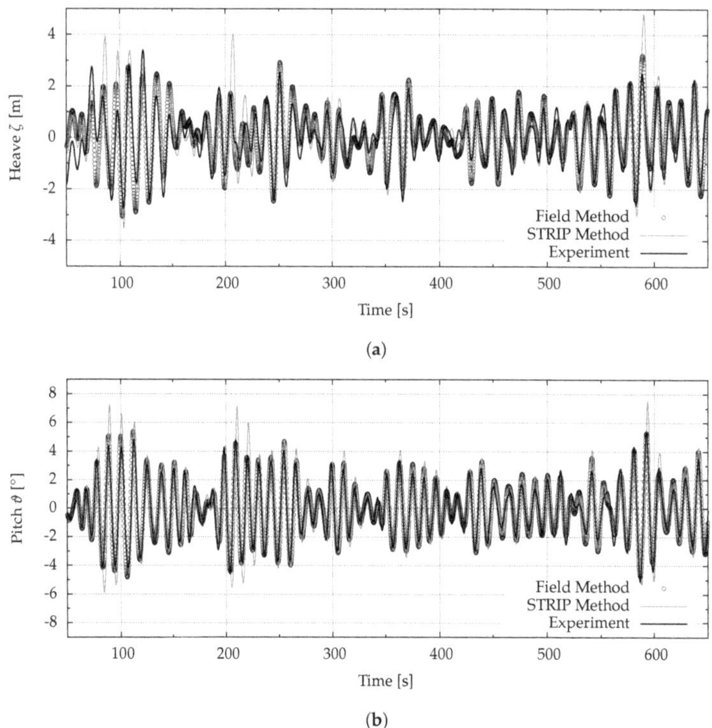

Figure 16. Cruise ship: comparison of time histories obtained with field method, strip method and experiments for (**a**) heave and (**b**) pitch motions.

LNG Carrier

For the LNG carrier, the field and Rankine source method were employed to simulate ship responses. Solely the field method, however, reconstructed the equivalent model test sea state realizations. Figures 17 and 18 show sample time histories of corresponding free surface elevations, ship motions and midship vertical bending moments in irregular waves. The selected time interval covered three severe wave groups. The field method captured the asymmetry between wave crests and troughs. Although differences of wave elevation and ship responses between model test and numerical results are most pronounced for heave, the overall agreement is satisfactory.

Pressures were measured using pressure sensors located at the ship's stern and bow as shown in Figure 19. Figure 20 presents two exemplary time histories of pressures obtained from sensor 34 at the stern and senor 10 at the bow. Pressures included the atmospheric pressure of one bar. Due to zero forward speed and relatively small bow and stern flares, pressures were moderate, and no distinct slamming peaks occurred during the time interval considered.

The green water column height above the weather deck was measured during experiments and monitored in numerical simulations. They agreed favorably over the time duration of such an event, as exemplarily shown in Figure 21. Here, the height of the computed water column is about 18% below the measured height. Considering the difficulties associated with determining the free surface elevation of a breaking wave and with a capacitance wave probe to accurately measure the water column height, the agreement was unexpectedly favorable. %endparacol

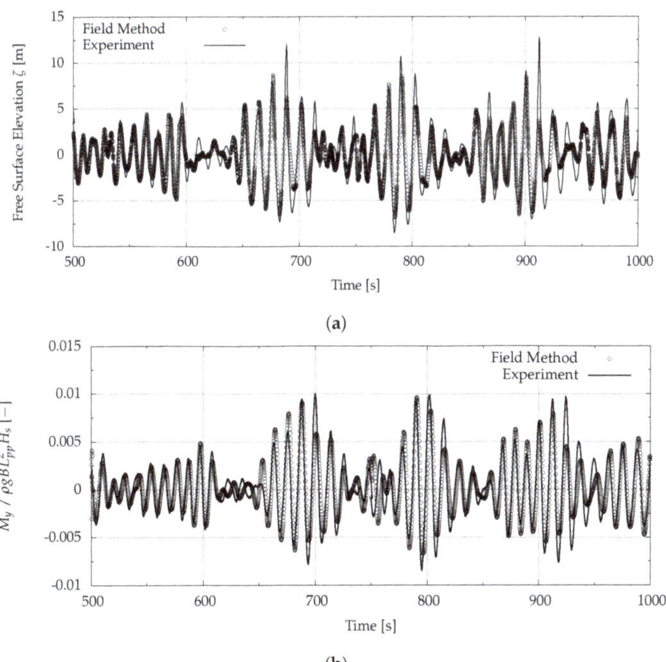

Figure 17. LNG carrier: field method computed and measured time histories of (**a**) free surface elevation and (**b**) midship vertical bending moment.

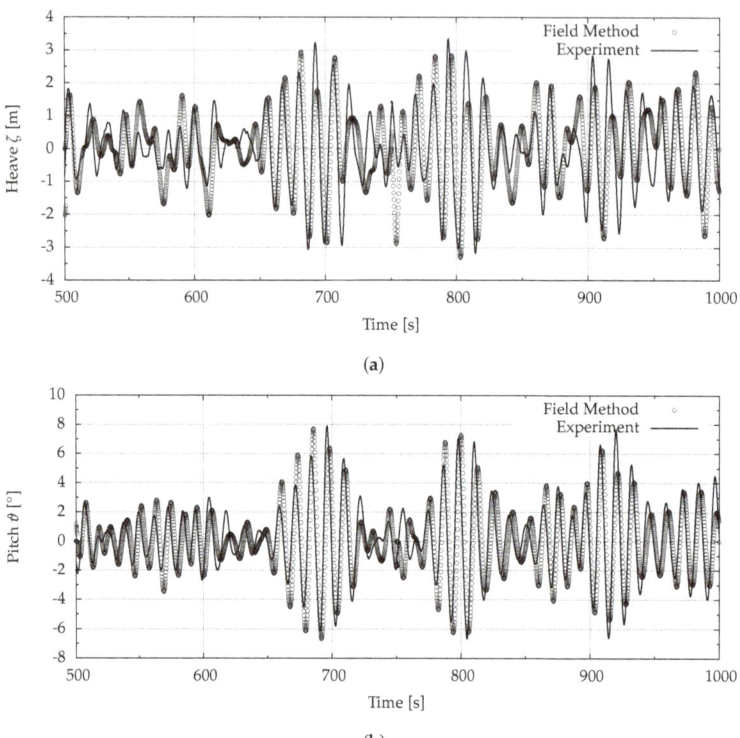

Figure 18. LNG carrier: field method computed and measured time histories of (**a**) heave and (**b**) pitch motions.

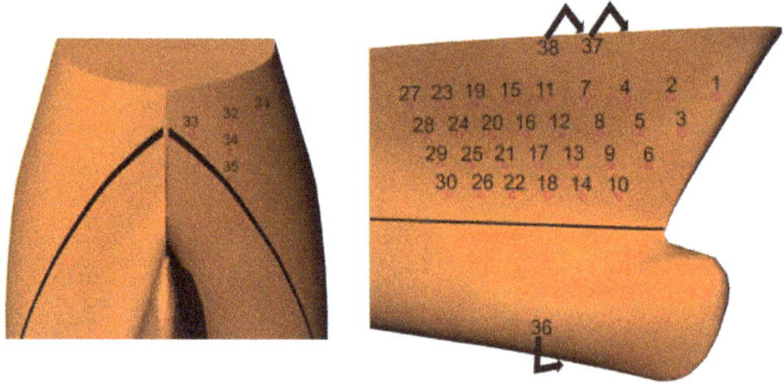

Figure 19. LNG carrier: pressure sensor locations at the ship's stern and bow [54].

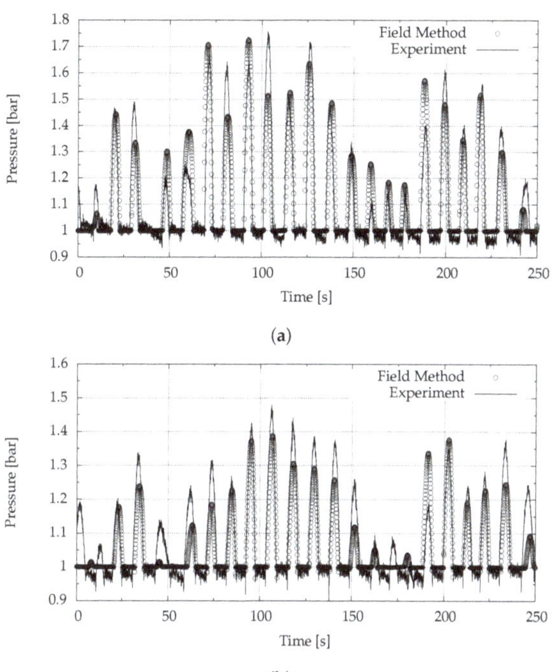

Figure 20. LNG carrier: field method computed time histories of pressures at (**a**) sensor 10 and (**b**) sensor 34.

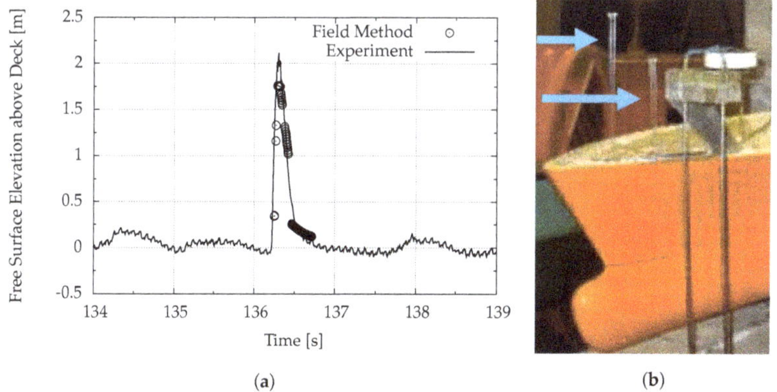

Figure 21. LNG carrier: (**a**) field method computed and measured time histories of free surface elevation above deck and (**b**) wave gauge arrangement on the physical model [54].

5.2.3. Short-Term Statistics

Response peaks were identified from time histories as maxima and minima between consecutive zero up-crossings of a response process. Rainflow counting yielded exceedance rate distributions of response cycles (double amplitudes). The exceedance rate, $\chi(r > r_C)$, is the average frequency (unit [1/s]) of r being larger than r_C. Evaluation was done for discrete response classes C_i. Assuming that T_z, the zero-upcrossing period of the process, is also the mean period of amplitudes due to the narrow-bandedness of the process, with probability P the rate of amplitudes outcrossing r_C reads

$$\chi(r > r_C) = \frac{P(r > r_C)}{T_z}. \tag{5}$$

Hence, $\chi(r > r_C) \cdot T_d$ gives the expected number of amplitudes larger than r_C during a time interval T_d.

Cruise Ship

Short-term statistics were obtained for the cruise ship in irregular sea states with each numerical method. Figure 22a shows exceedance rates of incident wave elevation obtained by the range-pair counting approach, based on a sampling interval of about 27 min in irregular waves, see Table 4. Maximum wave heights reached up to twice the significant wave height. Wave elevation statistics based on strip theory were measured amidships; wave elevation statistics based on the Rankine source BEM and the field method, ahead of the ship. Except for tail distributions, range pair counted wave heights from all codes compare well with experiments although this method provided no information about mean values. Figure 22b shows exceedance rates of wave-induced vertical bending moments. The field method replicates the trend for low, moderate and large response levels. Good agreement was obtained with the BEM for low and moderate response levels. Deviations at the tail were probably caused by the sampling variability. Strip method predictions for small amplitudes agreed favorably with measurements; moderate and large bending moment amplitudes, however, deviated significantly.

Figure 23 shows exceedance rates of heave and pitch, respectively. The field method obtains the best overall agreement with experiments, while the strip method overpredicts these motions already at moderate response levels although the midship vertical bending moment deviations are smaller. However, this is in agreement with previous comparisons. The BEM underpredicts these motions. Again, discrepancies are most significant in the tails of the distributions.

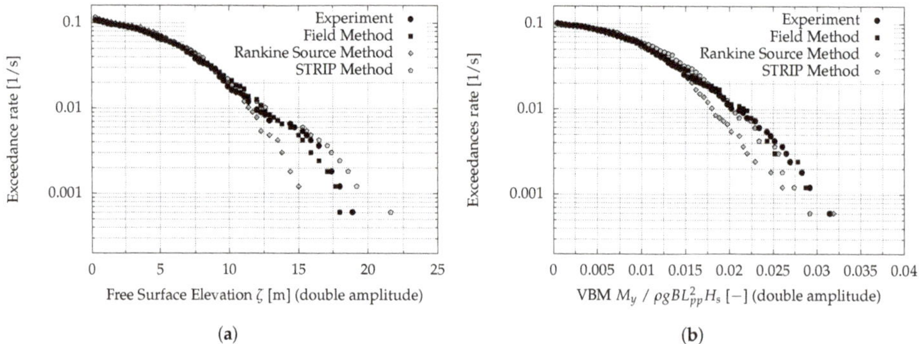

Figure 22. Cruise ship: short-term statistics of (**a**) free surface elevation and (**b**) midship vertical bending moment.

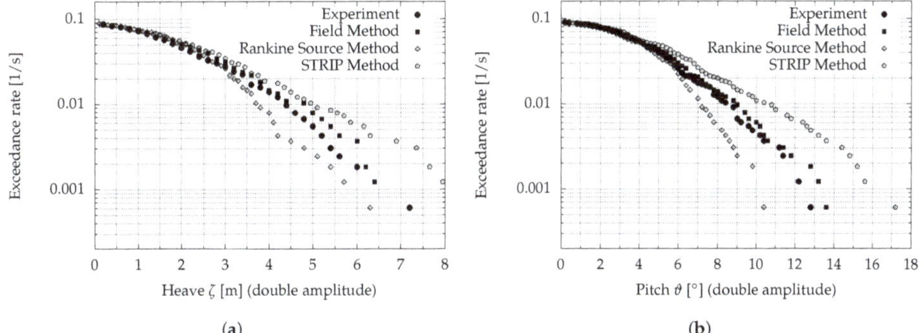

Figure 23. Cruise ship: short-term statistics of (**a**) heave and (**b**) pitch motions.

Chemical Tanker and LNG Carrier

The chemical tanker and LNG carrier were analyzed using the field method and the Rankine source method, see Table 4. The corresponding model tests employed equivalent sea state realizations which allow for direct comparisons of ship responses. In accordance with Table 4, the time duration of the irregular sea was 45 min. Simulation results for heave and pitch motions are compared with model test results for both vessels and shown in Figures 24 and 25. In both cases, field method computed heave and pitch motions fairly agree. Rankine source BEM computed and measured motions deviate notably.

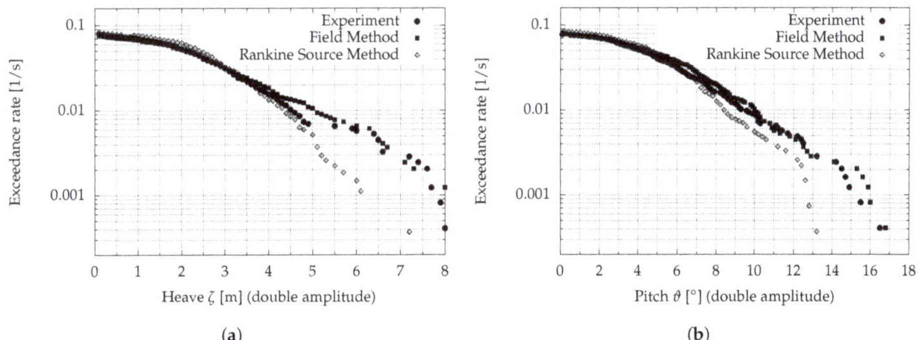

Figure 24. Chemical tanker: short-term statistics of (**a**) heave and (**b**) pitch motions.

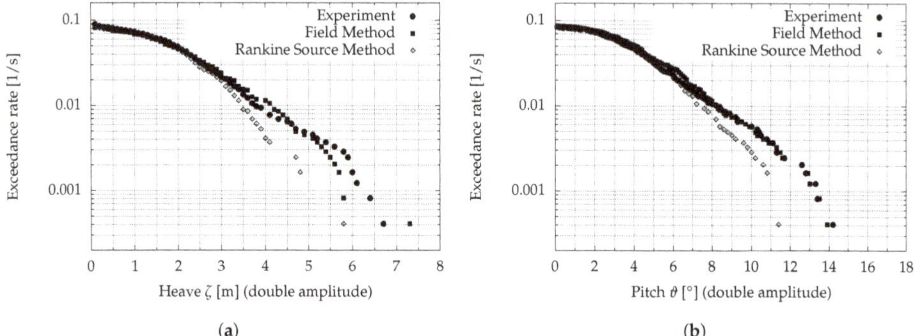

Figure 25. LNG carrier: short-term statistics of (**a**) heave and (**b**) pitch motions.

Figure 26 shows a direct comparison of ship responses for the same sea states. Both ships were tested in different campaigns, but experienced the same waves (sea state realization). The pitch and heave motions are presented with absolute values. As expected maximum heave and pitch motions are remarkably stronger pronounced for the smaller chemical tanker and differ by about 19.4% and 17.1%, respectively. This trend is significant and starts at low response levels.

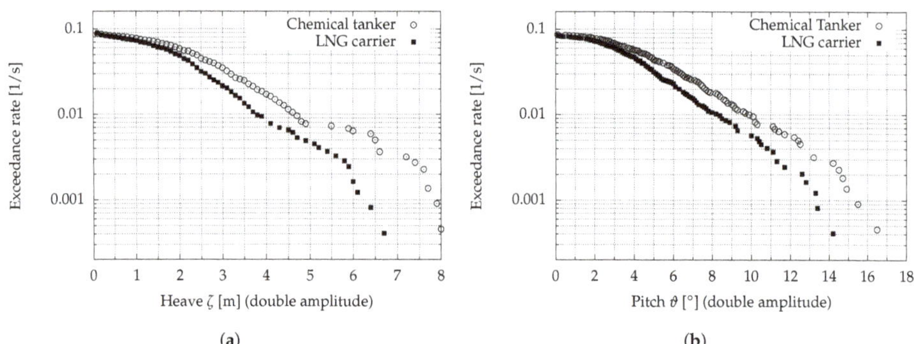

Figure 26. Chemical tanker vs. LNG carrier: short-term statistics of (**a**) heave and (**b**) pitch motions, model test results.

Containership

Short-term statistics were obtained for the containership in irregular sea states using field method and model tests. Range-pair counting yielded exceedance rates of ship responses both for the model tests and the field method numerical simulations. Computed and measured motions and the normalized vertical bending moments are compared.

Heave and pitch motions agree satisfyingly, see Figure 27. The selected ranges for the horizontal axes equal those from the chemical tanker and the LNG carrier. Figure 28 plots the exceedance rates of midship vertical bending moment amplitudes. To eliminate the high-frequency vibratory part from the time histories, the response was low-pass filtered with a cut-off frequency of 0.25 Hz. The remainder was assumed to correspond to the rigid-body midship vertical bending moment.

There is a significant difference in the slopes of exceedance rates of both filtered and unfiltered signals. First, the number of response cycles increases dramatically due to hull girder vibrations. This is favorably replicated in numerical simulations. Second, the maximum response cycle obtained from the unfiltered signal is about 60% larger than the maximum response from the rigid hull signal. Numerically and experimentally determined exceedance rates basically show the same vibratory amplification; however, numerical results underpredict the responses at low exceedance rates. In general, numerical predictions of unfiltered data compare favorably with experimental measurements. Apart from the tail, the slope for low response levels is well reproduced.

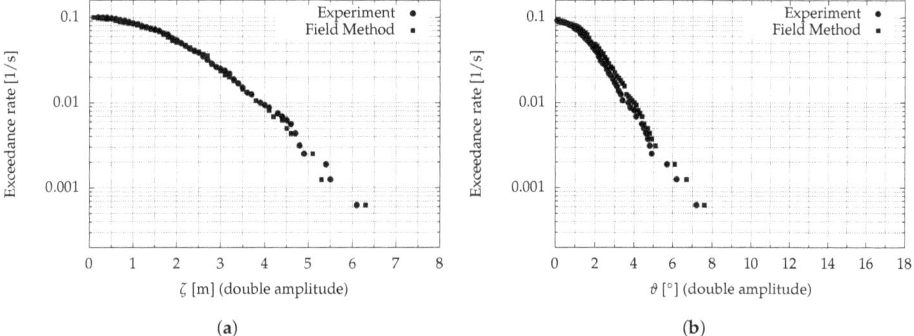

Figure 27. Containership: short-term statistics of (**a**) heave and (**b**) pitch motions.

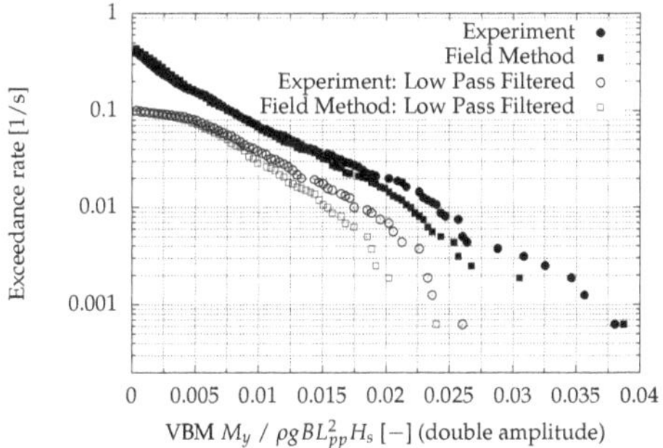

Figure 28. Containership: short-term statistics of midship vertical bending moment.

6. Summary and Conclusions

We compared results obtained with enhanced seakeeping codes that were applied to numerically simulate ship response in regular and irregular severe seas. These codes were based on the strip theory method, the Green function method, the Rankine source boundary element method and a field method. Systematic comparisons of numerical results and model test data were carried out, focusing on response amplitude operators, time histories and short-term statistics. The resulting time histories and short-term statistics comprised responses in severe irregular sea states. Midship vertical bending moment, heave and pitch motions and free-surface elevations as well as pressure distributions and green water columns above deck were addressed.

The purpose was to assess the suitability of numerical methods to predict ship responses in regular waves and their short-term statistical measures under severe sea conditions and to provide benchmark data (free surface elevations, ship motions, loads, hydroelastic effects, pressures) for different ship types. Extreme sea conditions meant that nonlinear effects associated with wave propagation and wave-induced ship responses as well as the occurrence of wave grouping and wave modulation instability had to be considered. Furthermore, hull girder ship loads were affected by green water pressures and slamming impacts. For strip theory and boundary element methods, this application was challenging because of nonlinearities. Field methods are rarely used to compute short-term statistics of ship responses and hydroelastic effects in extreme irregular seas of long duration.

The four different ship types we investigated comprised three medium size ships (a cruise ship, an LNG carrier, a chemical tanker) and a large modern containership. In general, RAOs of ship responses (motions and loads) obtained from the numerical methods compared favorably to model test results. Viscous field methods were considered to be too time consuming and inefficient for the determination of RAOs. However, using wave trains (instead of single regular waves) increased the efficiency of field methods substantially.

The conformity of numerically and experimentally predicted time histories for the ships sailing in extreme waves demonstrated the principle suitability of the numerical methods applied here, namely, the strip theory method and the field method. Corresponding statistical distributions of free surface elevations, heave and pitch motions and vertical bending moments amidships were presented as double amplitudes. Significant wave heights varied between 10.5 and 12.5 m, which meant that maximum wave heights reached values of up to about twice the significant wave height. Each numerical method mentioned above was employed for the cruise ship. As expected, most promising results were obtained with field methods. However, the computational effort greatly exceeded that of potential theory based methods. An underestimation of extreme wave heights did not necessarily yield an underestimation of motions and vertical bending moment as predicted from the exemplarily presented time histories. In compliance with observations from RAOs and from time histories, short-term statistics revealed that motion (double) amplitudes were overestimated by the strip method in comparison with model test results. The vertical bending moment (double) amplitudes, however, agreed favorably. For the Rankine source boundary element method, large ship motion amplitudes, in contrast, were underestimated, whereas the vertical bending moment amidships also agreed favorably to model test results. A comparison of tail distributions, however, was inconclusive as these distributions were affected by the sampling variability of the random sea state realization. Trends described above applied to three ship types, namely, the cruise ship, the chemical tanker and the LNG carrier.

Finally, short-term statistics were presented for the containership. Not only the numerical field method, but also the experiments accounted for hull girder elasticity. Unfiltered and low-pass filtered signals were evaluated. As expected, the number of response cycles increased dramatically due to hull girder vibrations. The maximum response cycle obtained from the unfiltered signal was about 60% larger than the maximum response from the rigid hull signal. Numerically and experimentally determined exceedance rates basically showed the same vibratory amplification. Numerical predictions of unfiltered data compared favorably to experimental measurements. Apart from the tail, the slope for low response levels was well reproduced.

Based on the results presented above, we conclude that the Rankine source boundary element method as well as the nonlinear strip method are suitable to predict small and moderate ship responses. While the boundary element method underestimated extreme ship responses, the nonlinear strip method overestimated these extremes. Accounting for strong nonlinearities associated with impact-related slamming and green water loads in extreme waves required the use of a field method coupled with the nonlinear rigid body equations of motions and the linear equations of elastic body motions.

The computational effort for this method was still high. However, it yielded the most promising ship responses over the entire range of wave conditions, extending from low amplitude regular waves to extremely large and steep irregular waves.

Author Contributions: Conceptualization, J.L. and O.e.M.; writing–original draft preparation, J.L. and O.e.M.; writing–review and editing, J.L. and O.e.M.; visualization, J.L. All authors have read and agreed to the published version of the manuscript.

Funding: The research leading to these results has received funding from European Union's Seventh Framework Programme FP7-SST-2008-RTD-1 under grant agreement No. 234175.

Acknowledgments: Model tests presented in this paper were carried out at Canal de Experencias Hidrodinámicas Del Pardo (CEHIPAR), Madrid, and Technical University of Berlin (TUB), Berlin. The computations with the Rankine Source boundary element method have been carried out by DNV GL, Oslo, the strip theory method was applied at Instituto Superior Técnico (IST), Lisbon. The authors thank all contributors.

Conflicts of Interest: The authors declare no conflict of interest.

References

1. International Union of Marine Insurance. 2016. Available online: http://www.iumi.com/ (accessed on 25 July 2020).
2. Bishop, R.E.D.; Price, W.G. *Hydroelasticity of Ships*; Cambridge University Press: Cambridge, UK, 1979; ISBN 780521017800.
3. Kahl, A.; Menzel, W. Full-Scale Measurements on a PanMax Containership. In Proceedings of the Ship Repair Technology Symposium Proc., Newcastle upon Tyne, UK, 1–2 September 2008; pp. 59–66.
4. Storhaug, G.; Vidic-Perunovic, J.; Rüdinger, F.; Holtsmark, G.; Helmers, J.B.; Gu, X. Springing/Whipping Response of a Large Ocean Going Vessel—A Comparison between Numerical Simulations an Full-Scale Measurements. In Proceedings of the 3rd International Conference on Hydroelasticity in Marine Technology, Oxford, UK, 15–17 September 2003; pp. 117–131.
5. Storhaug, G. Experimental Investigation of Wave Induced Vibrations and Their Effect on the Fatigue Loading of Ships. Ph.D. Thesis, Norwegian University of Science and Technology, Oslo, Norway, 2007.
6. Vidic-Perunovic, J.; Jensen, J.J. Non-Linear Springing Excitation due to a Bidirectional Wave Field. *Mar. Struct.* **2005**, *18*, 332–358. [CrossRef]
7. Hong, S.Y. Wave Induced Loads on Ships Joint Industry Project—II. In *First Year Model Test Report*; Technical Report No. BSPIS503 A-2112-2 (Confidential); MOERI: Daejeon, Korea, 2009.
8. Hong, S.Y. Wave Induced Loads on Ships Joint Industry Project—II. In *Technical Report No. BSPIS503 A-2207-2*; MOERI: Daejeon, Korea, 2010.
9. Hong, S.Y. Wave Induced Loads on Ships Joint Industry Project—III. In *Technical Report No. BSPIS7230-10306-6*; MOERI: Daejeon, Korea, 2013.
10. Hirdaris, S.E.; Miao, S.H.; Price, W.G.; Temarel, P. The Influence of Structural Modelling on the Dynamic Behaviour of a Bulker in Waves. In Proceedings of the 4th International Conference on Hydroelasticity in Marine Technology, Wuxi, China, 9–13 September 2006; Volume 1, pp. 25–33. ISBN 7-118-04728-7.
11. Hirdaris, S.E.; Temarel, P. Hydroelasticity of Ships—Recent Advances and Future Trends. *Proc. IMechE Part M J. Eng. Marit. Environ.* **2009**, *223*, 305–330. [CrossRef]
12. Hirdaris, S.E.; Lee, Y.; Mortola, G.; Incecik, A.; Turan, O.; Hong, S.Y.; Kim, B.W.; Kim, K.H.; Bennett, S.; Miao, S.H.; et al. The Influence of Nonlinearities on the Symmetric Hydrodynamic Response of a 10,000 TEU Container Ship. *Ocean. Eng.* **2016**, *111*, 166–178. [CrossRef]
13. El Moctar, O.; Oberhagemann, J.; Schellin, T.E. Free Surface RANS Method for Hull Girder Springing and Whipping. *SNAME Trans.* **2011**, *119*, 48–66.
14. Oberhagemann, J.; Ley, J.; el Moctar, O. Prediction of Ship Response Statistics in Severe Sea Conditions using RANSE. In Proceedings of the ASME 31th International Conf. on Ocean, Offshore and Arctic Engineering, Rio de Janeiro, Brazil, 1–6 July 2012; pp. 461–468.

15. El Moctar, O.; Ley, J.; Oberhagemann, J.; Schellin, T.E. Nonlinear Computational Methods for Hydroelastic Effects of Ships in Extreme Seas. *Ocean Eng.* **2017**, *130*, 659–673. [CrossRef]
16. Hirdaris, S.E.; Miao, S.H.; Temarel, P. The Effect of Structural Discontinuity on Antisymmetric Response of a Container Ship. In Proceedings of the 5th International Conference on Hydroelasticity in Marine Technology, Southampton, UK, 7–9 September 2009; Volume 1, pp. 57–68. ISBN 9780854329045.
17. International Ship and Offshore Structures Congress. In Proceedings of the 20th International Ship and Offshore Structures Congress, Amsterdam, The Netherlands, 9–14 September 2018; Volume 1.
18. Hirdaris, S.E.; Bai, W.; Dessi, D.; Ergin, A.; Gu, X.; Hermundstad, O.A.; Huijsmans, R.; Iijima, K.; Nielsen, U.D.; Parunov, J.; et al. Loads for Use in the Design of Ships and Offshore Structures. *Ocean Eng.* **2014**, *78*, 131–174. [CrossRef]
19. Newman, J.N. The Theory of Ship Motions. *Adv. Appl. Mech.* **1978**, *18*, 222–283.
20. Faltinsen, O. *Sea Loads on Ships and Offshore Structures*; Cambridge University Press: Cambridge, UK, 1990.
21. Söding, H. Ermittlung der Kentergefahr aus Bewegungssimulationen. *Ship Technol. Res.-Schiffstechnik* **1987**, *34*, 28–39.
22. Jensen, J.J. *Load and Global Response of Ships*; Elsevier Ocean Engineering Book Series; Elsevier: Amsterdam, The Netherlands, 2001; Volume 4.
23. Fonseca, N.; Guedes Soares, C. Time-Domain Analysis of Large-Amplitude Vertical Ship Motions and Wave Loads. *J. Ship Res.* **1998**, *42*, 139–153.
24. Söding, H.; Shigunov, V.; Schellin, T.E.; el Moctar, O. A Rankine Panel Method for Added Resistance of Ships in Waves. *J. Offshore Mech. Arct. Eng.* **2014**, *136*, 031601. [CrossRef]
25. Kim, Y.; Kim, K.-H.; Kim, Y. Springing Analysis of a Seagoing Vessel Using Fully Coupled BEM-FEM in the Time Domain. *Ocean Eng.* **2009**, 785–796. [CrossRef]
26. Sclavounos, P.D. Nonlinear Impulse of Ocean Waves on Floating Bodies. *J. Fluid Mech.* **2012**, *697*, 316–335. [CrossRef]
27. Riesner, M.; Chillcce, G.; el Moctar, O.; Schellin, T.E. Rankine Source Time Domain Method for Nonlinear Ship Motions in Steep Oblique Waves. *Ship Offshore Struct.* **2018**, *14*, 295–308. [CrossRef]
28. Riesner, M.; el Moctar, O. A Time Domain Boundary Element Method for Wave Added Resistance of Ships Taking into Account Viscous Effects. *Ocean Eng.* **2018**, *162*, 290–303. [CrossRef]
29. Shao, Y.L.; Faltinsen, O.M. Numerical Study of the Second-order Wave Loads on a Ship with Forward Speed. In Proceedings of the 26th International Workshop on Water Waves and Floating Bodies, Athens, Greece, 17–20 April 2011.
30. Papanikolaou, A.; Schellin, T.E. A Three-Dimensional Panel Method for Motions and Loads of Ships with Forward Speed. *J. Ship Technol. Res.* **1991**, *39*, 147–156.
31. El Moctar, O.; Brehm, A.; Schellin, T.E. Prediction of Slamming Loads for Ship Structural Design using Potential Flow and RANSE Codes. In Proceedings of the 25th Symposium on Naval Hydrodynamics, St. John's, NL, Cananda, 8–13 August 2004.
32. El Moctar, O.; Schellin, T.; Priebe, T. CFD and FE Methods to Predict Wave Loads and Ship Structure Response. In Proceedings of the 26th Symposium on Naval Hydrodynamics, Rome, Italy, 17–22 September 2006.
33. Oberhagemann, J.; el Moctar, O. Numerical and Experimental Investigations of Whipping and Springing of Ship Structures. *Int. J. Offshore Polar Eng.* **2012**, *22*, 108–114.
34. Oberhagemann, J.; Ley, J.; Shigunov, V.; el Moctar, O. Efficient Approaches for Ship Response Statistics using RANS. In Proceedings of the 22nd International Society of Offshore and Polar Engineers Conference, Rhodos, Greece, 17–23 June 2012.
35. Oberhagemann, J. On Prediction of Wave-Induced Loads and Vibration of Ship Structures with Finite Volume Fluid Dynamic Methods. Ph.D. Thesis, University of Duisburg-Essen, Duisburg, Germany, 2016.
36. Ley, J.; Oberhagemann, J.; Amian, C.; Langer, M.; Shigunov, V.; Rathje, H.; Schellin, T.E. Green Water Loads on a Cruise Ship. In Proceedings of the 32nd International Conference on Offshore Mechanics & Arctic Engineering, OMAE 2013-10132, Nantes, France, 9–14 June 2013.

37. Ley, J.; el Moctar, O.; Oberhagemann, J.; Schellin, T.E. Assessment of Loads and Structural Integrity of Ships in Extreme Seas. In Proceedings of the 30th Symposium on Naval Hydrodynamics, Hobart, Australia, 2–7 November 2014.
38. Ley, J.; el Moctar, O. An Enhanced 1-Way Coupling Method to Predict Elastic Global Hull Girder Loads. In Proceedings of the ASME 2014 33th International Conference on Ocean, Offshore and Arctic Engineering, OMAE, San Francisco, CA, USA, 8–13 June 2014.
39. Paik, K.J.; Carrica, P.M.; Lee, D.; Maki, K. Strongly Coupled Fluid-Structure Interaction Method for Structural Loads on Surface Ships. *J. Ocean Eng.* **2009**, *36*, 1346–1357. [CrossRef]
40. Seng, S.; Jensen, J.J. Slamming Simulations in a Conditional Wave. In Proceedings of the 31st International Conference on Ocean, Offshore and Arctic Engineering, OMAE2012-83310, Rio de Janeiro, Brazil, 1–6 July 2012.
41. Seng, S.; Vincent, I.; Jensen, J. On the Influence of Hull Girder Flexibility on the Wave induced Bending Moments. In Proceedings of the 6th International Conference on Hydroelasticity in Marine Technology, Tokyo, Japan, 19–21 September 2012; pp. 341–553.
42. Craig, M.; Piro, D.; Schambach, L.; Mesa, J.; Kring, D.; Maki, K. A Comparison of Fully-Coupled Hydroelastic Simulation Methods to Predict Slam-Induced Whipping. In Proceedings of the 7th International Conference on Hydroelasticity in Marine Technology, Split, Croatia, 16–19 September 2015.
43. Robert, M.; Monroy, C.; Reliquet, G.; Drouet, A.; Ducoin, A.; Guillerm, P.E.; Ferrant, P. Hydroelastic Response of a Flexible Barge Investigated with a Viscous Flow Solver. In Proceedings of the 7th International Conference on Hydroelasticity in Marine Technology, Split, Croatia, 16–19 September 2015.
44. Kim, Y.; Kim, J.H. Benchmark Study on Motions and Loads of a 6750-TEU Containership. *Ocean Eng.* **2016**, *119*, 262–273. [CrossRef]
45. Baarholm, G.S.; Moan, T. Estimation of Nonlinear Long-Term Extremes of Hull Girder Loads in Ships. *Mar. Struct.* **2000**, *13*, 495–516. [CrossRef]
46. Drummen, I.; Wu, M.; Moan, T. Numerical and Experimental Investigations into the Application of Response-Conditioned Waves for Long-Term Nonlinear Analyses. *J. Mar. Struct.* **2009**. [CrossRef]
47. Rajendran, S.; Fonseca, N.; Guedes Soares, C. *Extreme Seas Del. 4.4: Improvements in IST Non-Linear Strip Theory Method*; Technical Report; Instituto Superior Técnico: Brussels, Belgium, 2012.
48. Rajendran, S.; Guedes Soares, C. Numerical Investigation of the Vertical Response of a Containership in Large Amplitude Waves. *Ocean. Eng.* **2016**, *123*, 440–451. [CrossRef]
49. Rajendran, S.; Fonseca, N.; Guedes Soares, C. Simplified Body Nonlinear Time Domain Calculation of Vertical Ship Motions and Wave Loads in Large Amplitude Waves. *Ocean Eng.* **2015**, *107*, 157–177. [CrossRef]
50. Wang, S.; Zhang, H.D.; Guedes Soares, C. Slamming Occurrence for a Chemical Tanker Advancing in Extreme Waves Modelled with the Nonlinear Schrödinger Equation. *Ocean Eng.* **2016**, *119*, 135–142. [CrossRef]
51. Hui, S. *Extreme Seas Del. 4.1: A Method Based on WASIM to Simulate Ship Responses in Large Amplitude Incident Waves*; Technical Report; Det Norske Veritas: Brussels, Belgium, 2012.
52. Luo, Y.; Vada, T.; Greco, M. Numerical Investigation of Wave-Body Interaction in Shallow Water. In Proceedings of the 33rd International Conference on Ocean, Offshore and Arctic Engineering, OMAE2014-23042, San Francisco, CA, USA, 8–13 June 2014.
53. Pan, Z.; Vada, T.; Han, K. Computation of Wave Added Resistance by Control Surface Integration. In Proceedings of the 35th International Conference on Ocean, Offshore and Arctic Engineering, Busan, Korea, 19–24 June 2016; Paper OMAE2016-54353.
54. Klein, M.; Maron, A.; Clauss, G.; Dudek, M.; Miguel, F. *Extreme Seas Del. 5.4: Specifications of Models Tests*; Technical Report; Technical University of Berlin: Brussels, Belgium, 2012.
55. Gerritsma, J.; Beukelman, W.; Netherlands Ship Research Centre TNO; Shipbuilding Department. *Analysis of the Modified Strip Theory for the Calculation of Ship Motions and Wave Bending Moments*; Nederlands Scheeps-Studiecentrum TNO: Delft, The Netherlands, 1967.
56. Hachmann, D. Calculation of pressures on a ship's hull. *Ship Technol. Res.-Schiffstechnik* **1991**, *38*, 111–133.
57. Maron, A.; Kapsenberg, G. Design of a Ship Model For Hydroelastic Experiments in Waves. *Int. J. Nav. Archit. Ocean. Eng.* **2014**, *6*, 1130–1147. [CrossRef]

58. Pierson, W.; Moskowitz, L. Proposed Spectral Form for Fully Developed Wind Seas Based on the Similarity Theory of S. A. Kitaigorodskii. *J. Geophys. Res.* **1964**, *69*, 5181–5190. [CrossRef]
59. Hasselmann, K. Measurements of Wind-wave Growth and Swell Decay during the Joint North Sea Wave Project (JONSWAP). *Ergänzungsheft 8–12* **1973**, 1–95. Available online: http://resolver.tudelft.nl/uuid:f204e188-13b9-49d8-a6dc-4fb7c20562fc (accessed on 12 January 2021).
60. IACS. Recommendation No. 34 Standard Wave Data. 2001. Available online: http://www.iacs.org.uk/media/2604/rec_34_pdf186.pdf (accessed on 25 July 2020).

Journal of
Marine Science and Engineering

Article

Springing Analysis of a Passenger Ship in Waves

Jeremias Tilander [1,2], Matthew Patey [2] and Spyros Hirdaris [1,*]

1 Maritime Technology Group, Department of Mechanical Engineering, Aalto University, 02150 Espoo, Finland; Jeremias.Tilander@foreship.com
2 Foreship Ltd., 00810 Helsinki, Finland; Matthew.Patey@foreship.com
* Correspondence: spyros.hirdaris@aalto.fi

Received: 1 June 2020; Accepted: 1 July 2020; Published: 5 July 2020

Abstract: Traditionally, the evaluation of global loads experienced by passenger ships has been based on closed-form Classification Society Rule formulae or quasi direct analysis procedures. These approaches do not account for the combined influence of hull flexibility, slenderness, and environmental actions on global dynamic response. This paper presents a procedure for the prediction of the global wave-induced loads of a medium-size passenger ship using a potential flow Flexible Fluid Structure Interaction (FFSI) model. The study compares results from direct long-term hydro-structural computations against Classification Society Rules. It is demonstrated that for the specific vessel under consideration: (a) the elastic contributions of the responses on loads are negligible as springing effects occur outside of the wave energy spectrum, (b) deviations of the order of 28% arise by way of amidships when comparing direct hydrodynamic analysis predictions encompassing IACS UR S11A hog/sag nonlinear correction factors and the longitudinal strength standard, and (c) the interpretation of the wave scatter diagram influences predictions by approximately 20%. Based on these indications, it is recommended that further parametric studies over a range of passenger ship designs could help draw unified conclusions on the total influence of global and local hydrodynamic actions on passenger ship loads and dynamic response.

Keywords: ship dynamics; hydroelasticity of ships; flexible fluid-structure interactions (FFSI); long term wave loads; passenger ships

---

1. Introduction

Over the last ten years, the average size of passenger ships increased by approximately 30%. This trend reflects the demands of the economies of scale and tourist market expectations that consistently exceeded available supply [1]. Modern passenger vessels comprise of large effective superstructures and slender hulls. These unique design features imply that hull flexibility could be important in terms of predicting wave-induced loads [2]. Classification Rules and design procedures with direct application to passenger vessels do not account for the influence of hull flexibility on global loads in waves [3] and the possible influence of hydroelasticity on global response has not been investigated before.

Today, the global strength of these ships is assessed by semi-empirical rules or quasi direct analysis methods [4,5]. In direct analysis procedures, hydrodynamic and structural modeling are usually uncoupled [2,6]. Fluid structure interaction (FSI) is implemented by assuming that the ship is a rigid body that balances on a trochoidal waveform. The hydrodynamic pressures by way of the wetted part of the hull are extracted from a hull panel method and then applied to the finite element analysis (FEA) model. Hydroelasticity bypasses rigid hull assumptions by providing a coupled solution where the ship is treated as a flexible body [7]. Accordingly, the dynamic response in waves can be evaluated by combining structural dynamics and potential flow hydrodynamics [2,8].

The foundations of hydroelasticity theory subject to steady-state or transient wave-induced loads have been established by Bishop, Price and their collaborators [7–12]. Hydroelastic modeling consists of two parts, namely dry- and wet-analysis, and can be applied in two- or three-dimensional FSI domains. The two-dimensional form of the theory can model the dynamic response (motions, distortions, bending moments, shear forces, and torsional moments) of beam-like ships by combining Timoshenko beam or Vlasov beam dynamics with strip theory methods [9]. On the other hand, three-dimensional hydroelasticity applies to both beam-like and non-beam-like structures [7–11]. In this case, dry analysis uses FEA and the fluid actions associated with the distorting wet structure are determined from a panel method. To date, comparisons of two- and three-dimensional predictions have shown good overall agreement for symmetric responses [7–11]. However, in the antisymmetric plane, deviations may emerge due to the existence of large openings and the lack of ability of the beam like theories to model out of plane distortions [8].

Since the early 2000s, significant research efforts have focused on the validation of theoretical predictions by segmented model tests and the development of design procedures or methods that account for linear and weakly nonlinear effects on the dynamic response of slender, beam-like floating structures in waves. For example, Malenica et al. [13] reported on the global hydroelastic response of a barge to impulsive and non-impulsive wave loads. Their numerical model was validated against dedicated model tests for a barge modeled with 12 pontoons interconnected by means of two elastic plates. Shin et al. [14] and Bigot et al. [15] presented satisfactory comparisons of springing and whipping responses against model tests for the case of an ultra-large container ship (ULCS). Derbanne et al. [16] conducted springing tests on a very elastic box-like ship model and symmetric load-induced whipping tests on a passenger ship. Shin et al. [17] and Malenica and Tuitman [18] presented fully coupled hydroelastic models with nonlinear corrections for steady-state and slamming analysis. Once again, their numerical models were applied to a ULCS. Their frequency- and time-domain comparisons revealed that hull flexibility might have a significant influence on container ship loading. Im et al. [19] utilized a fully coupled symmetric springing and whipping assessment to question the influence of hydroelasticity on the fatigue loads experienced by a 19,000 TEU ULCS.

To date, research on the combined influence of irregular waves on a global rigid body or flexible dynamic response has been limited. For example, Rajendran et al. [20] confirmed the significant effects of bow flare variation on the rigid response of a bulker, a container ship and a passenger ship sailing in extreme waves and irregular seaways. Kim and Kim [21] and Jiao et al. [22] predicted extreme loads by time-domain hydroelastic methods. Their work shows that: (a) the combination of rigid and flexible ship dynamics may influence the long-term vertical sagging and hogging moments amidships and (b) Classification Society Rules for ultimate strength assessment may be dependent on sea state variations and the associated cycle times of alternate loads.

There are no publications where three-dimensional hydroelasticity theories have been applied for the prediction of global long-term wave-induced loads on passenger ships. To close this gap in the literature, this paper presents a method for the prediction of springing induced wave loads of a modern medium size passenger ship. The approach is based on the FFSI of Bishop et al. [7] and is applied along the lines of the modeling procedures introduced by Hirdaris et al. [2,8], Malenica et al. [23] and Classification Societies [24–26]. It is noted that the influence of equivalent design wave on total response or slamming induced whipping loads and associated effects on hull stresses is not considered. Instead, the main objective has been to compare itemized steady state, i.e., springing induced loads with Classification Rules and accordingly examine their influence on global long-term ship dynamic response in waves [27–29].

2. Theoretical Background

The principles of linear hydroelasticity theory are broadly discussed in the literature (e.g., [2,6,8,10,23]). This section highlights key theoretical and modeling items of direct relevance to three-dimensional FFSI idealizations.

An overview of the coupling procedure is presented in Figure 1. In 'dry analysis' the ship's structure can be modeled by the finite element method (FEM), and modal analysis effects (natural frequencies, mode shapes, and associated modal distortions) are evaluated in vacuo, i.e., in the absence of any damping effects, for mass and inertia properties corresponding to a typical load case. In 'wet analysis' ship hydrodynamics are modeled within the context of potential flow theory by a frequency domain 'Green function' panel method. Accordingly, hydrodynamic actions are evaluated by way of the wetted hull panels [10,20], and hydrodynamic pressures are integrated on the hull surface. FFSI is enabled by incorporating the influence of flexible ship distortions on hydrodynamic pressures. Structural responses are then separated into their rigid body and hydroelastic counterparts [18]. To predict the long-term loading of the ship, sea states the ship may encounter during her lifetime are considered. Thus, the long-term dynamic response accounting for cumulative short-term responses in sea states is defined by scatter diagrams [29] and transfer function (RAOs) are computed by spectral analysis.

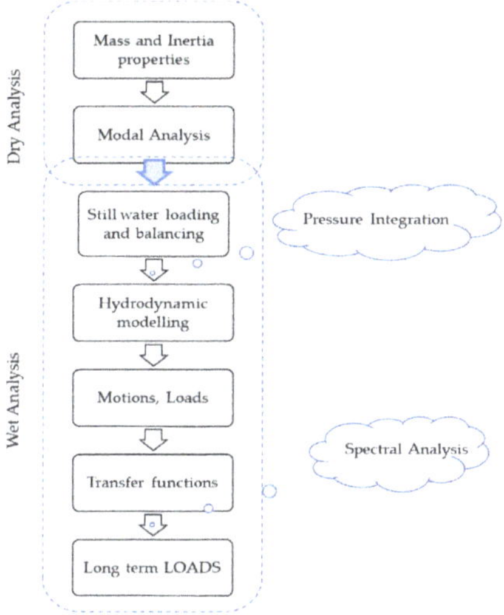

Figure 1. Hydroelastic fluid-structure interaction (FSI)-model analysis.

2.1. Frequency Domain Hydroelastic Seakeeping Model

A fundamental principle of hydroelasticity theory is that it accounts for the influence of hull distortions on dynamic response. These hydroelastic distortions represent the influence of continuous or discontinuous structural deflections, and are incorporated in the FFSI mathematical model according to Newman's generalized modal approach [30]. Within the context of linearity ship displacement is represented as the aggregate of modal displacements according to the Equation:

$$\boldsymbol{H}(x,y,z,\ t)=\sum_{i=1}^{N}\xi_i(t)\boldsymbol{h}^{i}(x,y,z)=\sum_{i=1}^{N}\xi_i(t)\ \left[h_x^i(x,y,z)\boldsymbol{i}+h_y^i(x,y,z)\boldsymbol{j}+h_z^i(x,y,z)\boldsymbol{k}\right] \tag{1}$$

where vector functions $\boldsymbol{h}^i$ (x, y, z) are the modal displacements of the rigid and elastic modes, ξ_i are the corresponding modal amplitudes, and N is the number of modes considered. Implementation of the modal approach leads to additional radiation potentials with the body boundary condition:

$$\frac{\partial \varphi_{Rj}}{\partial n}=\boldsymbol{h}^j\boldsymbol{n} \tag{2}$$

where $\boldsymbol{n}$ is the unit normal vector and $\boldsymbol{h}^j$ denotes the transferred modal displacements of rigid and elastic modes.

Hydrodynamic pressures are calculated after solving the boundary value problem. Integration of the pressures over the wetted body surface results in the Equation:

$$\left\{-\omega_e^2([\boldsymbol{m}]+[\boldsymbol{A}])-i\omega_e([\boldsymbol{B}]+[\boldsymbol{b}])+([\boldsymbol{k}]+[\boldsymbol{C}])\right\}\{\boldsymbol{\xi}\}=\left\{\boldsymbol{F}^{DI}\right\} \tag{3}$$

where ω_e is the wave encounter frequency; $\boldsymbol{m}$ represents the modal structural mass; $\boldsymbol{A}$,$\boldsymbol{B}$ are the hydrodynamic added mass and damping; $\boldsymbol{b}$ and$\boldsymbol{k}$ are the structural damping and stiffness; $\boldsymbol{C}$ is the hydrostatic restoring force; $\boldsymbol{\xi}$ is the modal amplitude; $\boldsymbol{F}^{DI}$ represents the vector of hydrodynamic excitations.

Equation (3) is essentially Newton's 2nd law of motion for the case of a flexible ship moving in waves. It accounts for linear springing and may be used to solve the principal coordinate amplitudes, i.e., modal amplitudes, and the corresponding phase angles for six rigid body motions and n distortion modes in regular waves. These modal amplitudes may then be used to define a corresponding number of modal internal actions, namely loads and stresses [6,31]. For example, the modal vertical bending moment (M_y) along the length of the ship can be defined as:

$$M_y(x,\ t)=\sum_{i=1}^{N}\xi_i(t)M_{yi}(x) \tag{4}$$

where M_{yi} is the modal vertical bending moment for the i_{th} mode of vibration. Respectively, the distribution of total stress can be defined as:

$$\sum(x,y,z,\omega)=\sum_{i=1}^{N}\xi_i(\omega)\sigma^i(x,y,z) \tag{5}$$

where σ^i (x, y, z) is the spatial distribution of the stresses (i.e., the stress tensor matrix) corresponding to each of the rigid and distortion modes. It is important to note that rigid body motions do not contribute to either the internal loads (see Equation (4)) or to the total stress (see Equation (5)). However, the number of considered distortion modes may influence the results. Generally, the first few lowest frequency global distortion modes may be considered enough to capture global responses [19]. Higher frequency dry distortion modes very often become localized, and it is believed that they do not significantly affect the global response [2,8].

2.2. Decomposition of Dynamic and Quasi-Static Responses

The influence of the flexibility or hydroelasticity of the hull can be captured by separating the quasi-static and dynamic part of the structural response through the decomposition method explained in this section.

By definition, the quasi-static part reflects the rigid body response. The dynamic part includes the influence of structural dynamics [18]. Therefore, the coupled dynamic equation can be expressed as:

$$\left(\begin{bmatrix}[\boldsymbol{RR}] & [\boldsymbol{RE}]\\ [\boldsymbol{ER}] & [\boldsymbol{EE}]\end{bmatrix}+\begin{bmatrix}[0] & [0]\\ [0] & [\boldsymbol{k}]\end{bmatrix}\right)\begin{Bmatrix}\xi^R\\ \xi^E\end{Bmatrix}=\begin{Bmatrix}\boldsymbol{F}^R\\ \boldsymbol{F}^E\end{Bmatrix} \tag{6}$$

where $\boldsymbol{R}$ represents the rigid body response and $\boldsymbol{E}$ the elastic structural dynamic response. In turn, the total response amplitudes for the rigid body responses are expressed as:

$$\xi^R=\xi_0^R+\xi_d^R \tag{7}$$

where subscripts *0* and *d* denote the quasi-static and dynamic parts. Similarly, the total response amplitudes for dynamic response are expressed as:

$$\xi^E = \xi_0^E + \xi_d^E \tag{8}$$

The quasi-static part of the response is defined as:

$$[\boldsymbol{RR}]\left\{\xi_0^R\right\} = \left\{\boldsymbol{F^R}\right\} \tag{9}$$

$$[\boldsymbol{k}]\left\{\xi_0^E\right\} = \left\{\boldsymbol{F^E}\right\} - [\boldsymbol{ER}]\left\{\xi_0^R\right\} \tag{10}$$

Substitution of Equations (7)–(10) into Equation (6), leads to the linear system of equations:

$$\left(\begin{bmatrix} [\boldsymbol{RR}] & [\boldsymbol{RE}] \\ [\boldsymbol{ER}] & [\boldsymbol{EE}] \end{bmatrix} + \begin{bmatrix} [0] & [0] \\ [0] & [\boldsymbol{k}] \end{bmatrix}\right)\left\{\begin{array}{c} \xi_d^R \\ \xi_d^E \end{array}\right\} = -\left\{\begin{array}{c} [\boldsymbol{RE}]\xi_0^R \\ [\boldsymbol{EE}]\xi_0^E \end{array}\right\} \tag{11}$$

The decomposition method separates the dynamic part and identifies it as a correction or amplification of the quasi-static part. Thus, hydroelasticity effects (e.g., springing) can be captured and simultaneously separated from the response in order to clearly point out their influence on the total wave-induced loading of the ship.

2.3. Linear Spectral Approach

The spectral analysis helps to compute long-term maxima for different response amplitude operators (RAOs) and hence may be used to describe ship responses in regular waves of unit amplitude. The first order spectral density of the response can be solved according to the Equation:

$$S_R(\omega) = RAO^2(\omega) * S_W(\omega, \beta) \tag{12}$$

where ω and β represent the wave frequency and wave direction, S_R represents the response spectrum, and S_W is the wave spectrum. In turn, spectral moments are used to calculate short-term responses that correspond to a duration (typically 3 h) of one stationary sea state:

$$m_n = \int_0^\infty \omega^n S_R(\omega) d\omega \tag{13}$$

where m_n is the *nth* order spectral moment. Assuming a narrow-banded process, the probability density of the response follows Rayleigh's distribution defined as [32]:

$$p(R) = \frac{R}{4m_0} e^{\left(\frac{-R^2}{8m_0}\right)} \tag{14}$$

where p is the probability density of the response, and R is a random variable that represents the range of the response. Respectively, the cumulative distribution function, P, is defined as [32]:

$$P(R) = 1 - e^{\left(\frac{-R^2}{8m_0}\right)} \tag{15}$$

Long-term responses can be obtained by summing up the results from the short-term analysis. Accordingly, the long-term analysis accounts for the maxima of all responses over all sea states with corresponding probabilities in the form of a wave scatter diagram [32].

2.4. Rule-Based Wave Loads

Traditionally, the vertical wave-induced dynamic response is assumed to play the most significant part in the longitudinal strength of ships [4]. Thus, the Rule envelope curves of the wave-induced vertical bending moments (VBM) and vertical shear forces (VSF) follow the International Association of Class Societies (IACS) Unified Requirements for ship longitudinal Strength assessment (URS) [28]. Antisymmetric dynamic responses (i.e., horizontal bending moments, HBM; torsional moments, TM) follow individual Classification Society Rules and, if applicable, direct analysis procedures suitably backed up by Classification Society Notations (e.g., [27]).

Rule-based wave loads are expressed in empirical formulae, including the ship's general particulars and empirical correction factors that allow for the influence of nonlinear hydrodynamic effects over a vessel's lifetime (i.e., 25 years). For example, see IACS UR S11, UR S11A standards comparisons presented in Table 1. UR S11A Rule formulae include nonlinear correction factors for hogging ($f_{NL\text{-}Hog}$) and sagging ($f_{NL\text{-}Sag}$) conditions. These are dependent on the block and waterplane area coefficients; $f_{NL\text{-}Hog}$ considers the draught of the ship and is not be taken greater than 1.1 while $f_{NL\text{-}Sag}$ accounts for the bow flare shape and is not to be taken less than 1.0 [28]. UR S11 does not include a nonlinear correction factor for hogging [28], and the sagging nonlinear correction factor can be considered as 1.2 maximum [33]. Because the nonlinear correction factors implemented in the latest version of UR S11A are the most universally accepted and were derived by extensive parametric seakeeping analysis studies over a range of modern ship designs, for the sake of completion this paper also presents comparisons against both UR S11 and UR S11A [33].

Table 1. Comparison of VBM rule load formulae according to IACS (L = ship length; B = ship breadth; C_B = ship block coefficient; C_W: water plane area coefficient; f_R = operational factor typically taken as 0.85 [28], C = ship length dependent wave parameter.

	IACS UR S11	IACS UR S11A
Hogging VBM	$190CL^2BC_B10^{-3}$	$1.5f_RL^3CC_W\left(\frac{B}{L}\right)^{0.8}f_{NL-Hog}$
Sagging VBM	$-110CL^2B(C_B+0.7)10^{-3}$	$-1.5f_RL^3CC_W\left(\frac{B}{L}\right)^{0.8}f_{NL-Sag}$

3. Case Study

The analysis presented in this paper is based on a medium-size passenger ship with the principal particulars shown in Table 2. This is a conventional passenger ship that has been professionally designed. The superstructure and the main hull are clearly separated, and key naval architecture information (e.g., hull form, key structural drawings, scantlings, basic CAD model etc.) was made available, thus enabling the development of a suitable FFSI model.

Table 2. Principal particulars of the ship used in the analysis.

Ship particular	Symbol	Value
Length overall (m)	L_{OA}	200
Gross tonnage	GT	41,500
Displacement (tonnes)	Δ	23,000

3.1. Structural Model

The global FEA model of the ship was developed by software FEMAP (see Figure 2). According to the Classification Society's structural strength assessment procedures for ship global strength analysis, a coarse mesh is considered adequate when FEA discretization accounts for the web frame spacing [24,27,34]. Based on these recommendations, the model included all primary longitudinal and transverse structural components contributing to the overall stiffness of the ship structure. It also accounted for ordinary stiffeners and large openings. Laminate elements were used to model the stiffened panels as equivalent shell elements [35]. Orthotropic plate elements were used to model the side shell openings [36]. Based on engineering experience, smaller openings and structures that do not contribute to longitudinal strength were omitted. This approach resulted in 34,545 elements and 16,239 nodes (see Table 3 and Figure 2).

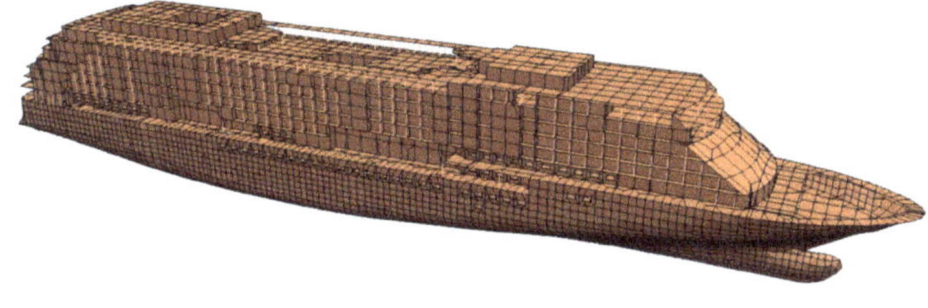

Figure 2. Structural mesh (34,545 elements and 16,239 nodes).

Table 3. FEA discretization

Element Type	Element Quantity
Laminate	17,463
Plate	6103
Beam	10,877
Mass	51
RBE3 (interpolation)	51
Total	34,545

The equivalent laminate elements consisted of three layers, namely: (1) deck plate, (2) stiffener web, and (3) stiffener flange [35]. Such layers have both iso- and orthotropic material properties and axial stiffness in the direction of the stiffeners. An isotropic material has equal Young's modulus and Poisson's ratio in all directions, whereas an orthotropic material has separate stiffness properties in different directions [36]. Accordingly, the deck plate layer was modeled with isotropic material properties, and the stiffener flange and web layers were suitably defined as orthotropic. The laminate elements also have out of plane shear stiffness [35]. Bulkhead structures without ordinary stiffening were modeled with plate elements having isotropic material properties. Plate elements with orthotropic material properties were used to represent superstructure side balcony bulkheads with large openings. The orthotropic plate elements followed the modeling approach presented in [36]. Thus, the side shell openings were homogenized by using equivalent stiffness properties.

Beam elements [37] were used to model the web frames and other primary structural elements such as pillars. These elements were uniaxial with tension, compression, and bending capabilities and associated properties specified for standard or arbitrary shapes. Liquids in tanks were modeled by FEMAP RBE3 elements connected to the tank boundaries' corners via massless strings (i.e., massless line elements) [37]. It is noted that such interpolation elements do not introduce any additional stiffness in the structural model. However, they help to correct the mass distribution of the model when they are attached to the mass elements.

The mass distribution of the FE model was adjusted to correspond to the departure loading condition of the ship. Accordingly, all consumable tanks were considered full and all passengers onboard. The coarse mesh led to the omission of some key masses and thus lowered the overall ship lightweight. To overcome this problem, missing masses were modeled as non-structural mass elements applied on decks. Then the mass distribution was fine-tuned by correcting the shear force along the length of the ship to correspond to that given for the selected loading condition in the loading manual.

3.2. Hydrodynamic Model

A hydrodynamic mesh representing the wetted hull geometry by flat quadrilateral or triangular panels with a normal vector pointing towards the fluid was developed using NAPA software [38].

Along the lines of BV hydro-structure computation guidelines, the model ignored any openings by way of bow thrusters or stabilizer fins [24]. Hydrodynamic discretization accounted for six panels by way of the shortest wavelength encountered by the ship. Based on modeling experience, such idealization may restrict artificial numerical instabilities (e.g., irregular frequency effects). The final mesh was port-starboard symmetric and comprised of 1596 panels per side (see Figure 3). Such a model may be considered adequate for rigid body wave encounter frequencies up to 2.5 rad/s [24].

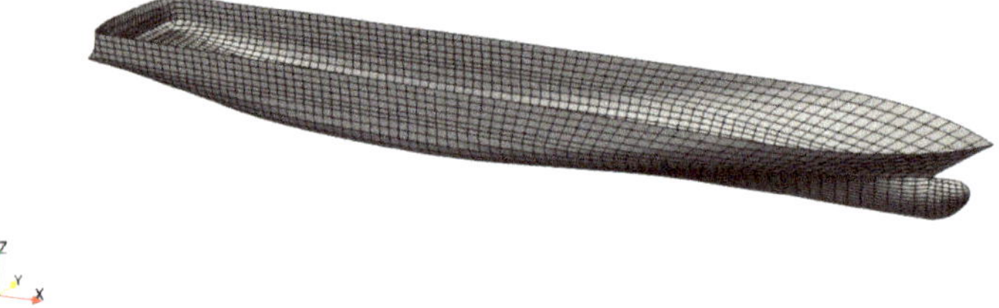

Figure 3. Hydrodynamic mesh (3192 hydrodynamic panels).

Ship balancing under still water conditions resulted in a trim angle of −0.184° and a heel angle of −0.089°. The resulting hull pressures under still water conditions are shown in Figure 4. Inconsistencies were checked by comparing the still water bending moments and shear forces with their counterparts defined in the loading manual of the ship for the loading condition shown in Figure 5.

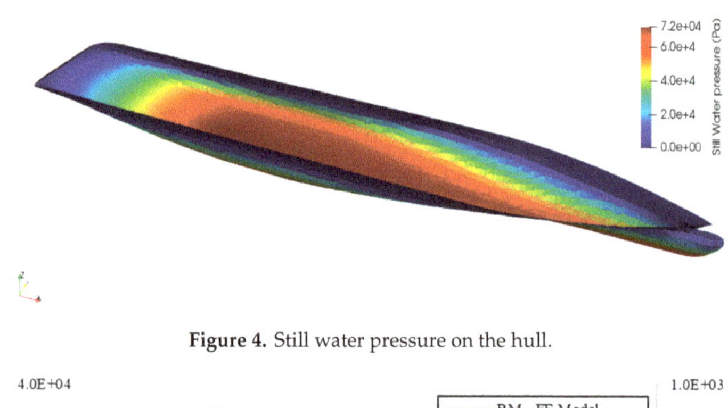

Figure 4. Still water pressure on the hull.

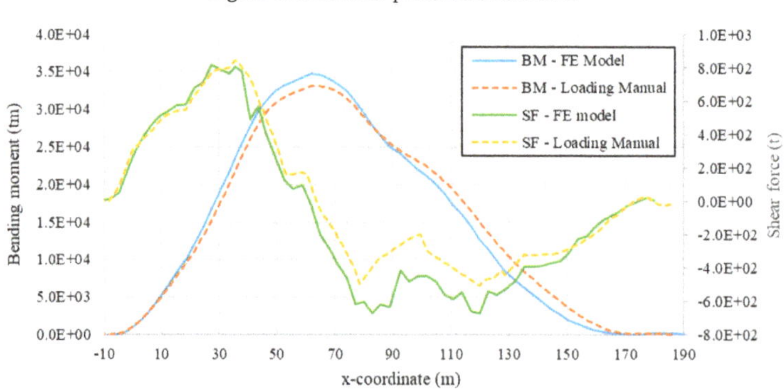

Figure 5. Still water bending moments and shear forces comparison.

4. Results

4.1. Dry Analysis

Six rigid body modes and four distortion modes were considered representative of the rigid body and hydroelasticity effects of the passenger vessel under consideration. Flexible distortions consisted of two symmetric and two antisymmetric modes (see Figure 6). A greater number of distortion modes could have been utilized, but the structural model of the ship introduced spurious higher-order modal behavior already at the fifth dry natural frequency. Based on hydroelastic modeling experience, such modes can be neglected as being non-representative of the real ship modal behavior, and four flexible distortion modes may be considered adequate in terms of accounting for the influence of flexible ship body dynamics on the response [11,24].

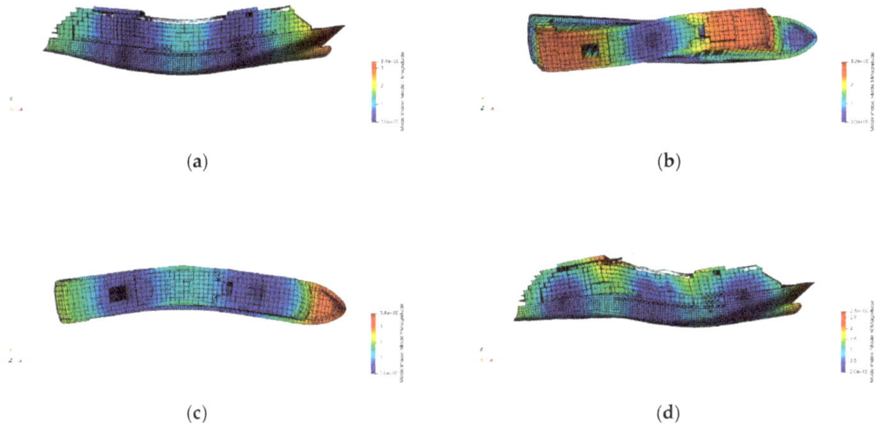

Figure 6. Dry mode shapes and eigenfrequencies. (**a**) 2-node VB mode at 12.56 rad/s (**b**) 1-node T+HB mode at 14.56 rad/s (**c**) 2-node HB+T mode at 14.92 rad/s (**d**) 3-node VB mode at 17.27 rad/s.

4.2. Wet Analysis

Hydrodynamic computations were carried out for zero speed, wave encounter frequencies from 0.0 to 20.0 rad/s in 0.1 rad/s steps (200 frequencies), and wave headings from 0 to 180 deg in 15 deg steps (13 headings in total). The zero-speed condition helped diminish possible numerical instability issues (e.g., irregular frequency effects). Before solving the wet resonance frequencies and the corresponding modal amplitudes, added hydrodynamic and structural damping were defined. The added hydrodynamic damping was defined for the roll motion to include the bilge keel roll damping effect. Based on modeling experience, roll damping was defined at a fraction corresponding to 7% of critical damping and was applied by way of 50% of the wetted hull nodes and 50% of the bilge keel nodes. Structural damping was defined according to Kumai empirical factors [39]. A summary of the dry and wet resonance characteristics is given in Table 4.

Table 4. Dry and wet resonance characteristics.

Mode	Dry Frequency (rad/s)	Wet Frequency (rad/s)	Ratio (Wet/Dry)	Structural Damping
7	12.56	7.94	0.63	0.002
8	14.56	11.75	0.81	0.003
9	14.92	12.79	0.86	0.007
10	17.27	14.29	0.83	0.005

Figure 7 presents the RAOs of the modal amplitude values for regular unit amplitude waves in head and oblique (135 deg.) seas. By way of the first distortion mode, the ship appears to experience significant resonance peaks between 8–12 rad/s. Minor rigid body dominant responses can be found at the lower frequencies of the 1st distortion mode. The second distortion mode resonance peaks occur between 13–14 rad/s. The third distortion mode resonance peaks are between 12.5–14 rad/s. By way of the fourth distortion mode, the ship experiences her first resonance peak at 10 rad/s. The highest resonance peak occurs at 12 rad/s.

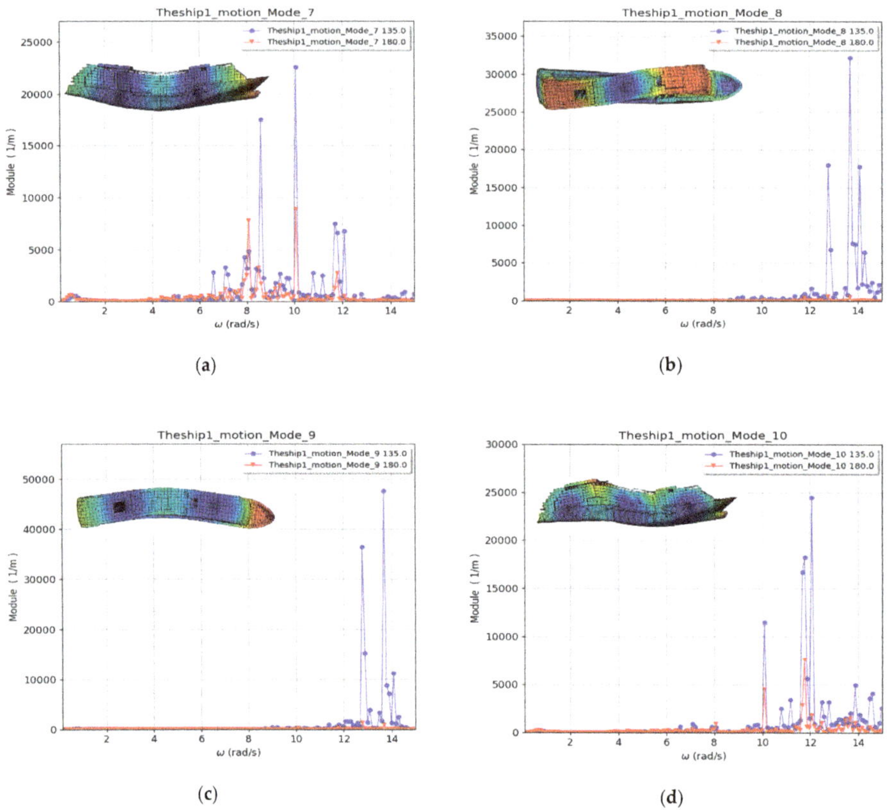

(a) (b) (c) (d)

Figure 7. Modal amplitudes of the distortion modes with the corresponding dry modes. (**a**) 2-node VBM mode at 12.56 rad/s (**b**) 1-node torsion mode at 14.56 rad/s (**c**) 2-node HBM mode at 14.92 rad/s (**d**) 3-node VBM mode at 17.27 rad/s.

4.3. Steady State Load RAOs

The VBM and HBM RAOs at amidships up to a rigid body frequency of 2.5 rad/s are presented in Figure 8. VBM reached a maximum in head waves at 0.6 rad/s. The response in oblique waves also seems to be of considerable magnitude for frequencies up to 1.0 rad/s. On the other hand, the HBM at amidships reached its maximum at 105 deg heading and frequency of 1.3 rad/s. Considerable responses can be found by way of wave headings ranging from 45–135 deg around the frequency of 1.0 rad/s.

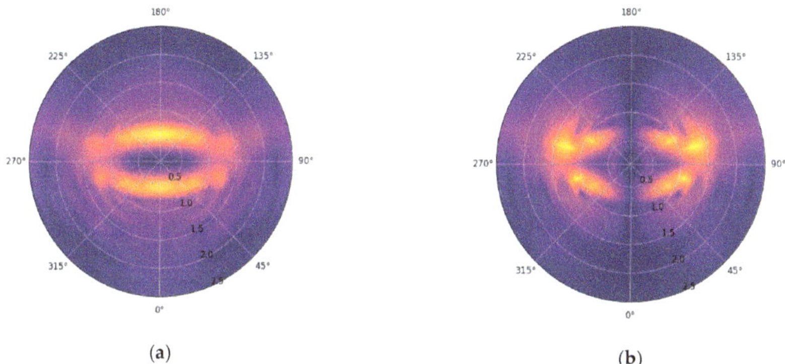

Figure 8. Bending moments at midship. (**a**) Vertical bending moment (**b**) Horizontal bending moment.

The TM RAOs, by way of the first and final quarter lengths of the hull (i.e., 0.25 *L* and 0.75 *L*) are shown in Figure 9. In both locations, they reach a maximum at a wave heading of 105 deg and frequency 1.3 rad/s. Significant torsional responses are also observed for wave headings 60 and 120 deg by way of the frequency of 1.0 rad/s.

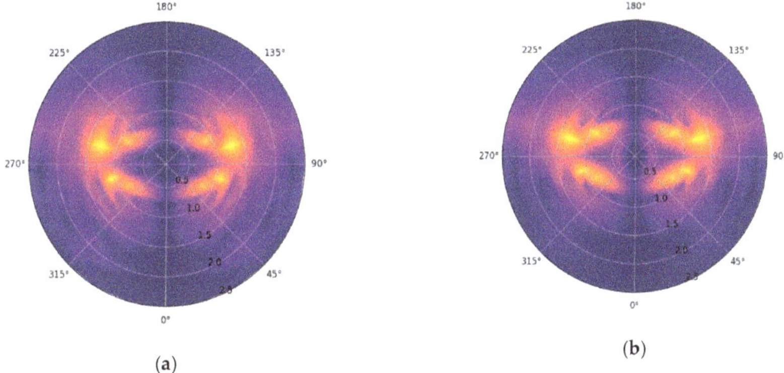

Figure 9. Torsional moment at (**a**) 0.25 *L*, and (**b**) 0.75 *L*.

The VBM in head waves and the HBM in oblique waves at amidships up to a frequency of 15 rad/s are shown in Figure 10. The VBM has three narrow peaks between 8–12 rad/s, and the total response is dominated by the elastic response. The locations of peak responses match very well with the modal amplitude peaks of the first distortion mode (Figure 7). The rigid body dominant response can be seen occurring at lower frequencies below 1.0 rad/s where the total steady-state response is entirely dominated by the quasi-static effects. The HBM has similar resonance peaks occurring at 12.5–14.0 rad/s by way of the same locations as the modal amplitude peaks of the third distortion mode (Figure 7). The total steady-state HBM response is dominated by the elastic contribution. The TM at 0.25 *L* and 0.75 *L* in 135 deg wave heading are shown in Figure 11. Resonance peaks occur between 12.5–14.0 rad/s. The locations of the peaks correspond well to the modal amplitude peaks of the second distortion mode (Figure 7).

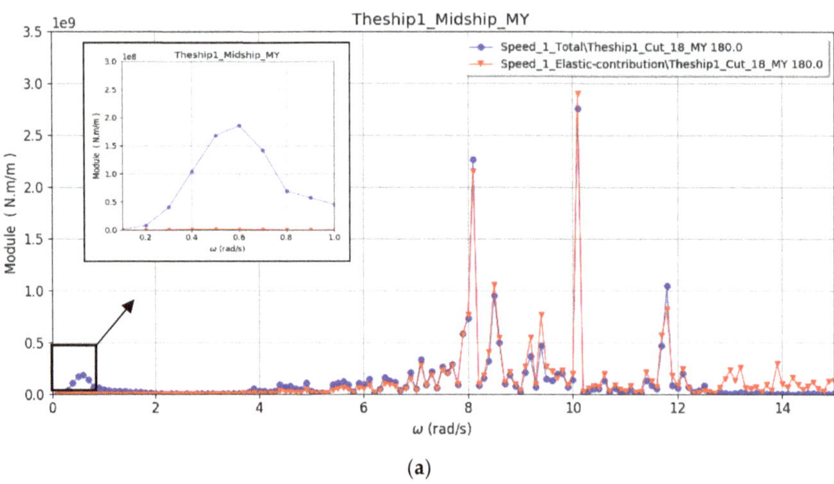

(a)

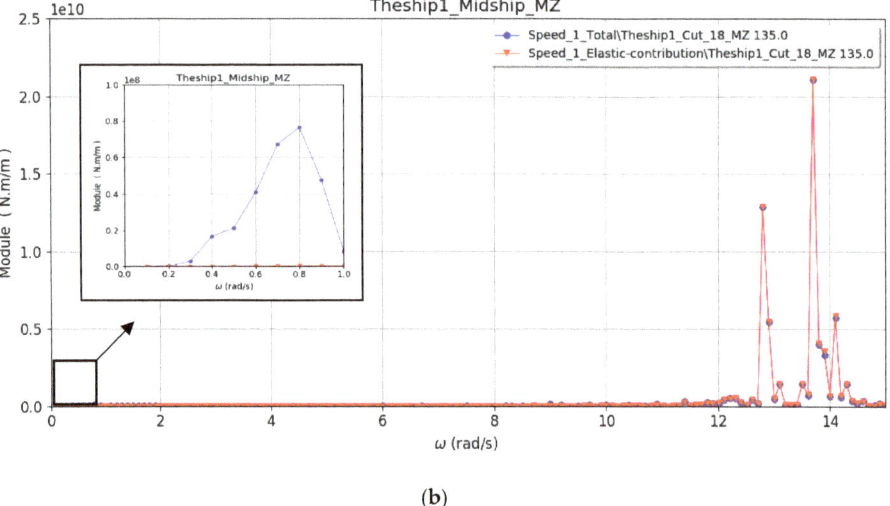

(b)

Figure 10. Bending moments at midship. (**a**) VBM in head waves (**b**) HBM in oblique waves (135 deg.).

4.4. Spectral Analysis

This section presents comparisons of load envelope curves against BV Rules [27] and the IACS longitudinal strength standard requirements [28]. Long-term spectral analysis of the internal load RAOs was conducted by BV STARSPEC [32] and assumed IACS Recommendation 34 assumptions; i.e., (1) North Atlantic wave scatter diagram, (2) Jonswap spectrum with unit peak enhancement function, (3) second-order wave spreading, (4) azimuth angles from 0° to 360° in steps of 5° (equal probability of occurrence), and (4) return period of 25 years (10^{-8} probability level) [40,41].

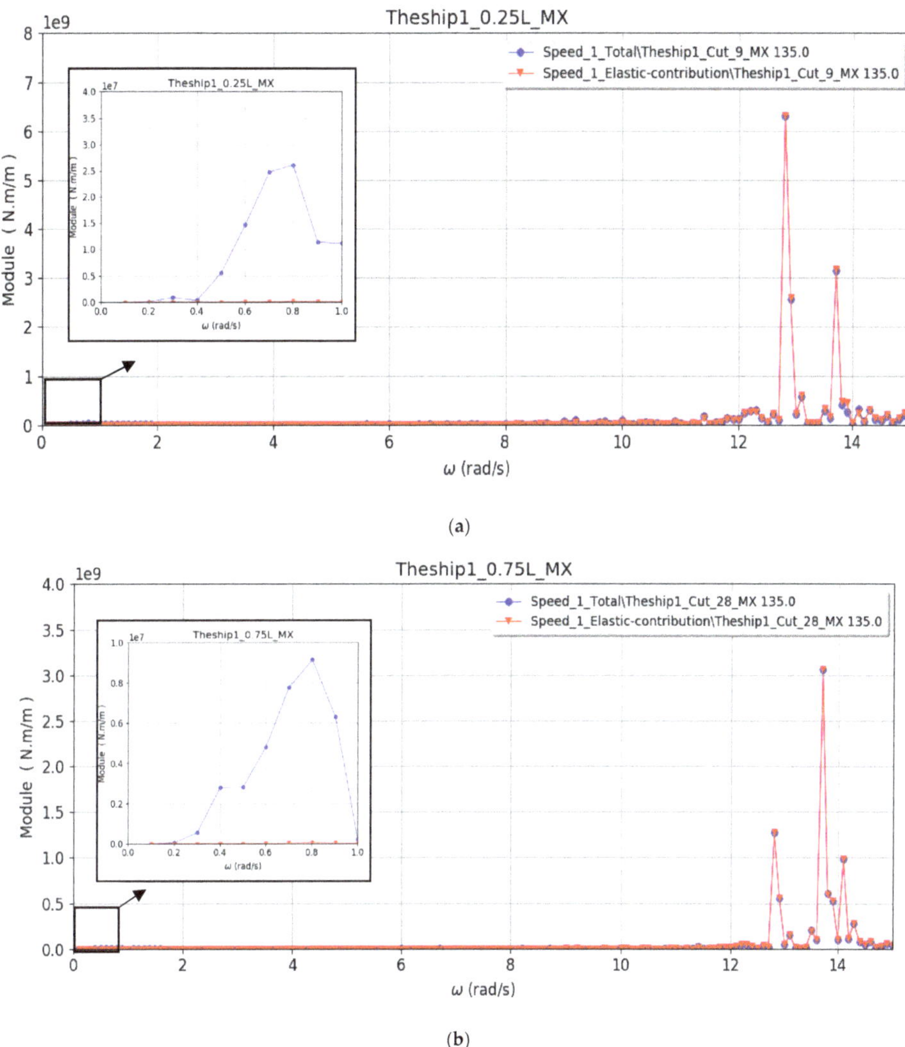

(a)

(b)

Figure 11. Torsional moment in oblique waves at (**a**) 0.25 *L* and (**b**) 0.75 *L*.

According to the decomposition method explained in Section 2.2, the contribution of elastic responses was investigated by using load RAOs containing the elastic part of the response and the total response. The analysis has shown that the influence of hydroelasticity is negligible as the contribution of elastic responses was less than 0.8% for all internal loads (Figure 12). Figure 13a,b demonstrate that symmetric load distributions along the length of the ship follow the trends of Rules [27–30]. Based on the nonlinear correction factors of 1.11 and 0.89 for sagging and hogging forces [27,28], the predicted hogging and sagging VBM distributions exceeded Rule values by up to 62% and 33% respectively at amidships. They also appear reduced by way of the extremities of the hull. The predicted hogging VSF exceeds rule values by 91% at 0.24 *L*, and the sagging shear force exceeded the rule value by 145% at 0.60 *L*. The predicted VSF longitudinal envelopes appear reduced by way of the forward end of the ship and very similar to Rule trends by way of the aft end. HBM and TM distributions follow the BV requirements [27] (see Figure 13c,d). However, the HBM envelope exceeded these Rule values by 20%

at 0.45 L and fell below the Rule envelope margins by way of the extremities of the hull. The predicted TM exceeded the rule values by 11% at 0.26 L.

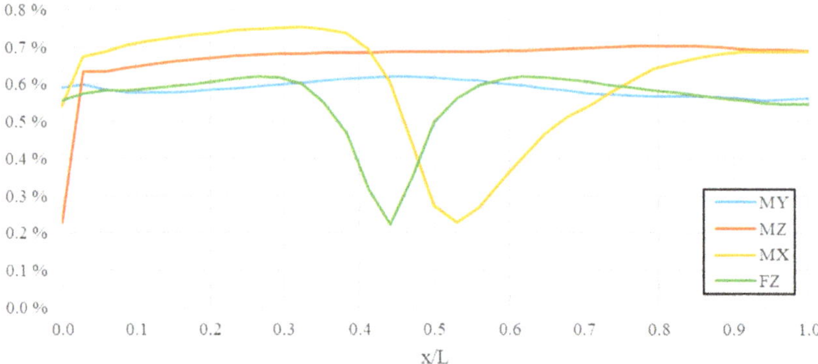

Figure 12. Elastic contribution on internal loads (MY, vertical bending moment; FZ, vertical shear force; MZ, horizontal bending moment; MX, torsional moment).

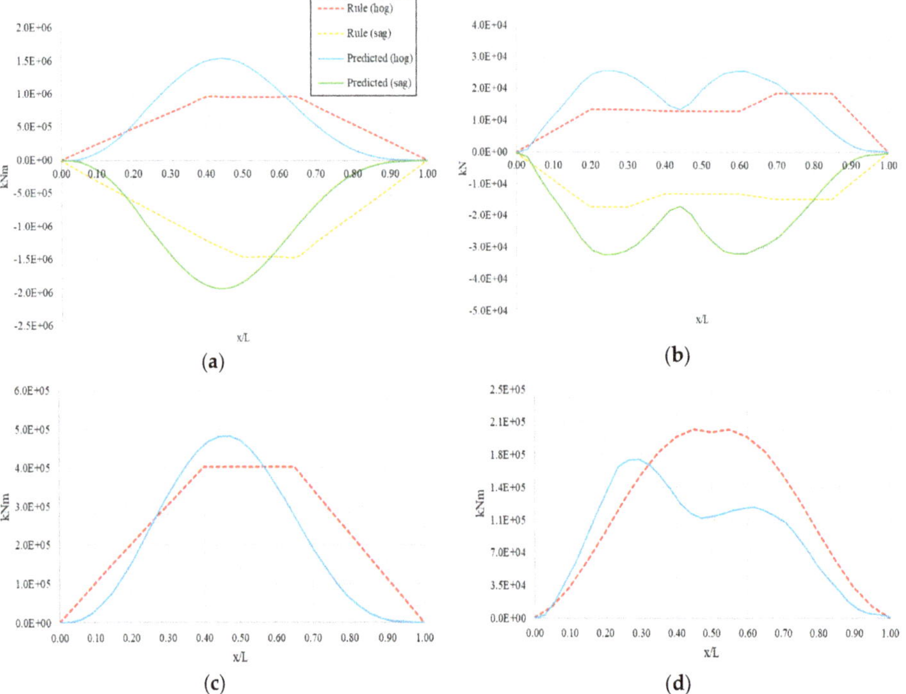

Figure 13. Comparison of load envelope curves between predictions and Rule-based values (10^{-8} probability level, 25 years return period); (**a**) Vertical Bending Moment—VBM; (**b**) Vertical Shear Force VSF; (**c**) Horizontal Bending Moment—HBM; (**d**) Torsional Moment—TM.

Comparisons of nonlinear and linearized long-term predictions of the hogging and sagging VBM and IACS UR S11 and S11A are shown in Figure 14 [28]. According to UR S11A requirements, the nonlinear correction factors were 0.52 for hogging and 1.94 for sagging. Figure 14a shows that predicted hogging and sagging VBM envelope curves exceeded the Rule values by 28% by way of amidships (0.47 *L*). However, towards the extremities of the hull the hogging VBM distribution was well contained by the Rule envelope curve with the predicted sagging envelope slightly shifted by way of the aft end of the ship. Figure 14b illustrates that total steady-state linear predictions in the North Atlantic exceeded amidships IACS UR S11 values by 61% and the renewed UR S11A requirements by 28%. The same figure shows that linear predictions for worldwide operation exceeded IACS UR S11 by 40% and UR S11A by 11%.

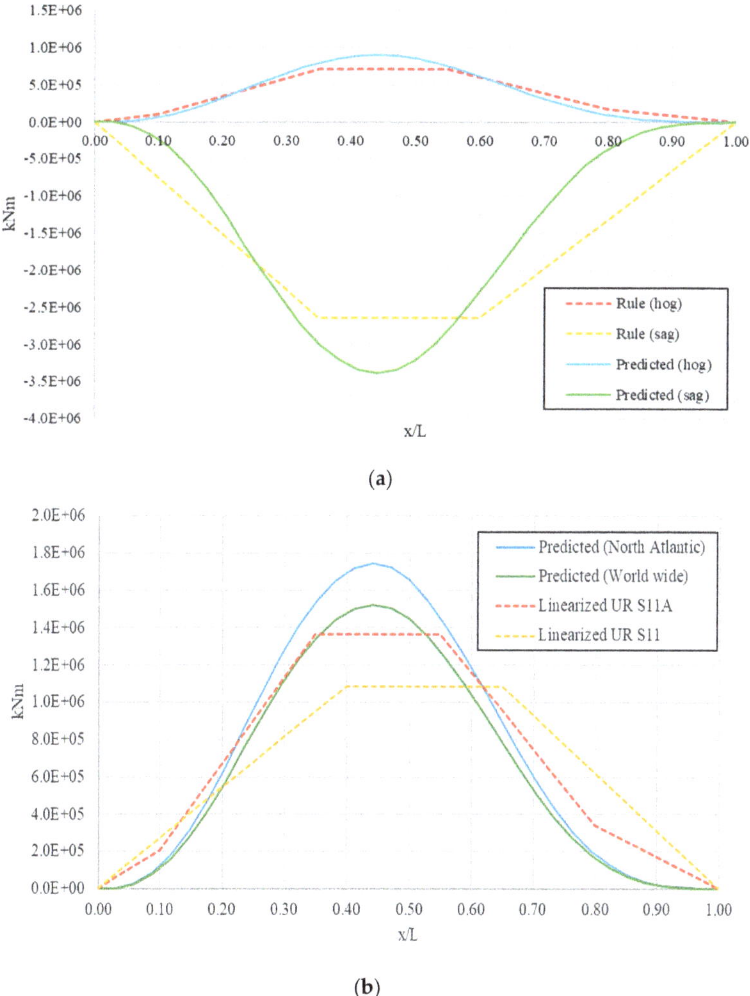

Figure 14. Long term load comparisons of (**a**) nonlinear VBM corrected predictions and IACS UR S11A requirement and (**b**) linear VBM predictions against linearized IACS UR S11 and S11A -requirements (10^{-8} probability level, 25 years return period).

5. Conclusions

The first four global symmetric and antisymmetric distortion modes captured well the influence of dry hull characteristics on steady-state global dynamic response; this agrees with past studies [9,17,42]. It appears that the modal amplitude peaks of the distortion modes shown in Figure 6 are in good agreement with the wet mode frequencies shown in Table 4. This practically means that the ship hull tends to resonate when the wave encounter frequency gets closer to the wet natural frequencies, and once again, the trends observed are similar to the ones presented in previous studies [9,12,17]. However, the elastic contributions of the responses were negligible, as hydroelasticity effects become significant only outside of the wave energy spectrum (see Figure 12).

Comparisons of the long-term direct calculation results and the Rule loads revealed significant discrepancies in VBM and VSF (see Figure 13a,b and 14). Similar load distribution trends may also be found in [43,44]. From an overall perspective, the predictions for antisymmetric responses appear to be in better compliance to the Rule envelope curves, yet differences are still evident (see Figure 14c,d and Section 5). Similar torsional moment distributions observed for the case of ULCS confirm in general the adequacy of results presented in this paper [45].

Comparisons of the VBM envelope curves demonstrated in Figure 14 show that the nonlinear correction factors implemented in IACS UR S11A offer improvements in comparison to IACS UR S11. For the ship under consideration the ratio between the UR S11 and UR S11A rules for hogging is 0.74 and the ratio for sagging was 1.82. However, significant deviations still existed over parts of the ship between the Rule requirements and the direct steady-state hydrodynamic analysis predictions corrected with nonlinear hog/sag correction factors introduced in the rules. For example, maximum deviations of the order of 28% arise by way of amidships when comparing VBM predictions encompassing IACS UR S11A hogging/sagging nonlinear correction factors and the longitudinal strength standard. The interpretation of the wave scatter diagram also influences predictions by approximately 20% (see Figure 14b).

In conclusion, the results presented show that for the vessel under consideration, hydroelasticity does not influence long term global loads. However, the total influence of springing on local stress tensors has not been evaluated. In addition, it is believed that hydroelasticity may have severe influence if the analysis is conducted on confined and detailed portions of the vessel in extreme waves where nonlinear hydrodynamic effects and impact loads may also be significant [22,46]. Accordingly, future work could concentrate more on understanding the combined influence of nonlinear hydrodynamic effects such as large amplitude motions, local slamming loads, and hull whipping on both global and local ship dynamics and strength over a broad range of passenger ship designs.

Author Contributions: Conceptualization, J.T. and S.H.; methodology, J.T., S.H., and M.P.; software, J.T.; validation, S.H. and M.P.; formal analysis, J.T.; investigation, J.T.; resources, J.T. and M.P.; data curation, S.H.; writing—original draft preparation, J.T.; writing, review, and editing, J.T., S.H. and M.P.; visualization, J.T.; supervision, S.H. and M.P.; project administration, J.T.; funding acquisition, J.T. All authors have read and agreed to the published version of the manuscript.

Funding: Part of this research was in-kind funded by Foreship Ltd. Aalto University funded the open-access Journal of Marine Science and Engineering fees.

Acknowledgments: The work presented here forms part of the MSc thesis of J.T. at the Marine Technology Group of Aalto University. Both J.T and M.P acknowledge the in kind support of Foreship Ltd. S.H. acknowledges financial support from the Academy of Finland University competitive funding award (SA Profi 2-T20404).

Conflicts of Interest: The authors declare no conflict of interest.

References

1. ILS. *Shipping Statistics and Market Review, 61-No. 5/6*; The Institute of Shipping Economics and Logistics (ILS): Bremen, Germany, 2017; ISSN 0947022.
2. Hirdaris, S.E.; Temarel, P. Hydroelasticity of ships: Recent advances and future trends. *Proc. Inst. Mech. Eng. Part M J. Eng. Marit. Environ.* **2009**, *223*, 305–330. [CrossRef]
3. Lloyd's Register. *Procedure for Strength Analysis of the Primary Structure of Passenger Ships*; ShipRight Design and Construction, Structural Design Assessment (SDA), Lloyd's Register Marine Business Stream: London, UK, 2017.
4. Bajic, D. Structural design of passenger cruise ships—An introduction to classification requirements. *Ships Offshore Struct.* **2015**, *10*, 232–238. [CrossRef]

5. Nitta, A.; Arai, H.; Magaino, A. Basis of IACS unified longitudinal strength standard. *Mar. Struct.* **1992**, *5*, 1–21. [CrossRef]
6. Malenica, S. Hydrostructure interaction in seakeeping. In Proceedings of the International Workshop on Coupled Methods in Numerical Dynamics, Dubrovnik, Croatia, 19–21 September 2007; pp. 1–25.
7. Bishop, R.E.D.; Price, W.G.; Tam, P.K.Y. A Unified Dynamic Analysis of Ship Response to Waves. *Trans. R. Inst. Naval Archit.* **1977**, *119*, 363–390.
8. Hirdaris, S.E.; Price, W.G.; Temarel, P. Two- and three-dimensional hydroelastic modelling of a bulker in regular waves. *Mar. Struct.* **2003**, *16*, 627–658. [CrossRef]
9. Bishop, R.E.D.; Price, W.G. An introduction to ship hydroelasticity. *J. Sound Vib.* **1983**, *87*, 391–407. [CrossRef]
10. Bishop, R.E.D.; Price, W.G.; Wu, Y. A general linear hydroelasticity theory of floating structures moving in a seaway. *Philos. Trans. R. Soc. Lond. Ser. A Math. Phys. Sci.* **1986**, *316*, 375–426.
11. Price, W.G.; Salas, M.; Temarel, P. The Dynamic Behavior of a Mono-hull in Oblique Waves Using Two- and Three-dimensional Fluid-structure Interaction Models. *Trans. R. Inst. Naval Archit.* **2002**, *144*, 1–26.
12. Basaran, I.; Belik, O.; Temarel, P. Dynamic Behaviour of a Container Ship Using Two- and Three-Dimensional Hydroelasticity Analyses. In Proceedings of the 27th International Conference on Offshore Mechanics and Arctic Engineering, Estoril, Portugal, 15–20 June 2008; Volume 6, pp. 219–228.
13. Malenica, S.; Molin, B.; Remy, F.; Senjanovic, I. Hydroelastic response of a barge to impulsive and non-impulsive wave loads. In Proceedings of the 3rd International Conference on Hydroelasticity in Marine Technology, The Oxford University, Oxford, UK, 15–17 September 2003; pp. 107–116, ISBN 0-952-62081-2.
14. Lee, Y.; White, N.; Wang, Z.; Zhang, S.; Hirdaris, S.E. Comparison of springing and whipping responses of model tests with predicted nonlinear hydroelastic analyses. *Int. J. Offshore Polar Eng.* **2012**, *22*, 1–8.
15. Bigot, F.; Derbanne, Q.; Sireta, F.X.; Malenica, S.; Tuitman, J.T. Global hydroelastic ship response comparison of numerical model and WILS model tests. In Proceedings of the 21st International Offshore and Polar Engineering Conference, Maui, HI, USA, 19–24 June 2011; ISOPE-I-11-138.
16. Derbanne, Q.; Malenica, Š.; Tuitman, J.T.; Bigot, F.; Chen, X.B. Validation of the global hydroelastic model for springing & whipping of ships. In Proceedings of the 11th International Symposium on Practical Design of Ships and Other Floating Structures, Rio de Janeiro, Brazil, 19–24 September 2010; pp. 331–340, ISBN 9788528501407.
17. Shin, K.; Jo, J.; Hirdaris, S.E.; Jeong, S.; Park, J.B.; Lin, F.; Wang, Z.; White, N. Two- and three-dimensional springing analysis of a 16,000 TEU container ship in regular waves. *Ships Offshore Struct. Loads Ships Offshore Struct.* **2015**, *10*, 498–509.
18. Malenica, S.; Tuitman, J.T. 3DFEM-3DBEM Model for springing and whipping analyses of ships. In Proceedings of the International Conference on Design and Operation of Container Ships, The Royal Institution of Naval Architects, London, UK, 3–4 July 2008; pp. 13–22, ISBN 9781510885578.
19. Im, H.I.; Vladimir, N.; Malenica, S.; Cho, D.S. Hydroelastic response of 19,000 TEU class ultra large container ship with novel mobile deckhouse for maximizing cargo capacity. *Int. J. Naval Archit. Ocean Eng.* **2017**, *9*, 339–349. [CrossRef]
20. Rajendran, S.; Vasquez, G.; Soares, C.G. Effect of bow flare on the vertical ship responses in abnormal waves and extreme seas. *Ocean Eng.* **2016**, *124*, 49–436. [CrossRef]
21. Kim, J.H.; Kim, Y. Prediction of extreme loads on ultra-large containerships with structural hydroelasticity. *J. Mar. Sci. Technol.* **2018**, *23*, 253–266. [CrossRef]
22. Jiao, J.; Jiang, Y.; Zhang, H.; Li, C.; Chen, C. Predictions of Ship Extreme Hydroelastic Load Responses in Harsh Irregular Waves and Hull Girder Ultimate Strength Assessment. *Appl. Sci.* **2019**, *9*, 240. [CrossRef]
23. Malenica, S.; Derbanne, Q. Hydro-structural issues in the design of ultra large container ships. *Int. J. Naval Archit. Ocean Eng.* **2014**, *6*, 983–999. [CrossRef]
24. BV. *Homer 2.1.9 User Guide*; Bureau Veritas VeriSTAR: Paris, France, 2019.
25. BV. *NR 583. Whipping and Springing Assessment*; Bureau Veritas: Paris, France, 2015; Available online: http://erules.veristar.com/dy/data/bv/pdf/583-NR_2015-07.pdf (accessed on 3 July 2020).
26. Lloyd's Register. *Global Design Loads of Container Ships and Other Ships Prone to Whipping and Springing*; ShipRight Design and Construction—Structural Design Assessment (SDA), Lloyd's Register Marine Business Stream: London, UK, 2018; Available online: https://www.lr.org/en/shipright-procedures/#accordion-structuraldesignassessment(sda) (accessed on 3 July 2020).

27. BV. *Rules for the Classification of Steel Ships. Part B—Hull and Stability. Chapter 5 Design Loads, Section 2 Hull Girder Loads*; Bureau Veritas: Paris, France, 2019.
28. IACS. Longitudinal Strength Standard S11. 2015. Available online: http://www.iacs.org.uk/publications/unified-requirements/ur-s/ (accessed on 3 July 2020).
29. BV. *NI 638. Guidance for Long-term Hydro-structure Calculations*; Bureau Veritas VeriSTAR: Paris, France, 2019; Available online: http://erules.veristar.com/dy/data/bv/pdf/638-NI_2019-02.pdf (accessed on 3 July 2020).
30. Newman, J.N. Wave effects on deformable bodies. *Appl. Ocean Res.* **1994**, *16*, 47–59. [CrossRef]
31. Harding, R.D.; Hirdaris, S.E.; Miao, S.H.; Pittilo, M.; Temarel, P. Use of Hydroelasticity analysis in design. In Proceedings of the 4th International Conference on Hydroelasticity in Marine Technology, Wuxi, China, 10–14 September 2006; pp. 1–12.
32. BV. *StarSpec User Guide*; Bureau Veritas VeriSTAR: Paris, France, 2016.
33. Derbanne, Q.; Storhaug, G.; Shigunov, V.; Xie, G.; Zheng, G. Rule formulation of vertical hull girder wave loads based on direct computation. In Proceedings of the 13th International Symposium on Practical design of Ships and other Floating Structures, Copenhagen, Denmark, 4–8 September 2016; ISBN 9788774754732.
34. DNV GL. *Classification Guideline CG-0127. Finite Element Analysis*; DNV GL AS Rules and Standards: Oslo, Norway, 2016; Available online: https://rules.dnvgl.com/docs/pdf/DNVGL/CG/2016-02/DNVGL-CG-0127.pdf (accessed on 3 July 2020).
35. Avi, E.; Lillemäe, I.; Romanoff, J.; Niemelä, A. Equivalent shell element for ship structural design. *Ships Offshore Struct.* **2015**, *10*, 239–255. [CrossRef]
36. Kaldoja, M. Modeling of Ship's Side Shell Openings in Global Finite Element Models. Master's Thesis, Aalto University, Espoo, Finland, 25 September 2017. Available online: http://urn.fi/URN:NBN:fi:aalto-201710307320 (accessed on 3 July 2020).
37. *Siemens. Femap User Guide*; Version 2019.1; Siemens Product Lifecycle Management Software Inc.: Plano, TX, USA, 2019.
38. *NAPA Manuals*; Release 2019.2; NAPA: Helsinki, Finland, 2019.
39. Kumai, T. Damping factors in higher modes of vibrations. *Eur. Shipbuild. Prog.* **1958**, *111*, 29–34. [CrossRef]
40. IACS. Recommendation No.34 on Standard Wave Data. 2001. Available online: http://www.iacs.org.uk/publications/recommendations/21-40/ (accessed on 3 July 2020).
41. DNV GL. *Classification Guideline CG-0130. Wave Loads*; DNV GL AS Rules and Standards: Oslo, Norway, 2018; Available online: https://rules.dnvgl.com/docs/pdf/DNVGL/CG/2018-01/DNVGL-CG-0130.pdf (accessed on 3 July 2020).
42. Malenica, S.; Tuitman, J.T.; Bigot, F.; Sireta, F.X. Some aspects of 3D linear hydroelastic models of springing. In Proceedings of the 8th International Conference on Hydrodynamics, Nantes, France, 30 September–3 October 2008; pp. 529–538.
43. Parunov, J.; Senjanovi, I.; Paviaeeviae, M. Use of Vertical Wave Bending Moments from Hydrodynamic Analysis in Design of Oil Tankers. *Trans. R. Inst. Naval Archit.* **2004**, *146*, 10. [CrossRef]
44. Jensen, J.J.; Pedersen, P.; Shi, B.; Wang, S.; Petricic, M.; Mansour, A. Wave induced extreme hull girder loads on containerships. *Trans. Soc. Naval Archit. Mar. Eng.* **2009**, *116*, 128–151.
45. Rörup, J.; Rathje, H.; Schellin, T. Status of Analysis Methods for Hull Girder Torsion of Containerships. In Proceedings of the 4th PAAMES and AMEC Conference, Singapore, 6–8 December 2010; pp. 1–6.
46. Bennett, S.S.; Hudson, D.A.; Temarel, P. The influence of forward speed on ship motions in abnormal waves: Experimental measurements and numerical predictions. *J. Fluids Struct.* **2013**, *39*, 154–172. [CrossRef]

Article

3D Numerical Simulations of Green Water Impact on Forward-Speed Wigley Hull Using Open Source Codes

Linfeng Chen [1], Yitao Wang [2], Xueliang Wang [2,*] and Xueshen Cao [1]

[1] School of Naval Architecture and Ocean Engineering, Jiangsu University of Science and Technology, Zhenjiang 212000, China; chenlinfeng@just.edu.cn (L.C.); 182010010@stu.just.edu.cn (X.C.)
[2] China Ship Scientific Research Center, Wuxi 214082, China; wangyt@cssrc.com.cn
* Correspondence: wangxl@cssrc.com.cn

Received: 19 March 2020; Accepted: 28 April 2020; Published: 6 May 2020

Abstract: A series of CFD RANS simulations are presented for Wigley hulls of two freeboard heights progressing with forward speed in waves. Free surface effects are captured using the Volume of Fluid (VOF) method embedded in open source software OpenFOAM. Comparisons of heave, pitch motions and added resistance of the first Wigley model against the experiments of Kashiwagi (2013) confirm the numerical validity of the hydrodynamic modelling approach. Further simulations for the lower-freeboard Wigley model reveal that the highest green water impact on decks appears in way of $\lambda/L = 1.3$ and at the highest instantaneous pitch amplitude where the water propagates far downstream and across the deck. The simulations also demonstrate that the green water events are associated with air bubble entrapment.

Keywords: ship motions; green water impact; computational fluid dynamics; volume of fluid method; RANS

1. Introduction

Under harsh sea conditions, vessels or offshore platforms endure large wave loads, such as bottom and bow slamming, green water on decks, etc. Severe slamming may lead to serious structural damage and may affect the safety and stability of the vessel or offshore platform. Green water may not pose a direct threat to integrity or survivability but may result in increasing expenses for repairs and renewal of damaged topside modules or outfitting equipment [1,2].

Over the years the development of non linear numerical methods for the assessment of risks associated with impact of green water on decks has been challenging. Measuring freeboard exceedance or free surface elevation relative to the deck using 3D linear diffraction theory has been extensively applied for the prediction and statistical analysis of green water risk due to its relative simplicity [3–5]. In reality, a green water event is highly nonlinear. This implies that linear potential flow theory and the assumption of a Rayleigh distribution for the statistics may be inaccurate [3,6–9].

Another method used is the so-called composite method. In this numerical idealisation the green water event is divided into four different stages and each of them is modeled separately using distinct methodologies. The freeboard exceedance is used as input for a dam-break model [10–14] or as a solution to the shallow water equations [15,16] that can simulate hydrodynamic behaviour of on-deck flows. The green water loading on deck structures is approximated either analytically by the similarity method [17] and Wagner-type analysis [18] or empirically by utilizing extensive scale model tests [3]. Computational Fluid Dynamics (CFD) is then adopted to idealise the influence of local fluid flow in way of the localized regions around the bow or to simulate local interactions between green water flows and deck structures [3,19,20]. Uncertainties associated with modeling assumptions constrain the accuracy of this method that, in any case, could be considered more suitable for preliminary design stage.

With the successful development of high performance computing, CFD simulations using the volume of fluid (VOF) method may be considered as a valuable alternative [2,21–28]. However, to capture the nonlinear evolution of wave groups and wave-structure interactions, high mesh resolution is required in such simulations. For example, for the simulation of green water events, finer mesh is required to accommodate for the influence of shallow water hydrodynamics on deck.

This paper presents a series of CFD RANS simulations for Wigley hulls of two freeboard heights progressing with forward speed in waves. Free surface effects are captured using the Volume of Fluid (VOF) method embedded in open source software OpenFOAM. Heave, pitch motions and added resistance of the first Wigley hull model are compared against the experiments of Kashiwagi (2013) [29]. Pressures on the deck bow of the lower-freeboard Wigley hull model are measured and flow observations of green water are carried out. The study provides complementary understanding of green water dynamics on deck and key physics of the green water events.

2. Mathematical Model and Numerical Method

The flow is governed by the incompressible Navier–Stokes equations

$$\frac{\partial u_i}{\partial x_i} = 0, \tag{1}$$

$$\frac{\partial(\rho u_i)}{\partial t} + u_j\frac{\partial(\rho u_i)}{\partial x_j} = -\frac{\partial p}{\partial x_i} + \mu\frac{\partial^2 u_i}{\partial x_j \partial x_j} + g, \tag{2}$$

where u_i ($i = 1, 2, 3$) denotes flow velocity at the x-, y- and z-axes respectively, ρ represents the mixture density of the two phases separately considered, p denotes pressure, μ is dynamic viscosity of the fluid, g represents the gravitational acceleration.

A Reynolds averaging approach is adopted to smear out the fluctuating components of the solution variables in the instantaneous Navier–Stokes equations so as to alleviate the computational cost. Averaging the continuity and Navier–Stokes equations yields

$$\frac{\partial \bar{u}_i}{\partial x_i} = 0, \tag{3}$$

$$\frac{\partial(\rho \bar{u}_i)}{\partial t} + \bar{u}_j\frac{\partial(\rho \bar{u}_i)}{\partial x_j} = -\frac{\partial \bar{p}}{\partial x_i} + \mu\frac{\partial^2 \bar{u}_i}{\partial x_j \partial x_j} - \frac{\partial(\rho\overline{u_i'u_j'})}{\partial x_j} + g, \tag{4}$$

where $\bar{u}_i$ and $\bar{p}$ represent the ensemble-averaged velocity and pressure; the product $\rho\overline{u_i'u_j'}$ denotes the Reynolds stress component. Thereafter Reynolds-averaged Navier–Stokes (RANS) equations govern the transport of flow. The Reynolds stress in Equation (4) is modeled as a function of turbulent viscosity μ_T and the mean velocity gradients according to Boussinesq hypothesis

$$-\rho\overline{u_i'u_j'} = \mu_T\left(\frac{\partial \bar{u}_i}{\partial x_j} + \frac{\partial \bar{u}_j}{\partial x_i}\right) - \frac{2}{3}\rho k\delta_{ij}, \tag{5}$$

where k is is the turbulent kinetic energy and δ_{ij} is the Kronecker delta.

To obtain the turbulent viscosity, the shear-stress transport (SST) k-ω two-equation model proposed by Menter [30] is used as follows

$$\frac{\partial(\rho k)}{\partial t} + \frac{\partial(\rho k \bar{u}_i)}{\partial x_i} = \frac{\partial}{\partial x_i}\left[(\mu + \mu_T)\frac{\partial k}{\partial x_i}\right] + \min(\mu_T S^2, 10\beta^* k\omega) - \rho\beta^* k\omega, \tag{6}$$

$$\frac{\partial(\rho \omega)}{\partial t} + \frac{\partial(\rho \omega \bar{u}_i)}{\partial x_i} = \frac{\partial}{\partial x_i}\left[(\mu + \mu_T)\frac{\partial \omega}{\partial x_i}\right] + \alpha\rho S^2 - \rho\beta^*\omega^2 + 2(1 - F_1)\rho\sigma_{\omega,2}\frac{\partial k}{\partial x_i}\frac{\partial \omega}{\partial x_i}, \tag{7}$$

where ω represents the specific dissipation rate,

$$\mu_T = \frac{\rho k}{\omega} / \max\left[\frac{1}{\alpha^*}, \frac{SF_2}{\alpha_1 \omega}\right], \tag{8}$$

and F_1 and F_2 are the blending functions [30].

The volume of fluid (VOF) method with artificial bounded compression technique proposed by Hirt and Nichols (1981) [31] was used to capture the free surface. The free surface was considered as a mixture of water and air and accordingly the VOF transport equation was expressed as

$$\frac{\partial \alpha}{\partial t} + \nabla \cdot (\boldsymbol{U}_r (1-\alpha)\alpha) = 0, \tag{9}$$

where $\boldsymbol{U}_r = \boldsymbol{U}_{water} - \boldsymbol{U}_{air}$ is the relative velocity between the water and air and termed as "compression velocity"; α represents the volume fraction defined as the relative volume proportion of water in a cell, namely

$$\begin{cases} \alpha = 0 & \text{air} \\ 0 < \alpha < 1 & \text{interface} \\ \alpha = 1 & \text{water} \end{cases} \tag{10}$$

The spatial variation of fluid density and dynamic viscosity can be then expressed as

$$\rho = \alpha \rho_{water} + (1-\alpha)\rho_{air}, \tag{11}$$

$$\mu = \alpha \mu_{water} + (1-\alpha)\mu_{air}, \tag{12}$$

where the subscripts "water" and "air" represent the corresponding fluid property of water and air respectively.

Rigid body motion equations, in which all other motions are constrained apart from heave and pitch, were used to model the dynamic response of the ship as follows

$$m\ddot{z} = F_z, \tag{13}$$

$$I_{yy}\ddot{\theta} = N, \tag{14}$$

where m denotes the ship mass; z is displacement in heave; F_z represents the total force in the vertical direction; I_{yy} denotes moment of inertia in pitch; θ represents angle in pitch, N is moment of force in pitch. The Newmark-β method [32] with $\gamma = 0.25$ and $\beta = 0.5$ was used to solve Equations (13) and (14).

Mesh deformation was achieved by moving the node points of the mesh according to the ship motion and without changing the mesh topology. The displacements of the node points were then calculated by using the equation

$$\nabla \cdot (\frac{1}{r} D_i) = 0 \tag{15}$$

where D_i represents the displacements of the node points; r denotes the distance of the cell centre to the nearest moving wall boundary.

3. Problem Setup

At first instance Equation (9) was solved for a fraction of volume and then integrated across the entire flow field. Then, the Reynolds-averaged Navier–Stokes equations using the SST $k-\omega$ model were solved by the OpenFOAM interFoam subroutine as follows

$$\frac{2y}{B} = \left[1-\left(\frac{z}{d}\right)^2\right]\left[1-\left(\frac{2x}{L}\right)^2\right]\left[1+0.6\left(\frac{2x}{L}\right)^2+\left(\frac{2x}{L}\right)^4\right]+\left(\frac{z}{d}\right)^2\left[1-\left(\frac{z}{d}\right)^8\right]\left[1-\left(\frac{2x}{L}\right)^2\right]^4, \tag{16}$$

where B denotes the waterline breadth; L denotes the length between perpendiculars; d represents the draft. To observe the green water occurrence, two blunt Wigley hull models with different freeboard heights were considered (see Figure 1 and Table 1).

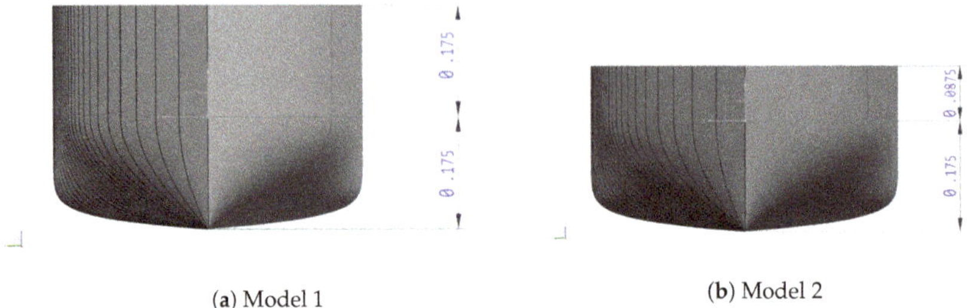

(a) Model 1 (b) Model 2

Figure 1. Forward views of the blunt Wigley hull models.

Table 1. Principle parameters of the ship model.

Length between perpendiculars, L (m)	2.5
Waterline breadth, B (m)	0.5
Draft, d (m)	0.175
Displacement volume, V (m^3)	0.139
Height of gravity center, KG (m)	0.145
Gyration radius in pitch, k_{yy}/L	0.236
Freeboard height, f (m)	0.175 (model 1), 0.0875 (model 2)

A top view of the layout of the computational domain according to Chen et al. (2019) [2] is shown in Figure 2. The first-order Stokes regular head wave was generated in the wave tank for Froude number (Fn) of 0.2. Assuming that the wave propagates along the negative x-axis direction, the wave elevation ζ, the encounter frequency ω_e and the incident wave potential φ were written as

$$\zeta = A cos(kx + \omega_e t), \tag{17}$$

$$\omega_e = \omega + kU, \tag{18}$$

$$\varphi = \frac{gA}{2\omega}\frac{cosh[k(z+H)]}{sinh(kH)}(kx + \omega_e t), \tag{19}$$

the velocity components of the water were then defined as follows

$$u_x(x,y,z,t) = U + A\omega_e e^{kz} cos(kx - \omega_e t + \theta), \tag{20}$$

$$u_y(x,y,z,t) = 0, \tag{21}$$

$$u_z(x,y,z,t) = A\omega_e e^{kz} sin(kx - \omega_e t + \theta), \tag{22}$$

where A represents the wave amplitude; k is the wave number; ω denotes the natural frequency; U represents the ship speed; H is the water depth.

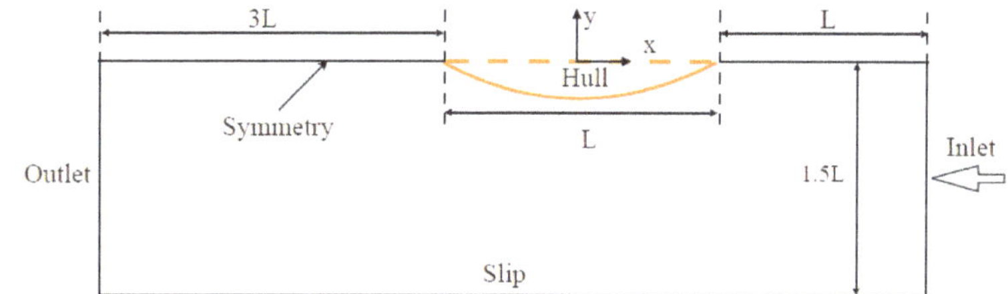

Figure 2. Top view of the layout ($z = 0$) of the computational domain with $-1.5L < x < 3.5L$, $0 < y < 1.5L$, $-1.5L < z < 0.5L$. The direction of x axis is opposite to the incoming flow, y axis is parallel to the free surface, the direction of z axis is opposite to the gravity.

For all cases, the free surface was set at $z = 0$ with the draft $d = 0.175$. Equations (20)–(22) were then used to define the boundary condition at the inlet. At the outlet, a damping zone of wave length λ was set to damp the vertical motions of an interface by applying a force ($d(mu_z)/dt = -\kappa m u_z$) to the momentum equation proportional to the momentum of the flow in the direction of gravity. Fluid damping effects lead to exponential decay of the vertical velocity ($u_z = u_{z0}e^{-\kappa t}$), where κ is set based on the desired level of damping and the residence time of a perturbation through the damping zone. At the mid plane of the hull, a symmetry boundary condition was used.

Figure 3. Top and side view of the mesh for the computational domain. Upper one: top view; lower one: side view.

Unstructured mesh including hexahedral and tetrahedral cells was generated using snappyHexMesh tool of OpenFOAM. Top view and side view of the mesh in the computational domain are shown in Figure 3. Refinements with 5 different levels were used to refine the mesh around the hull and free surface. The finest mesh size was set to satisfy $\lambda/\Delta x > 250$ in the longitudinal direction and $d/\Delta z = 20$ in the vertical direction. In addition, five layers of thin cells were added for the boundary layer around the ship hull. The first layer of cells around the ship hull was set to satisfy the nondimensional thickness of the cell $\eta^+ \approx 1000$, where $\eta^+ = U\eta/\nu$ (U is the ship speed, η denotes the thickness of the first layer of cell,ν represents kinematic viscosity of the fluid). Meanwhile, standard wall functions were applied to computing ν_T at the first off-wall nodes ($\nu_T = \mu_T/\rho$ represents turbulence kinematic viscosity).

4. Results and Discussions

4.1. Code Validation

To carry out a validation for the wave model, the same wave parameters to Kashiwagi (2013) [29] were used (see Table 2). Seven cases with different wave lengths were then implemented for model 1. The nondimensional total resistance of the hull was defined as

$$C_t = \frac{\overline{R}_{wave}}{\rho g A^2 B^2 / L}, \tag{23}$$

and was computed using 1.51, 2.33 and 3.02 million cells of meshes and three time steps (Δt) to test the grid and time step convergence (see Figure 4). In Equation (23) $\overline{R}_{wave}$ denotes the mean resistance of the hull in waves. The figure shows the convergence and a mesh with about 2.33 million cells and $\Delta t = 0.00025$ was adopted for all the further simulations.

Table 2. Wave parameters for the simulations.

Froude Number, Fr	Wave Amplitude, A/L	Wave Length, λ/L	Wave Steepness, H/λ	Encounter Frequency, ω_e
		0.5	0.048	11.998
		0.7	0.0343	9.49
		0.9	0.02667	7.996
0.2	0.012	1.1	0.0218	6.994
		1.3	0.01846	6.267
		1.6	0.015	5.48
		2.0	0.012	4.75

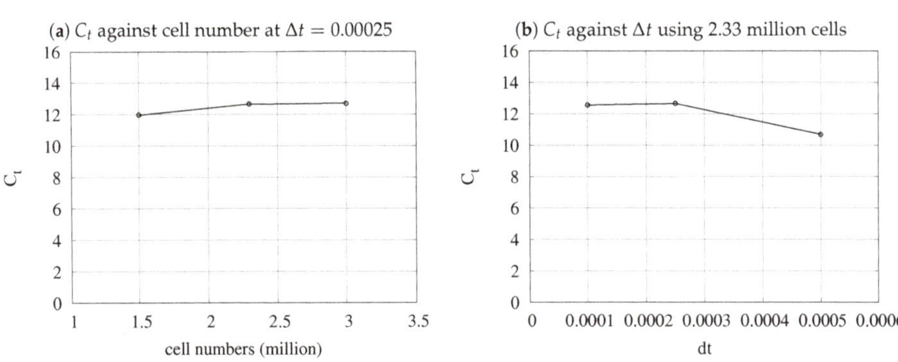

Figure 4. Total resistance (C_t) computation of model 1 at $\lambda/L = 1.1$ for testing grid and time step convergence.

In the simulation of each case, 15 encounter periods ($T_e = 2\pi\omega_e$) of computation, which takes about 50 CPU hours of high performance computing with eight Intel Xeon Gold processors, were undertaken and 10 periods of data were used to calculate the results of the wave-induced motion and resistance after the system reached a steadily periodic state. Time histories of heave and pitch were expanded into Fourier series

$$\zeta(t) = \frac{a_0}{2} + \sum_{n=1}^{N} a_n cos(n\omega_e t + \theta_n) \tag{24}$$

where a_n is the nth harmonic amplitude and θ_n is the corresponding phase angle. The nth dimensionless harmonic amplitudes of the heave ($\zeta_3^{(n)} = a_3^{(n)}/A$) and pitch ($\zeta_5^{(n)} = a_5^{(n)}/(kA)$) were then obtained using Fourier transformation. Profiles of the first harmonic amplitude of heave $\zeta_3^{(1)}$ over a range of dimensionless wave length (λ/L) are shown in Figure 5. Experimental data, computed results by Enhanced Unified Theory (EUT), New Strip Method (NSM) and Rankine Panel Method (RPM) originally introduced by Kashiwagi (1995 [33], 2013 [29]), Iwashita and Ito (1998) [34] are also included in this comparison. Both EUT and RPM are frequency-domain linear calculation methods, in which the disturbed flow is divided into the steady (ϕ_s) and unsteady (ϕ_u) flows. In EUT, the steady disturbance potential ϕ_s is ignored. In RPM it is computed numerically as double-body flow and its effects on the body and free-surface boundary conditions are considered. In NSM hydrodynamic actions on and motions of the ship are determined by integrating the results from two-dimensional hydrodynamic coefficients over the ship length. As shown in Figure 5, EUT and NSM are not in good agreement with the experimental results. In general comparisons against RPM are are good with the exception of $\lambda/L = 1.3$.

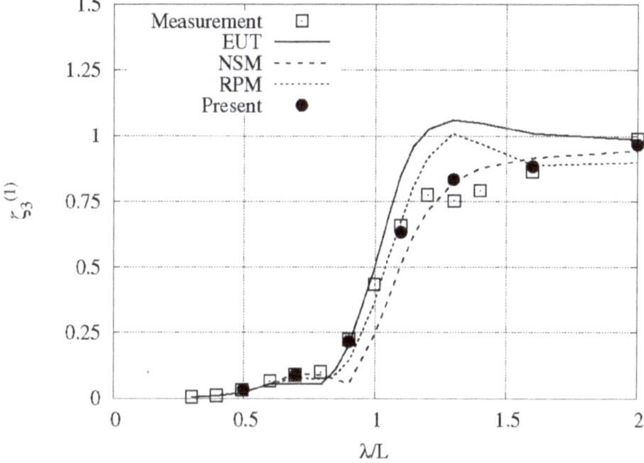

Figure 5. The first harmonic amplitude of the heave against the wave length.

Variations of the first harmonic amplitude of pitch $\zeta_5^{(1)}$ with the wave length are shown in Figure 6. Both EUT and NSM results do not coincide well with the experiments and visible underpredictions appear at large λ/L. Results obtained by the RPM model are less accurate than those obtained by the novel CFD method presented in this paper for $\lambda/L < 1.3$.

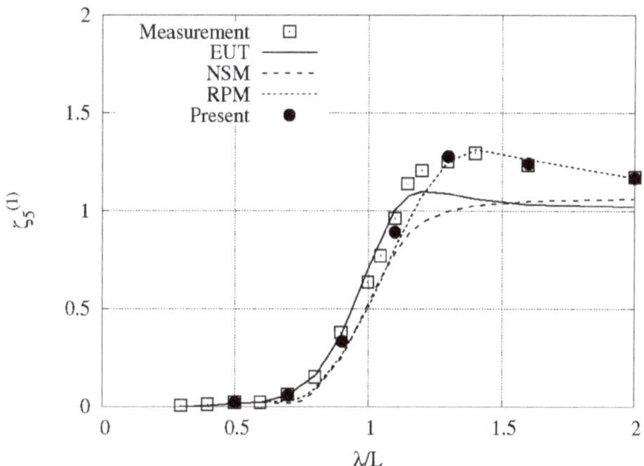

Figure 6. The first harmonic amplitude of the pitch against the wave length.

Added resistance is defined as a difference between the mean (time-averaged) resistance in waves and in calm water at the same speed U. The nondimensional added resistance was defined as

$$C_{aw} = \frac{\overline{R}_{wave} - \overline{R}_{cw}}{\rho g A^2 B^2 / L}, \tag{25}$$

where $\overline{R}_{wave}$ and $\overline{R}_{cw}$ represent the mean resistance of the hull in waves and in calm water respectively; ρ is the density of fluid; g is the gravitational acceleration; A is the wave amplitude; B and L are waterline breadth and length between perpendiculars of the ship respectively. Comparisons of the distributions of the nondimensional added resistance obtained by experiments, EUT and CFD are shown in Figure 7. It is shown that EUT underpredicts added resistance. CFD results are generally in better agreement with experiment. However, differences at low λ/L become more obvious.

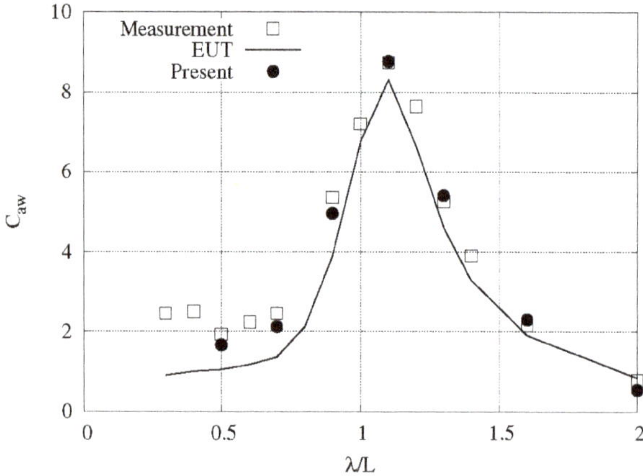

Figure 7. Added resistance of model 1 versus λ/L.

4.2. Green Water on Decks

To observe the green water events, simulations associated with seven wave lengths were implemented in Model 2 (see Figure 1). The impact of green water on decks was examined by measuring the pressure at five monitoring points with coordinates (1.25, 0), (1.1, 0), (1.0, 0), (0.9, 0) and (0.8, 0) on the deck along the centreline of the bow (see Figure 8).

Figure 8. Monitoring points (Point 1 to 5 starting from the bow) on the deck for the pressure measurement.

In green water events free surface exceeds the ship freeboard and hydrodynamic flow propagates over the deck with high-velocity. In way of impact the fluid flow pressure sharply increases. Figure 9 shows the variation of time histories of pressure coefficients over nondimensional time (t/T_e) in way of points 1 and 2 (see Figure 8). For each case the pressure coefficient was defined as

$$C_p = \frac{p}{0.5\rho U^2 LB}. \quad (26)$$

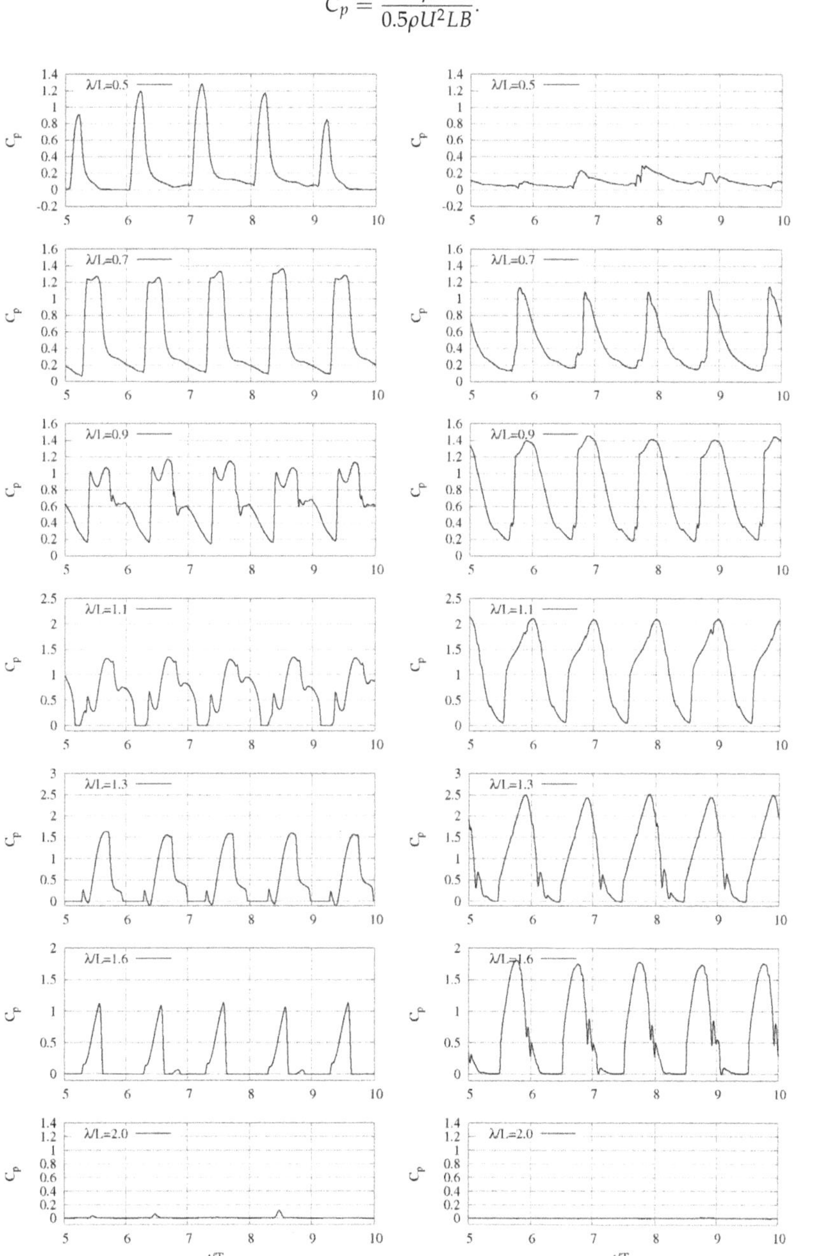

Figure 9. Time histories of the pressure coefficients at point 1 (left column) and point 2 (right column).

It was found that the green water impact at point 1 increased from $\lambda/L = 0.5$ to 0.7 and decreased from $\lambda/L = 0.7$ to 0.9. On the other hand, the pressure at point 2 is larger than that in way of point 1 at $\lambda/L = 0.9$. This is because at $\lambda/L = 0.9$ green water becomes stronger and the water flow over the freeboard rapidly propagates downstream while an air layer is formed underneath. Eventually, an air bubble is entrapped inside the on-deck water. The pressure at point 2 increases from $\lambda/L = 0.9$ to 1.3 and reaches its highest value at $\lambda/L = 1.3$. This implies that the highest influence of green water on decks occurs at $\lambda/L = 1.3$. The green water event then starts to decline until it disappears at $\lambda/L = 2.0$. Occurrence of the strongest green water impact is associated with the highest 1st harmonic amplitudes of the pitch.

Figure 10 shows the pressure coefficient distributions over the time at five points for $\lambda/L = 0.9$ and 1.3. The profiles show that for $\lambda/L = 0.9$ pressures at all the monitoring points are smaller than those of $\lambda/L = 1.3$. This confirms that the green water at $\lambda/L = 0.9$ is relatively weaker than at $\lambda/L = 1.3$. This demonstrates that while a large amount of water exceeds the freeboard and keeps propagating along the deck, pressures remain high at the five monitored points. The time lag in pressure peaks experienced between point 1 and points 2 to 5 reflects that water first impacts point 1, then impacts point 2 as it climbs up the freeboard. The water rapidly spreads on deck (points 2–5) without major pressure fluctuations.

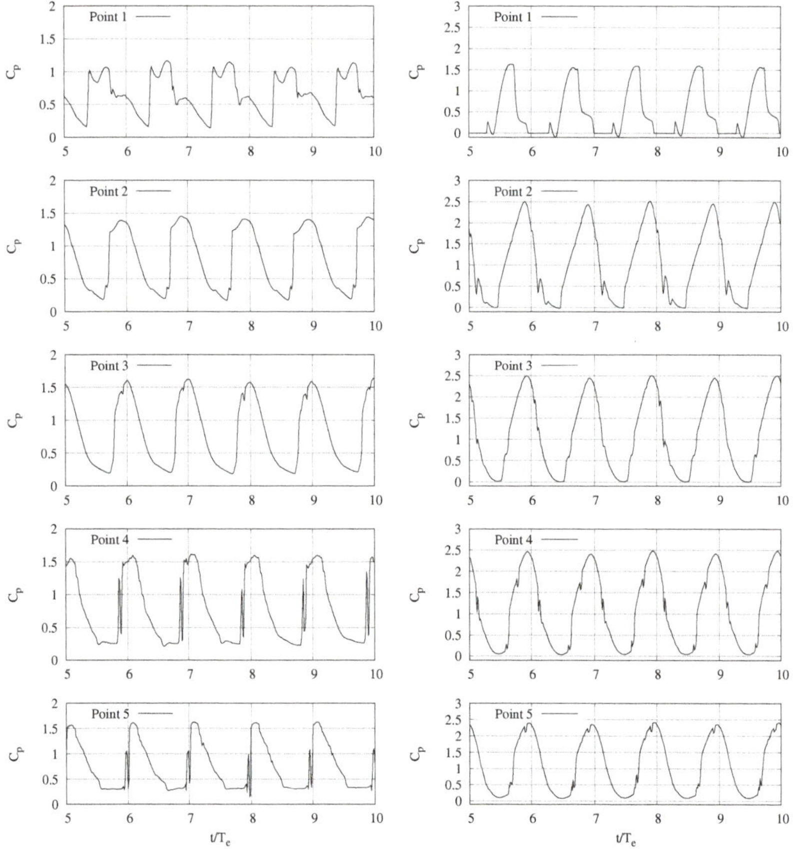

Figure 10. Time histories of the pressure at five monitoring points for $\lambda/L = 0.9$ (left column) and 1.3 (right column).

4.3. Flow Visualization

Flow visualization was achieved by plotting velocity contours on the free surface and pressure contours in way of the x-z plane at $y = 0$ at seven instants in one motion period corresponding to $\lambda/L = 1.1$. Free surface was defined at $\alpha = 0.5$. As shown in Figures 11and 12 on the top left side of each flow velocity contour, the pressure profile at point 1 is attached to highlight the pressure magnitude. At first instance, the hull is almost at a horizontal state and the pressure at point 1 is low (see Figure 11a,b). At the second instant, a trim by the bow starts and wave-structure interaction causes climb up of the water above the bow height (see Figure 11c,d). Consequently, water impact with high-velocity flow causes sharp increase of pressure at point 1. At the third time instant, it can be seen that the pressure at point 1 decreases (see Figure 11e,f). This is because the water flow has very high velocity. As the vessel trims by the bow, a low pressure air layer grows between the water and the deck (see Figure 11f). At the fourth time instant, a trim by the aft starts and the wave-structure interaction becomes stronger (see Figure 11g,h). This causes the pressure at point 1 to reach its highest magnitude. When the trim by aft starts, the water over the deck rolls down and entraps the air underneath to form an air bubble in a form of a low-pressure zone on the deck bow (see Figure 11h). At the fifth time instant, the air bubble propagates downstream along the deck and air bubbles are formed downstream (see Figure 12a,b). At the sixth time instant, the hull trims by the aft, the bow is no longer submerged in waves and the on-deck water is starting to decline, thus leading to decrease of the pressure at point 1 (see Figure 12c,d). At the seventh instant, the volume of water at the deck bow continues to decline and some of on-deck water propagates downstream on the deck. This leads to lower pressure at the other four monitoring points (see Figure 12e,f). The free surface ahead of the bow is falling and the pressure at point 1 approaches its lowest magnitude.

4.4. Motion Quantities and Added Resistance

During the green water event, the wetted surface also decrease at the moment of occurrence of a trim by aft. This may lead to changes in the hydrodynamic forces on the hull. The green water impact would also have an effect on the hydrodynamic forces on the hull, variations in motions and added resistance. Figure 13 illustrates variations of heave, pitch and added resistance for both Wigley models (see Figure 1). It may be concluded that because of the higher free-board height of Model 2 in comparison to Model 1 the impact of green water on decks is low and does not influence that much the first harmonic heave/pitch amplitudes at $\lambda/L = 0.5$, 0.7 and 1.6. For both models from $\lambda/L = 0.9$ to 1.3, the green water event remains strong, thus both the decrease of the wetted surface and the green water impact cause decreases of the first harmonic amplitudes of the heave and pitch as well as added resistance.

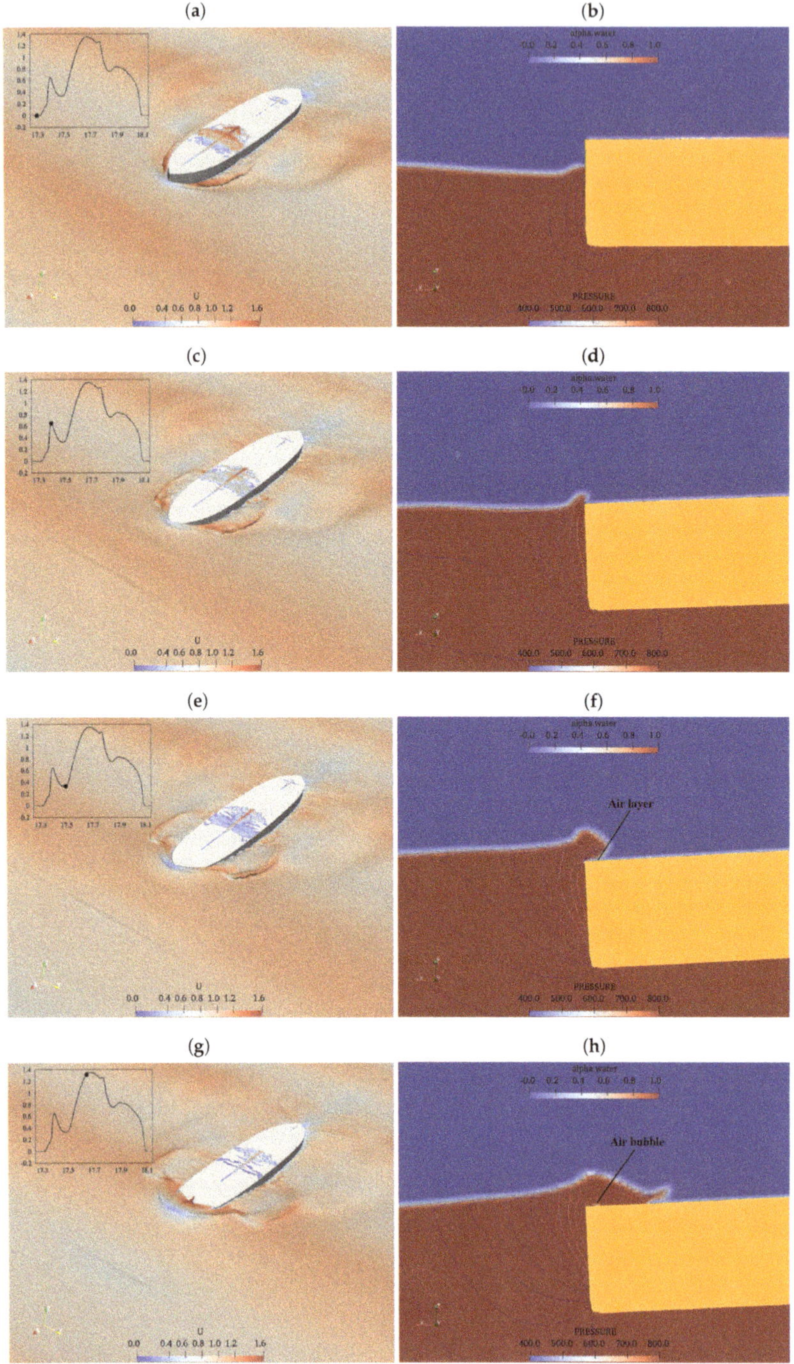

Figure 11. Flow velocity contours on the free surface (**a**,**c**,**e**,**g**) and pressure contours on the x–z plane at $y = 0$ (**b**,**d**,**f**,**h**) at time instants 1–4 for $\lambda/L = 1.1$.

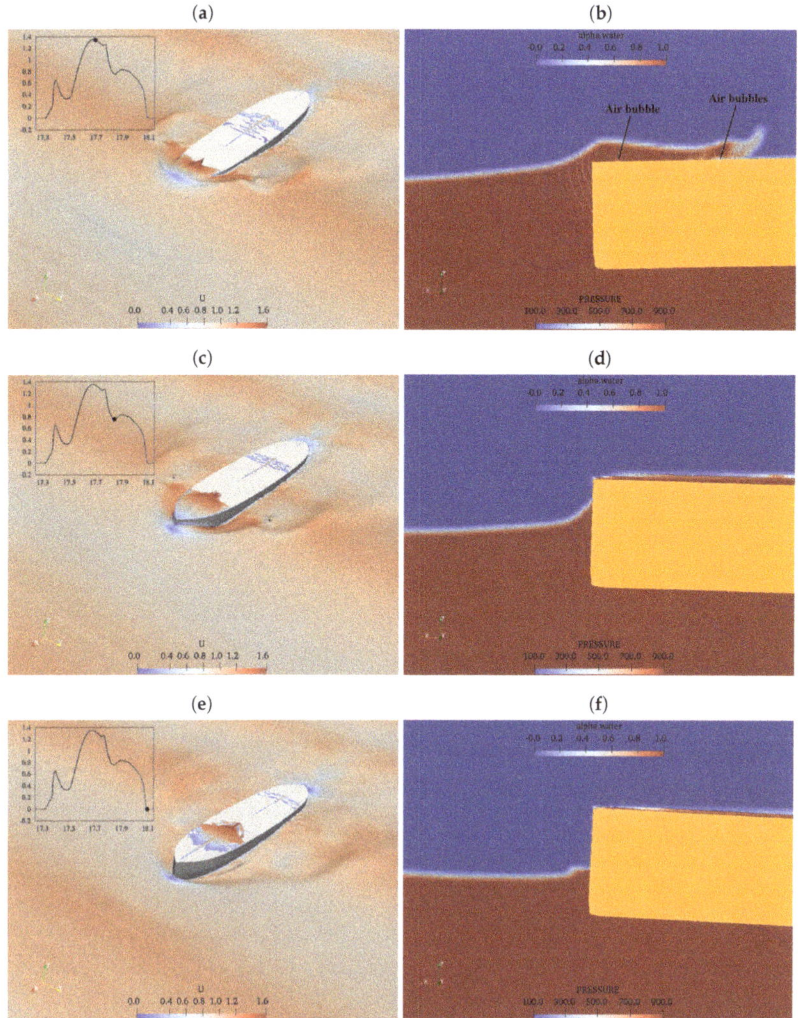

Figure 12. Flow velocity contours on the free surface (**a**,**c**,**e**) and pressure contours on the x–z plane at $y = 0$ (**b**,**d**,**f**) at time instants 5–7 for $\lambda/L = 1.1$.

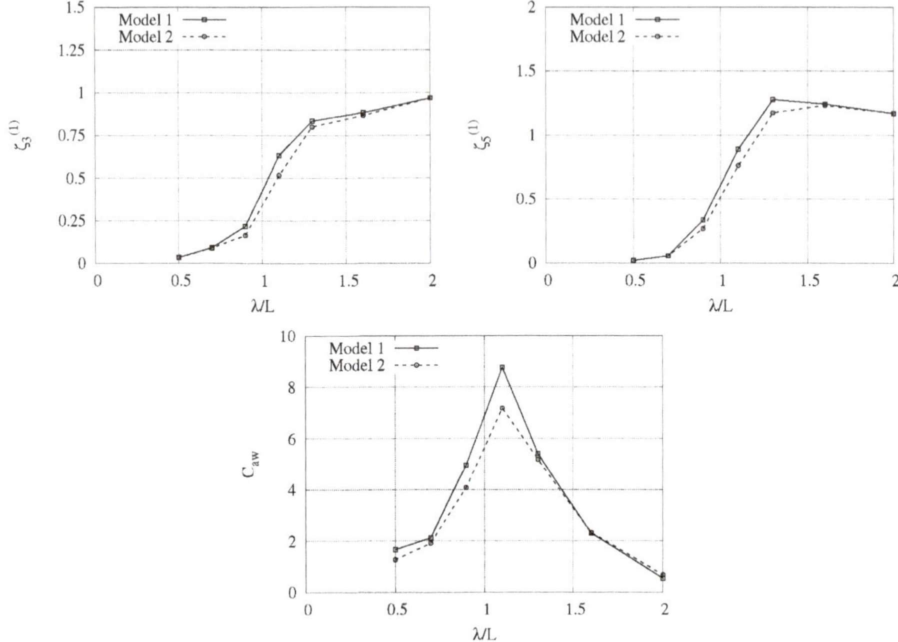

Figure 13. Profiles of heave, pitch and added resistance of the two models against λ/L.

5. Conclusions

This paper presented 3D RANS CFD simulations for low and high freeboard Wigley hull models. Free surface effects were solved by the VOF method in the openFOAM numerical wave tank. Comparisons between computations and experimental results helped to validate fluid structure interaction modeling procedures, realise fluid flow phenomena and pressure peaks of relevance to the phenomenon of green water on decks. Results revealed that CFD simulations provide improved physics representation. The highest green water impact appears at $\lambda/L = 1.3$ at which the highest pitch is also present. For $\lambda/L = 1.1$ to 1.3, the water propagates far downstream and an air bubble is entrapped in a form of a small low-pressure zone.

Author Contributions: Conceptualization, L.C. and X.W.; methodology, L.C. and X.W.; validation, L.C. and X.C.; formal analysis, L.C. and X.C.; investigation, L.C.; writing–original draft preparation, L.C. and Y.W.; writing–review and editing, L.C. Y.W. and X.W.; visualization, L.C., Y.W. and X.C.; supervision, X.W.; project administration, L.C.; funding acquisition, L.C. All authors have read and agreed to the published version of the manuscript.

Funding: This research was funded by the Natural Science Foundation of Jiangsu Province (grant no. SBK2018040999) and Major Basic Research Project of the Natural Science Foundation of the Jiangsu Higher Education Institutions (18KJB570001).

Conflicts of Interest: The authors declare no conflict of interest.

References

1. He, G.; Zhang, Z.; Tian, N.; Wang, Z. Nonlinear analysis of green water impact on forward-speed Wigley hull. In Proceedings of the 27th International Offshore and Polar Engineering Conference, San Francisco, CA, USA, 25–30 June 2017.
2. Chen, L.; Taylor, P.H.; Draper, S.; Wolgamot, H. 3-D numerical modelling of greenwater loading on fixed ship-shaped FPSOs. *J. Fluids Struct.* **2019**, *84*, 283–301. [CrossRef]

3. Buchner, B. Green Water on Ship-Type Offshore Structures. Ph.D. Thesis, Delft University of Technology, Delft, The Netherlands, 2002.
4. Schiller, R.; Pakozdi, C.; Stansberg, C.; Yuba, D.; Carvalho, D. Green water on FPSO predicted by a practical engineering method and validated agianst model test data for irregular waves. In Proceedings of the ASME International Conference on Ocean, Offshore and Arctic Engineering—OMAE, San Francisco, CA, USA, 8–13 June 2014.
5. Wang, S.; Wang, X.; Woo, W.L.; Seow, T.H. Study on green water prediction for FPSOs by a practical numerical approach. *Ocean Eng.* **2017**, *143*, 88–96. [CrossRef]
6. Cox, D.T.; Scott, C.P. Exceedance probability for wave overtopping on a fixed deck. *Ocean Eng.* **2001**, *28*, 707–721. [CrossRef]
7. Ogawa, Y.; Minami, M.; Tanizawa, K.; Kumano, A.; Matsunami, R.; Hayashi, T. Shipping water load due to deck wetness. In Proceedings of the Twelfth International Offshore and Polar Engineering Conference, Kitakyushu, Japan, 26–31 May 2002.
8. Soares, C.G.; Fonseca, N.; Pascoal, R. Experimental and numerical study of the motions of a turret moored FPSO in waves. *J. Offshore Mech. Arct. Eng.* **2004**, *127*, 197–204. [CrossRef]
9. Werter, R. Green Water Along the Side of an FPSO. Ph.D. Thesis, Delft University of Technology, Delft, The Netherlands, 2016.
10. Buchner, B. The impact of green water on FPSO design. In Proceedings of the 27th Offshore Technology Conference, Houston, TX, USA, 1–4 May 1995.
11. Schonberg, T.; Rainey, R.C.T. A hydrodynamic model of green water incidents. *Appl. Ocean Res.* **2002**, *24*, 299–307. [CrossRef]
12. Yilmaz, O.; Incecik, A.; Han, J. Simulation of green water flow on deck using non-linear dam breaking theory. *Ocean Eng.* **2003**, *30*, 601–610. [CrossRef]
13. Ryu, Y.; Chang, K.A.; Mercier, R. Application of dam-break flow to green water prediction. *Appl. Ocean Res.* **2007**, *29*, 128 – 136. [CrossRef]
14. Pakozdi, C.; de Carvalho e Silva, D.; Ostman, A.; Stansberg, C. Green water on FPSO analyzed by a coupled potential-flow-NS-VOF method. In Proceedings of the International Conference on Offshore Mechanics and Arctic Engineering—OMAE, San Francisco, CA, USA, 8–13 June 2014.
15. Zhou, Z.; Kat, J.; Buchner, B. A nonlinear 3-D approach to simulate green water dynamics on deck. In Proceddings of the Seventh International Conference on Numerical Ship Hydrodynamics, Nantes, France, 19–22 July 1999.
16. Greco, M.; Lugni, C. 3-D seakeeping analysis with water on deck and slamming. Part 1: Numerical solver. *J. Fluids Struct.* **2012**, *33*, 127–147. [CrossRef]
17. Zhang, S.; Yue, D.K.P.; Tanizawa, K. Simulation of plunging wave impact on a vertical wall. *J. Fluid Mech.* **1996**, *327*, 221–254. [CrossRef]
18. Greco, M. A Two-Dimensional Study of Green-Water Loading. Ph.D. Thesis, Norwegian University of Science and Technology, Trondheim, Norway, 2001.
19. Pham, X.P.; Varyani, K.S. Evaluation of green water loads on high-speed containership using CFD. *Ocean Eng.* **2005**, *32*, 571–585. [CrossRef]
20. Greco, M.; Colicchio, G.; Faltinsen, O.M. Shipping of water on a two-dimensional structure. Part 2. *J. Fluid Mech.* **2007**, *581*, 371–399. [CrossRef]
21. Shen, Z.; Wan, D. RANS computations of added resistance and motions of a ship in head waves. *Int. J. Offshore Polar Eng.* **2013**, *23*, 263–271.
22. Nielsen, K.B.; Mayer, S. Numerical prediction of green water incidents. *Ocean Eng.* **2004**, *31*, 363–399. [CrossRef]
23. Lu, H.; Yang, C.; L?hner, R. Numerical studies of green water impact on fixed and moving bodies. *Int. J. Offshore Polar Eng.* **2012**, *22*, 123–132.
24. Zhao, X.; Hu, C. Numerical and experimental study on a 2-D floating body under extreme wave conditions. *Appl. Ocean Res.* **2012**, *35*, 1–13. [CrossRef]
25. Zha, R.; YE, H.; Shen, Z.; Wan, D. Numerical computations of resistance of high speed catamaran in calm water. *J. Hydrodyn.* **2014**, *26*, 930–938. [CrossRef]

26. Ostman, A.; Sileo, C.; Stansberg, C.; Fonseca de Carvalho e Silva, D. A fully nonlinear RANS-VOF numerical wave tank applied in the analysis of green water on FPSO in waves. In Proceedings of the International Conference on Offshore Mechanics and Arctic Engineering—OMAE, San Francisco, CA, USA, 8–13 June 2014.
27. Kim, M.; Hizir, O.; Turan, O.; Incecik, A. Numerical studies on added resistance and motions of KVLCC2 in head seas for various ship speeds. *Ocean Eng.* **2017**, *140*, 466–476. [CrossRef]
28. Chen, S.; Hino, T.; Ma, N.; Gu, X. RANS investigation of influence of wave steepness on ship motions and added resistance in regular waves. *J. Mar. Sci. Technol.* **2018**, *23*, 991–1003. [CrossRef]
29. Kashiwagi, M. Hydrodynamic study on added resistance using unsteady wave analysis. *J. Mar. Sci. Technol.* **2018**, *23*, 991–1003. [CrossRef]
30. Menter, F.R. Two-equation eddy-viscosity turbulence models for engineering applications. *AIAA J.* **1994**, *32*, 1598–1605. [CrossRef]
31. Hirt, C.; Nichols, B. Volume of fluid (VOF) method for the dynamics of free boundaries. *J. Comput. Phys.* **1981**, *39*, 201–225. [CrossRef]
32. Newmark, N.M. A Method of Computation for Structural Dynamics. *J. Eng. Mech. Div.* **1959**, *85*, 67–94.
33. Kashiwagi, M. Prediction of surge and its effect on added resistance by means of the enhanced unified theory. *Trans. West-Jpn. Soc. Naval Archit.* **1995**, *89*, 77–89.
34. Iwashita, H.; Ito, A. Seakeeping computations of a blunt ship capturing the influence of the steady flow. *J. Ship Technol. Res.* **1998**, *45*, 159–171.

Journal of Marine Science and Engineering

Article

Experimental Determination of Non-Linear Roll Damping of an FPSO Pure Roll Coupled with Liquid Sloshing in Two-Row Tanks

Jane-Frances Igbadumhe [1], Omar Sallam [1,2], Mirjam Fürth [1,2,*] and Rihui Feng [1,2]

1 Department of Civil, Environmental and Ocean Engineering, Stevens Institute of Technology, Castle Point on Hudson, Hoboken, NJ 07030, USA; jigbadum@stevens.edu (J.-F.I.); osallam@stevens.edu (O.S.); rfeng1@stevens.edu (R.F.)

2 Department of Ocean Engineering, Texas A & M University, College Station, TX 77843, USA

* Correspondence: mfurth@stevens.edu

Received: 16 June 2020; Accepted: 24 July 2020; Published: 3 August 2020

Abstract: Wave excited roll motion poses danger for moored offshore vessels such as Floating Production Storage and Offloading (FPSO) because they cannot divert to avoid bad weather. Furthermore, slack cargo tanks are almost always present in FPSOs by design. These pose an increased risk of roll instability due to the presence of free surfaces. The most common method of determining roll damping is roll decay tests, yet very few test have been performed with liquid cargo, and most liquid cargo experiments use tanks that span the entire width of the vessel; which is seldom the case for full scale FPSO vessels during normal operations. This paper presents a series of roll decay test carried out on a FPSO model with two two-row-prismatic tanks with different filling levels. To directly investigate the coupling between the liquid sloshing and the vessel motion, without modifying the damping, tests were performed at a constant draft. The equivalent linear roll damping coefficients consisting of linear, quadratic and cubic damping terms are analyzed for each loading condition using four established methods, the Quasi-linear method, Froude Energy method, Averaging method and the Perturbation method. The results show that the cubic damping term is paramount for FPSOs and at low filling levels, were the FPSO is more damped. Recommendations regarding the applicability of the methods, their accuracy and computational effort is given and the effect of the liquid motion on the vessel motion is discussed.

Keywords: nonlinear roll motion; FPSO; roll damping; roll decay; liquid cargo motion

---

1. Introduction

Roll motion is the most dangerous motion amongst the six degrees of freedom because it can lead to capsizing [1,2], causes crew discomfort, and thus reduce the vessel's efficiency [3]. Roll motion is especially critical for offshore floating vessels such as Floating Production Storage and Offloading (FPSO) units because they are expected to operate at a location for extended duration and as such cannot avoid severe weather conditions. In addition, FPSOs experience continuous cargo loading and offloading [4], hence slack tanks cannot be avoided. Oscillations in slack cargo tanks can effect the roll motion characteristics of the vessel, it is therefore important to consider the free surface effect on roll damping [5].

Damping effects can be classified into linear and nonlinear components. Linear damping does not consider the viscosity of the liquid and it is described using linear radiation/diffraction theory [6]. Nonlinear damping includes the effects of liquid viscosity and appendages [6,7]. The vessel's roll damping can be described as linear for vessels with small roll amplitudes [8] but linear models are insufficient for vessels with large roll amplitudes, therefore, nonlinearities need to be accounted for when the roll amplitude is large enough to capsize the vessel [9]. Nonlinearities are further important when considering the effect of bilge keels [10] or liquid cargo motion [5].

Froude [11] was the earliest to predict nonlinear roll motion effects on floating vessels. Subsequently, Ikeda et al. [12] developed an empirical method that divided roll damping into five components (skin friction on the hull, eddy making, wave, lift and appendages). This empirical method provided low accuracy for damping at large angles, cannot handle attributes of complex flows [13] and are only applicable at the design stage [14]. Beyond design stage, excited roll motion, forced roll motion and the free roll decay test [15] are three methods for determining the nonlinear roll motion. These three methods can be done by experiment or by Computational Fluid Dynamics (CFD).

For forced roll motion, the model is rotated with the aid of a mechanical device and it is maintained in a fixed axis [15,16]. Kinnas [17] simulated a FPSO with and without bilge keels in harmonic forced motion and he observed a linear relationship between roll moment and roll amplitude in inviscid flow and a nonlinear relationship between the roll moment and roll angle in viscous flow. Thiagarajan et al. [18] did forced roll motion studies numerically and experimentally for an FPSO and concluded that the amplitude of the damping was influenced by the roll angular velocity and the bilge keel widths.

Using the excited roll motion method, the model is freely floating and it is excited by regular beam waves or an internal mechanical device such as a gyro roll exciter or a contra-rotating mass or an internal lateral shifting mass [14]. Blume [19] obtained roll damping coefficients using this method and he showed that roll damping coefficient was dependent on the maximum roll angle, metacentric height and heel angle of the vessel but Blume's method had long measurement time. Using same method, Handschel and Maksoud [20] improved on Blumes work by reducing the measuring time, improving the estimation accuracy of roll damping coefficient over a wide frequency range and provided possibilities of determining roll damping with nonlinear stability moment curve. Wasserman et al. [21] compared experimental and numerical results obtained from excited roll motion and free roll decay test and concluded that the excited roll motion was superior to the roll decay test for larger damping values.

Despite the findings by Wasserman et al. [21], free roll decay test is the most widely used method [5,22–27] because it is the recommended technique by the International Maritime Organization (IMO) [28]. There are five methods to determine the damping coefficients for the vessel roll motion from roll decay test [29]: Quasi-linear method, Froude Energy method, Roberts Energy method, Averaging method and the Perturbation method.

The Quasi-linear method is the simplest method because the equivalent linearised damping coefficient with respect to the roll angle amplitude can be directly obtained without the need for any curve fitting. It further, permits aggregating data from different decay tests as long as it is the same loading condition [14], this is beneficial for vessels with higher dampening since they will experience fewer oscillations before reaching steady state. The roll decrement is calculated from successive peaks or successive troughs or the double amplitudes of the roll decay curve (see Figure 1) [21]. This method is sometimes referred to as the logarithmic decrement method [30]. Though widely used [5,14,28,30–32], this method can be highly inconsistent because it sometimes shows poor correlation between the roll decrements and the roll angle amplitudes, leading to a incorrect damping coefficient [33].

The two main energy methods based on energy conservation are: Froude and Roberts [21,29]. Froude energy method assumes that for each half cycle, the energy lost by damping is equal to the work done by the hydrostatic restoring moment that reduces the roll amplitude. Roberts energy method [34] equates roll damping to an energy loss function. Froude energy method does not necessarily require a curve fitting for the roll angle series, it is weakly sensitive to shift in the origin of roll angle measurement in the vertical and horizontal direction due to roll period, errors in righting arm estimation, and roll amplitude definition. Roberts method on the other hand requires curve fitting and it is strongly sensitive to a shift of the origin of roll angle measurement [21]. A comparative studies have been carried out to compare the Quasi-linear and Energy methods and they agree that the Quasi-linear method is more susceptible to errors [29] and the energy methods are suitable for larger amplitudes based on curve fitting of the energy envelop [15,21]. Since both Roberts and Froude method are both based on conservation of energy, only Froude energy method will be treated in this paper.

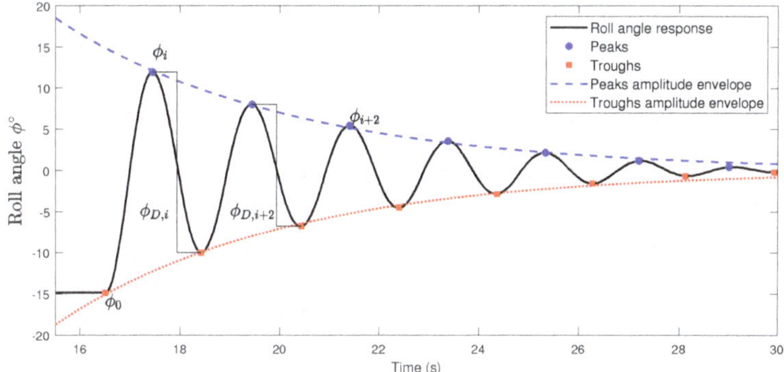

Figure 1. Roll angle decay response with peaks and double amplitudes annotations test case 8, see Section 3.3.

The Averaging method by Kryloff and Bgoliuboff [35] assumes the profile of the roll decay is sinusoidal and the rate of change of the amplitude and phase is constant at their average values for each cycle, that make this method to be more applicable for light damped systems [35]. It is more complex than the other methods because it produces a complex expression for the decay envelop. Though it is reliable, it is sensitive to the first peak of the decay motion hence, a distortion in the first peak of the decay would affect results obtained by this method [29].

The Perturbation method assumes that the nonlinear damping terms are small compared to the linear damping terms. This assumption permits the solution of the roll motion to be treated as a linear motion with some perturbations. This makes this method more complicated than the other methods as the roll decay angle amplitude envelop needs to be fitted with a higher order nonlinear equation with three unknown variables [29].

The damping coefficient may be dependant on the chosen analyzing method [28] or it may be affected by the different definitions of the average roll amplitude (ϕ_m) (Equations (11) and (12)) [21]. Other factors that can affect the damping coefficients obtained are the estimation of the righting moment (GZ) of the vessel [15]. Therefore, choosing which method, mathematical model and the average roll amplitude to use to determine the damping coefficients is challenging, and more research on the topic is needed.

This paper investigates the equivalent linear damping coefficients obtained using four methods and two mathematical models for roll damping from a roll decay test of a model scale FPSO. A series of roll decay test on the FPSO model with two two-row prismatic tanks with different filling levels was carried out at the Davidson Laboratory of Stevens Institute of Technology. The test were performed at constant draft, constant displacement and only roll motion was applied. The obtained damping is dependant on the chosen analyzing method, by comparing the effect of the number of terms included in the curve fit needed to obtain the dampening coefficient, guidance regarding method selection can be provided. The effect of the liquid motion on the vessel motion and the effect of change in draft are further discussed.

2. Theoretical Background

The model was free to roll, but fixed in all other degrees of freedom (DOFs), the dynamic behaviour of the vessel is therefore described by the single degree of freedom (DOF) equation:

$$(I + A)\ddot{\phi} + B(\phi)\dot{\phi} + C(\phi)\phi = F(t) \tag{1}$$

where I is the roll mass moment of inertia (Kg m^2), A is the added mass moment inertia (Kg m^2), $B(\phi)$ is the roll damping coefficient (Nm.s), $C(\phi)$ is the restoring stiffness coefficient (Nm) and F is the time dependent forcing moment which is zero for calm water tests. ϕ, $\dot{\phi}$ and $\ddot{\phi}$ are the roll angle, angular velocity and angular acceleration respectively. $C = \rho g \nabla GZ$ where $\rho\nabla$ is the model mass (Kg) and GZ is the angle dependent restoring lever (m) [36]. The normalised form of the roll motion is:

$$\ddot{\phi} + b(\phi)\dot{\phi} + c(\phi)\phi = f(t) \tag{2}$$

where $b(\phi)$ is the roll damping coefficient per mass moment of inertia = $\frac{B(\phi)}{(I+A)}$, $c(\phi)$ is the roll stiffness coefficient per unit mass moment of inertia = $\frac{C(\phi)}{(I+A)}$ and $f(t)$ is the applied moment per mass moment of inertia = $\frac{F(t)}{(I+A)}$.

The nonlinear damping moment per mass moment of inertia $b(\phi)$ can be expressed by a series expansion consisting of a linear term and higher order nonlinear terms, as a function of ϕ and $\dot{\phi}$ [37]:

$$\overbrace{b(\phi)\dot{\phi}}^{Total\ damping} = \overbrace{b_1\dot{\phi}}^{Linear\ damping\ term} + \overbrace{b_2|\dot{\phi}|\dot{\phi}}^{Quadratic\ damping\ term} + \overbrace{b_3\dot{\phi}^3}^{Cubic\ damping\ term} + \overbrace{O(\phi^4)}^{Higher\ 4^{th}\ order\ terms} \tag{3}$$

Inserting Equation (3) in Equation (2) for a calm water roll decay test, where $f(t) = 0$, the normalised equation of motion for the vessel becomes:

$$\ddot{\phi} + b_1\dot{\phi} + b_2|\dot{\phi}|\dot{\phi} + b_3\dot{\phi}^3 + c(\phi)\phi = 0 \tag{4}$$

Equation (4) is an accurate mathematical model of dynamic ship rolling in calm water, since it includes the effects of the resistance due to the linear damping coefficient b_1 and nonlinear damping coefficients b_2 and b_3. However, this nonlinear equation is rarely solved [7], since it requires iterative numerical ODE solvers to overcome the non-linearity in the equation represented by terms $|\dot{\phi}|\dot{\phi}$ and $\dot{\phi}^3$. Hence using a linearized equivalent equation is beneficial. In the linearized form the total damping coefficient $b(\phi)$ is replaced by an equivalent linear damping coefficient $b_e(\phi_a)$ [7,29] which is dependent on the vessel roll angle amplitude and the roll period. The linear, quadratic and cubic damping coefficients are also kept constant in the equivalent linear damping. The linearized one DOF equation of motion with equivalent damping coefficient is:

$$\ddot{\phi} + b_e(\phi_a)\dot{\phi} + c_e(\phi_a)\phi = 0 \tag{5}$$

where $c_e(\phi_a)$ is the amplitude dependent stiffness and ϕ_a is the roll angle amplitude at certain time. The relationship between the total damping and the equivalent linear damping is:

$$b(\phi) = b_e(\phi_a) \tag{6}$$

$$b_e(\phi_a) = b_1 + \frac{8}{3\pi}\omega_d\phi_a b_2 + \frac{3}{4}\omega_d^2\phi_a^2 b_3 \tag{7}$$

2.1. Quasi-Linear Method (Logarithmic Decrement)

The Quasi-linear or the Logarithmic decrement method is considered to be the simplest technique to directly compute the equivalent linear damping coefficient b_e with respect to the roll angle amplitude obtained through roll decay tests [29]. The transient roll angle $\phi(t)$ is assumed to be a sinusoidal motion with decaying amplitude as shown in Equation (8), where ϕ_0 is the initial roll angle amplitude and $\phi_a(t)$ is the decaying amplitudes for peaks or troughs as shown in Figure 1 and Equation (9).

By applying Equation (9) for two successive peaks i and $i+1$ (or troughs), of the decaying envelope, the logarithmic decay can be determined (Equation (10)) and then the equivalent linear damping coefficient (Equation (11)).

When measuring the roll angle, a measurement offset may occur causing irregularity of peaks and troughs amplitudes envelopes leading to different predictions of the damping coefficient by peak based or trough based envelopes. The double amplitude based logarithmic decay technique [21] is used to compensate this possible offset, instead of analysing the peaks' envelope and the troughs' envelope separately, the absolute difference between each peak and following trough is considered as shown in Figure 1, and the equivalent linear damping coefficient is obtained using Equation (12) [36]. Figure 2 shows the equivalent linear damping coefficient b_e (12) as a function of the average roll angle amplitude ϕ_m, the plot is a combination of four roll decay tests with different initial roll angles for loading condition 8 (see Section 3.3).

$$\phi(t) = \phi_0 \, e^{\frac{-b_e t}{2}} \cos(\omega_d t) \tag{8}$$

$$\phi_a(t) = \phi_0 \, e^{\frac{-b_e t}{2}} \tag{9}$$

$$\frac{\phi_{i+1}}{\phi_i} = e^{\frac{-b_e}{2}(t_{i+1}-t_i)} \tag{10}$$

$$b_e(\phi_m) = \frac{\omega_d}{\pi} \ln \frac{\phi_i}{\phi_{i+1}}, for \ \phi_m = \frac{\phi_i + \phi_{i+1}}{2} \tag{11}$$

$$b_e(\phi_m) = \frac{\omega_d}{\pi} \ln \frac{\phi_{D,i}}{\phi_{D,i+2}}, for \ \phi_m = \frac{\phi_{D,i} + \phi_{D,i+2}}{4} \tag{12}$$

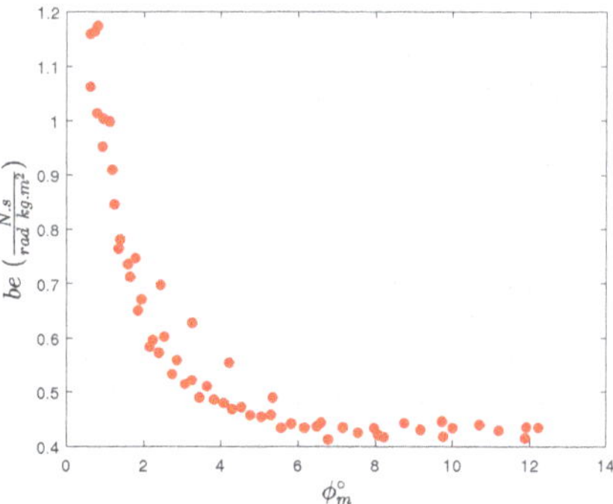

Figure 2. Equivalent linear damping coefficient estimation using quasi-linear method Equation (12), test case 8 by combining results of 4 tests with different initial roll angles, see Section 3.3.

2.2. Froude Energy Method

Froude energy method [7,36] is based on the assumption that the energy lost due to the damping $E_{D\,\frac{1}{2}}$ in a half cycle is equivalent to the hydrostatic stiffness restoring energy applied by the restoring lever arm $E_{r\,\frac{1}{2}}$:

$$E_{D\,\frac{1}{2}} = E_{r\,\frac{1}{2}} \tag{13}$$

The energy lost by damping in a half cycle and restored by stiffness are calculated by assuming that the motion is sinusoidal through each quarter cycle as shown in Equation (14), compared with the

Quasi-linear method that assumes that the motion is sinusoidal with an exponential decaying term (Equation (8)).

$$\phi = \phi_i sin(w_d t) \tag{14}$$

ϕ_1 is the amplitude of the roll motion in the first quarter cycle quarter $\phi_1 \leq \phi \leq 0$, ϕ_2 is the amplitude of the roll motion over the second quarter cycle $0 \leq \phi \leq \phi_2$ as shown in Figure 3, ω_d is the natural damped angular frequency. Equation (14) is only valid for a quarter cycle, hence the half cycled energy terms must be calculated by the summation of energy terms in each cycle quarter as:

$$E_{D1\,\frac{1}{4}} + E_{D2\,\frac{1}{4}} = E_{r1\,\frac{1}{4}} + E_{r2\,\frac{1}{4}} \tag{15}$$

The energy dissipation including linear, quadratic and cubic damping in a quarter cycle is described by Equation (16), hence the total energy dissipation during the half cycle between amplitudes ϕ_1 and ϕ_2 can be described using Equation (17).

$$E_{Di\,\frac{1}{4}} = \int_0^{\frac{\pi}{2w_d}} (b_1\dot{\phi}_i^2 + b_2|\dot{\phi}_i|\dot{\phi}_i^3 + b_3\dot{\phi}_i^4) \cdot dt \tag{16}$$

$$E_{D\,\frac{1}{2}} = \int_0^{\frac{\pi}{2w_d}} (b_1(\dot{\phi}_1^2 + \dot{\phi}_2^2) + b_2(|\dot{\phi}_1 + \dot{\phi}_2|(\dot{\phi}_1^3 + \dot{\phi}_2^3)) + b_3(\dot{\phi}_1^4 + \dot{\phi}_2^4)) \cdot dt \tag{17}$$

Integrating Equation (17) gives Equation (18) which is the energy dissipated in the half cycle from ϕ_1 to ϕ_2. For $\phi_1 \simeq \phi_2$ and $\phi_a = (\frac{\phi_1+\phi_2}{2})$ the total energy dissipation can be simplified as shown in Equation (19).

$$E_{D\,\frac{1}{2}} = \frac{1}{4}\pi b_1\omega_d(\phi_1^2 + \phi_1^2) + \frac{2}{3}b_2\omega_d^2(\phi_1^3 + \phi_2^3) + \frac{3}{16}b_3\pi\omega_d^3(\phi_1^4 + \phi_2^4) \tag{18}$$

$$E_{D\,\frac{1}{2}} = \frac{1}{2}\pi b_1\omega_d\phi_a^2 + \frac{4}{3}b_2\omega_d^2\phi_a^3 + \frac{3}{8}b_3\omega_d^3\pi\phi_a^4 \tag{19}$$

The energy done by the restoring moment lever arm during the half cycle between ϕ_1 and ϕ_2 is described by Equation (20). By averaging the the successive amplitudes the restoring energy can be represented using Equation (21), where $\phi_a = \frac{\phi_1+\phi_2}{2}$, $c = \omega_d^2$ and $\frac{d\phi}{dr} = \phi_1 - \phi_2$.

$$E_{r\,\frac{1}{2}} = \int_{\phi_2}^{\phi_1} c\phi d\phi = \frac{c}{2}[\phi_1^2 - \phi_2^2] \tag{20}$$

$$E_{r\,\frac{1}{2}} = \omega_d^2\phi_a\frac{d\phi}{dr} \tag{21}$$

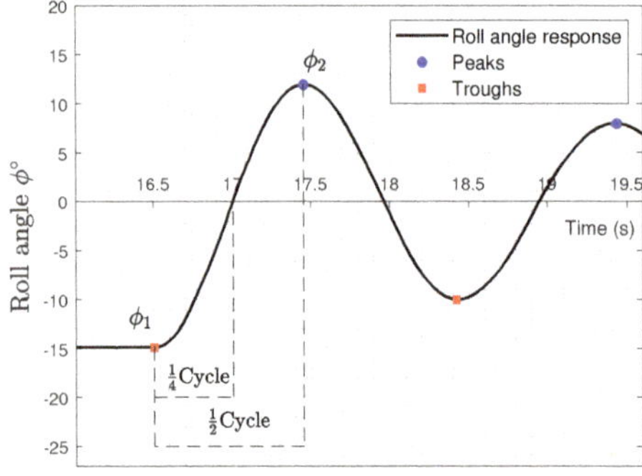

Figure 3. Annotation of peak and trough amplitudes for half cycle, test case 8, see Section 3.3.

By equating Equations (19) and (21) the decrease of amplitudes per half cycle $\frac{d\phi_a}{dr}$ can be represented by a third order polynomial as shown in Equation (22). This equation can be solved by determining the decay extinction coefficients: a, b and c. They can be determined by a third order polynomial fit or second order polynomial fit if only the linear (a) and quadratic (b) damping are considered.

$$\frac{d\phi_a}{dr} = \underbrace{\frac{\pi b_1}{2\omega_d}\phi_a}_{a\phi_a} + \underbrace{\frac{4}{3}b_2\phi_a^2}_{b\phi_a^2} + \underbrace{\frac{3}{8}\omega_d\pi b_3\phi_a^3}_{c\phi_a^3} \tag{22}$$

$$a_e = a + b\phi_a + c\phi_a^2 = \frac{\pi b_e}{2\omega_d} \tag{23}$$

Figures 4 and 5 shows two examples of the Froude energy method used to determine the equivalent damping coefficient per mass moment of inertia b_e (Equation (23)). In Figure 4a the decrease in half cycle amplitudes is plotted as a function of the mean peak roll angle amplitude then fitted with a quadratic polynomial to estimate the first two coefficients a and b (see Equation (22)), then the equivalent damping coefficient is determined for the range of roll peak amplitudes using Equation (23), see Figure 4b. In Figure 5a the results are fitted by a third order degree polynomial to compute the three extinction coefficients a, b and c, the equivalent damping coefficient including the cubic damping term are plotted in Figure 5b. By comparing Figures 4b and 5b It is obvious that after including the cubic damping term, the behaviour of the equivalent damping with respect to the roll angle peak amplitude is changed as an inflection point appears near 10° after which the the damping is increasing.

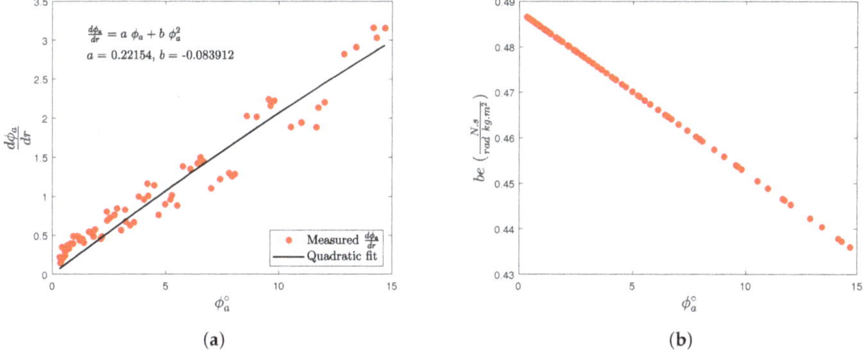

Figure 4. Equivalent linear damping coefficient estimation using Froude energy method with linear and quadratic damping terms, test case 8 by combining results of 4 tests with different initial roll angles, see Section 3.3. (**a**) Estimation of extinction coefficients a and b using quadratic curve fitting, Equation (22); (**b**) Equivalent linear damping coefficient be estimation based on the extinction coefficients a and b using Equation (23).

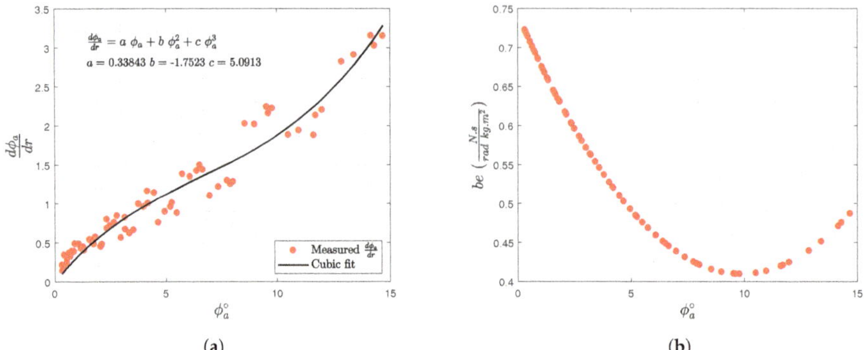

Figure 5. Equivalent linear damping coefficient estimation using Froude energy method with linear, quadratic and cubic damping terms, test case 8 by combining results of 4 tests with different initial roll angles, see Section 3.3. (**a**) Estimation of extinction coefficients a, b and c using cubic curve fitting, Equation (22); (**b**) Equivalent linear damping coefficient be estimation based on the extinction coefficients a, b and c using Equation (23).

2.3. Averaging Method

The Averaging method developed by Kryloff and Bgoliuboff [29,35] is designed for lightly damped systems. The Froude energy method assumes a sinusoidal motion of only half of the cycle, the Averaging method on the other hand assumes the roll angle to be in a sinusoidal form for the whole cycle, with a slight amplitude decay with respect to the time, as expressed in Equation (25). By inserting $\phi(t)$ (Equation (25)) in the equation of motion (24) and integrating over the whole cycle [35], an expression for the rate change of amplitude $\frac{d\phi_a}{dt}$ is generated as shown in Equations (26) and (27). Equation (27) represents the relation between the amplitude rate change $\frac{d\phi_a}{dt}$ with the roll angle amplitude ϕ_a by a second order equation which is fitted by the nonlinear least square algorithm to evaluate the linear and nonlinear damping coefficients b_1 and b_2 (Equation (28)) as shown in Figure 6a. Figure 6b shows the estimated equivalent linear damping coefficient per mass moment of inertia b_e, Equation (28).

$$\ddot{\phi} + b_1\dot{\phi} + b_2\dot{\phi}|\dot{\phi}| + c\phi = 0 \tag{24}$$

$$\phi(t) = \phi_a(t)\cos(\psi)\ ,\ \psi = \omega_d t + \theta \tag{25}$$

$$\frac{d\phi_a}{dt} = -\frac{1}{2\pi\omega_d}\int_0^{2\pi}\left[b_1\phi_a\omega_d cos^2(\psi) + b_2(\phi_a\omega_d)^2|cos(\psi)|cos^2(\psi)\right]d(\psi) \tag{26}$$

$$\frac{d\phi_a}{dt} = -\underbrace{\left(\frac{4b_2\omega_d}{3\pi}\right)\left(\frac{3\pi b_1}{8b_2\omega_d}\right)\phi_a}_{A\ B\ \phi_a} - \underbrace{\left(\frac{4b_2\omega_d}{3\pi}\right)\phi_a^2}_{A\ \phi_a^2} \tag{27}$$

$$b_e = b_1 + \frac{8}{3\pi}b_2\phi_a\omega_d \tag{28}$$

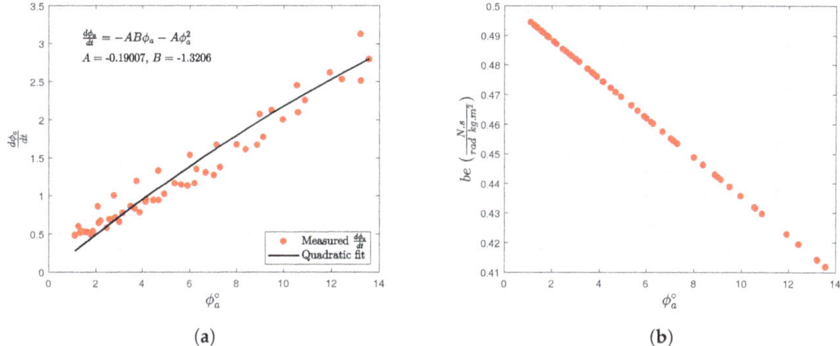

Figure 6. Equivalent linear damping coefficient estimation using Averaging method, test case 8 by combining results of 4 tests with different initial roll angles, see Section 3.3. (**a**) Estimation of A, B coefficients using quadratic curve fitting between $\frac{d\phi_a}{dt}$ and ϕ_a Equation (27); (**b**) Equivalent linear damping coefficient *be* estimation based on b_1 and b_2 using Equations (28).

2.4. Perturbation Method

The Perturbation method is considered the most complicated method for analyzing roll decay test data [29], because the roll decay angle amplitude envelope (Figure 7a) is fitted with a high order nonlinear equation with three unknown variables.

The Perturbation method assumes the nonlinear damping coefficients are small compared to the linear damping coefficient. The motion is described using Equation (29), where ϵ is the perturbation expansion parameter. This equation is assumed to have a solution represented by a power series of ϵ, Equation (30). By inserting the zero, first and second order expansions (Equation (30)) in (29), a new equation of motion is obtained for each expansion. The equations are Equations (31)–(33), respectively. Equation (31) represents the linear damping case and has an analytic solution representing the standard roll angle decay profile with linear damping, Equation (34). By inserting ϕ_0 from Equation (34) in the Equation of motion for the ϵ^1 expansion (Equation (32)), ϕ_1 can be estimated as shown in Equation (35), this process can be repeated to obtain an estimation of ϕ_n, where n is the order of the perturbation expansion. The roll angle $\phi(t)$ is the geometric sum of the all solutions, as seen in Equation (30), with $\epsilon = 1$ resulting in the roll angle profile, $\phi(t)$ as shown in Equation (36) and with an amplitude envelope profile ϕ_r shown in Equation (37).

$$\ddot{\phi} + b_1\dot{\phi} + \epsilon b_2 \dot{\phi}|\dot{\phi}| + c\phi = 0 \tag{29}$$

$$\phi(t) = \overbrace{\underbrace{\underbrace{\phi_0(t)}_{\epsilon^0 expansion} + \epsilon\phi_1(t)}_{\epsilon^1 expansion} + \epsilon^2\phi_2(t)}^{\epsilon^2 expansion} + \overbrace{O(\epsilon^4)}^{Higher\ 4^{th}\ order\ terms} \tag{30}$$

$$Equation\ of\ motion\ with\ \epsilon^0\ expansion: \ \ddot{\phi}_0 + b_1\dot{\phi}_0 + c\phi_0 = 0 \tag{31}$$

$$Equation\ of\ motion\ with\ \epsilon^1\ expansion: \ \ddot{\phi}_1 + b_1\dot{\phi}_1 + c\phi_1 = -b_2\ \dot{\phi}_0^2\ sgn(\dot{\phi}_0) \tag{32}$$

$$Equation\ of\ motion\ with\ \epsilon^2\ expansion: \ \ddot{\phi}_2 + b_1\dot{\phi}_2 + c\phi_2 = -b_2\ \dot{\phi}_0\ \dot{\phi}_1\ sgn(\dot{\phi}_0 + \epsilon\dot{\phi}_1) \tag{33}$$

$$\phi_0 = \phi_{01}\, e^{\frac{-b_1 t}{2}} \cos\left(\omega_d t + \theta_{01}\right) \tag{34}$$

$$\phi_1 = \frac{8 b_2 \omega_d}{3\pi b_1}\, \phi_{01}^2\, e^{\frac{-b_1 t}{2}} \cos\left(\omega_d t + \theta_{01}\right) \tag{35}$$

$$\phi(t) = \frac{3\pi b_1 \phi_{01}}{3\pi b_1 e^{\frac{-b_1 t}{2}} - 8\, b_2\, \phi_{01}\, \omega_d} cos(\omega_d t) \tag{36}$$

$$\phi_r = \left[\frac{1}{\phi_{01}} e^{\frac{-b_1 \pi r}{2\omega_d}} - \frac{8\, b_2 \omega_d}{3\pi b_1}\right]^{-1} \tag{37}$$

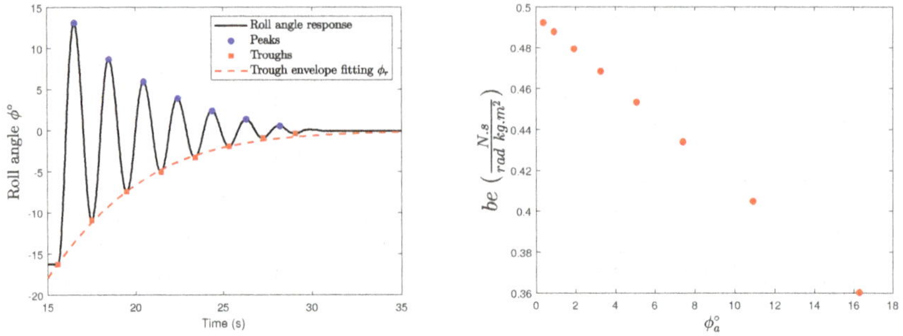

(**a**) (**b**)

Figure 7. Equivalent linear damping coefficient estimation using the perturbation method, test case 8, see Section 3.3. (**a**) Roll angle decay test envelope nonlinear curve fitting with Equation (37); (**b**) Equivalent linear damping coefficient using the estimated coefficients b_1 and b_2 Equation (38).

The perturbation derived roll angle amplitude envelope (Equation (37)) is fitted with the measured roll angle amplitude envelope ϕ_r from the roll decay test (see Figure 7a). The fitted curve is nonlinear with three unknown parameters (b_1, $b2$ and ϕ_{01}) where ϕ_{01} is the initial amplitude parameter or amplitude of zero order solution, and obtained by the nonlinear curve fitting of Equation (37). Figure 7a shows a roll angle decay test where the trough measured envelope is fitted with Equation (37) using the nonlinear least square algorithm to compute b_1 and b_2, the equivalent linear damping coefficient per unit mass moment of inertia b_e (Equation (38)) as shown in Figure 7b is then obtained.

$$b_e = b_1 + \frac{8}{3\pi} b_2 \phi_a \omega_d \tag{38}$$

3. Experimental Set Up

3.1. Model and Instrumentation

An existing tanker hull model made of wood is fitted with two two-row cargo tanks made of acrylic plates. The cargo tanks were placed in the model such that the sloshing occurs in the longitudinal axis of the tank during the roll motion. The model attached to the heave post and tow carriage is shown in Figure 8a. A roll pivot box is connected to the tow carriage to measure the roll amplitude and the walls of the port stern cargo tank is fitted with four water level gauges see Figure 8b to measure liquid water level in the tanks. All instrumentation used are calibrated prior to the test, see Appendix A. The main particulars of the model and cargo tanks are shown in Table 1. The draft of the vessel was measured at three locations; stern, amidships and the bow during the free floating inclining test and the roll decay tests and they are shown in Table A2.

Table 1. FPSO and cargo tank characteristics.

Parameter	Value	Unit
Length overall LOA	2.18	m
Beam B	0.341	m
Depth D	0.19	m
Unballasted vessel weight	24.96	kg
Ballasted vessel weight	40	kg
Displacement (ballast condition) Δ_m	0.029	ton
Kyy	1.09	m
Kzz	1.09	m
Kxx	0.12	m
Cargo Tank Outer Dimensions		
Length L_T	0.49	m
Width B_T	0.26	m
Width of one tank B_{T1}	0.13	m
Depth D_T	0.17	m
Thickness t	0.01	m

(**a**) Model attached to tow carriage

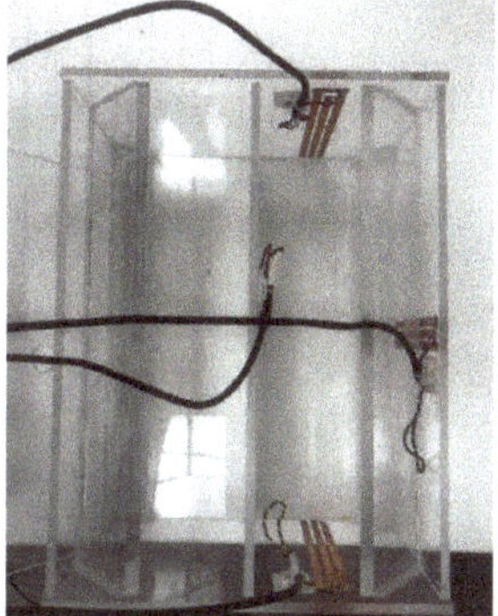

(**b**) Cargo tank with water level gauges

Figure 8. Model attached to heave post and cargo tank.

3.2. Inclination Test

To obtain the model's vertical center of gravity KG, an inclining test was carried out in calm water by horizontally shifting an inclination mass of w_i = 2.04 kg to the port and starboard sides with a horizontal shift d causing heeling angles [38]. The weight displacement were kept constant (inclination weight was taken from the loaded ballast) as shown in Figure 9, Figure A1 and Table A1. The Metacentric Height GM, was obtained using Equation (39) [38]. Table 2 shows the measured distances, the load moment per displacement, and the resulting roll angle. To get the vertical distance from the keel to the Metacenter KM, a hydrostatic analysis was conducted using ORCA3D [39] where the KM is evaluated at various vessel displacements. The vertical center of gravity KG 91.6 mm, was determined using Equation (40) .

$$GM = \frac{w_i\, d}{\Delta\, \tan\varphi} \tag{39}$$

$$KG = KM - GM \tag{40}$$

Figure 10a shows the inclining moment for the inclining tests with respect to the heeling angles φ, where a linear curve fitted to the dataset showed a slope of 70.0 mm representing the metacentric height GM as shown in Equation (39). Where Δ is the total displacement of the vessel in kg, see Table 1.

To validate the experimental inclining test, a reverse numerical inclining test was implemented in ORCA3D using KG from the experimental test and the same loading conditions. As shown in Figure 10b the numerical metacentric height for this loading case is 70.7 mm, confirming the experimental results.

Figure 9. Model during Incline Test.

Table 2. Inclination test.

Side	Load Horizontal Shift d (mm)	Load Moment per Displacement $\frac{wd}{\Delta}$ (mm)	Roll Angle $\phi°$
Port side	−150	−78	−5.9
	−100	−52	−3.9
	−50	−26	−1.9
	0	0	0.1
Starboard side	50	26	2.0
	100	52	4.0
	150	78	6.0

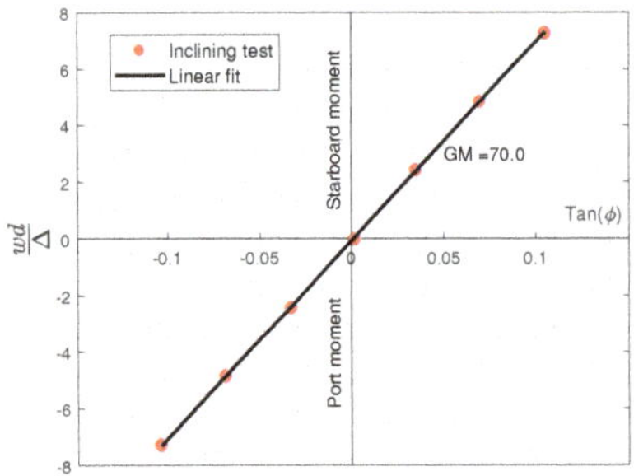

(**a**) Experimental determination of Metacentric Height GM from incling test

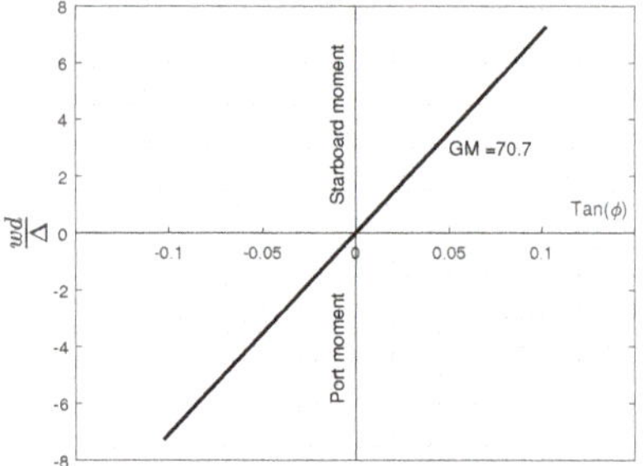

(**b**) Numerical determination of Metacentric Height GM using ORCA3D software

Figure 10. Determination of Metacentric Height GM from experimental inclining test data and numerical modelling using ORCA3D software.

3.3. Roll Decay Test

The vessel roll decay experiment followed the guidelines and procedures by the International Marine Organisation (IMO) [40] and the International Towing Tank Conference (ITTC) [7] which advised that for highly damped vessels with low oscillation number, several roll tests (at least four) should be implemented with different initial roll angles, so the estimated damping coefficient would be continuous through the whole range of roll angle amplitudes. During the test, both two-row cargo tanks (4 compartments in total), were filled with water as shown in Table 3. The initial angle of the test was imposed by a system of pulley and strings as shown in Figures 8a and A1. The volume displacement was kept constant (constant draft) for loading conditions 1–8 and the draft was increased for loading condition 9. The initial disturbance was between 7.3° and 17.7°, ITTC [7] recommends a disturbance larger than 25°, however it was not possible to achieve with the current experimental set up.

Table 3. Water Tank Loading Conditions.

Condition	Forward				Aft			
	Port		Starboard		Port		Starboard	
	Volume (L)	Percentage (%)	Volume (L)	Percentage (%)	Volume (L)	Percentage (%)	Volume (L)	Percentage (%)
1	0	0	0	0	0	0	0	0
2	0	0	3	50	5.5	95	0	0
3	0	0	0.6	10	3	50	5.5	95
4	0.6	10	5.5	95	3	50	0.6	10
5	3	50	0	0	0.6	10	0.6	10
6	3	50	0	0	0	0	0.6	10
7	5.5	95	0	0	0	0	0	0
8	5.5	95	5.5	95	5.5	95	5.5	95
9 *	5.5	95	5.5	95	5.5	95	5.5	95

* Sinkage was allowed for Condition 9 as shown in Table A2 and Figure A2.

Roll amplitude was recorded by a rotational potentiometer installed inside the roll pivot box as shown in Figure A1. The model was allowed to roll freely after the release of the pulley-string system for all conditions, while sway, surge, yaw, pitch and heave were restricted.

To ensure no initial angular velocity nor unintended sloshing effect of the loaded water were induced, each test were performed for at least 3 minutes after the initial angle was set. Data collection was started before the release of pulley-string system to show the setup was at steady state prior to the test, in accordance with IMO and ITTC procedures [7,40].

To evaluate the repeatably of the the roll decay tests, some loading conditions were repeated with the same initial roll angle. The number of peaks and troughs before steady state for each repeated pairs are the same, however a deviation in the natural decay period is observed. Table 4 shows the percentage difference between each two measured natural periods, no dramatic deviation except for condition 8 where the cargo volume is at its maximum and the liquid sloshing on tanks wall is random in nature.

Table 4. Natural decay period for conditions with the same initial roll angle.

Condition	Initial Roll Angle $\varphi \pm 0.1°$	Natural Period (s)	Percentage Difference %
3	16.6	1.68 1.58	6.13
5	12.6	1.26 1.14	10
7	13.6	1.30 1.22	6.35
8	16.6	1.62 2.09	25.3

4. Results and Discussion

4.1. *Effect of Analyzing Method*

The equivalent linear damping coefficient per total mass moment of inertia (be) is studied using the roll decay analysis methods presented in Section 2 for the 9 filling conditions described in

Tables 3 and A2. For most conditions and analysis methods, the equivalent linear damping coefficient is observed to have the same trend; it is inversely proportional to the roll angle amplitude as shown in Figure 11. As expected using the Averaging method, quadratic Froude energy method and Perturbation methods yields almost a linear relation between the equivalent linear damping coefficient and roll angle amplitude. However, for the Quasi-linear and cubic Froude energy methods where the cubic damping term is included, a nonlinear relation is observed, that implies that the cubic damping term can not be ignored for the tested vessel type.

A high deviation of the equivalent linear damping coefficient is observed when comparing the Averaging method and Froude energy method with quadratic and cubic terms at small roll angle amplitudes, across all loading conditions, see Figure 11. The Averaging method assumes that the motion is sinusoidal through the whole cycle, therefore it does not produce accurate results at lower angles where damping is high. The Froude energy method on the other hand assumes the motion is sinusoidal for only the cycle quarter and therefore produces more reliable results at low rolling angles [29]. At higher roll angle amplitudes, where the damping decreases, the Averaging method converges with the results from the other methods, this is expected and consistent across all loading conditions, since the Averaging method is recommended for lightly damped systems.

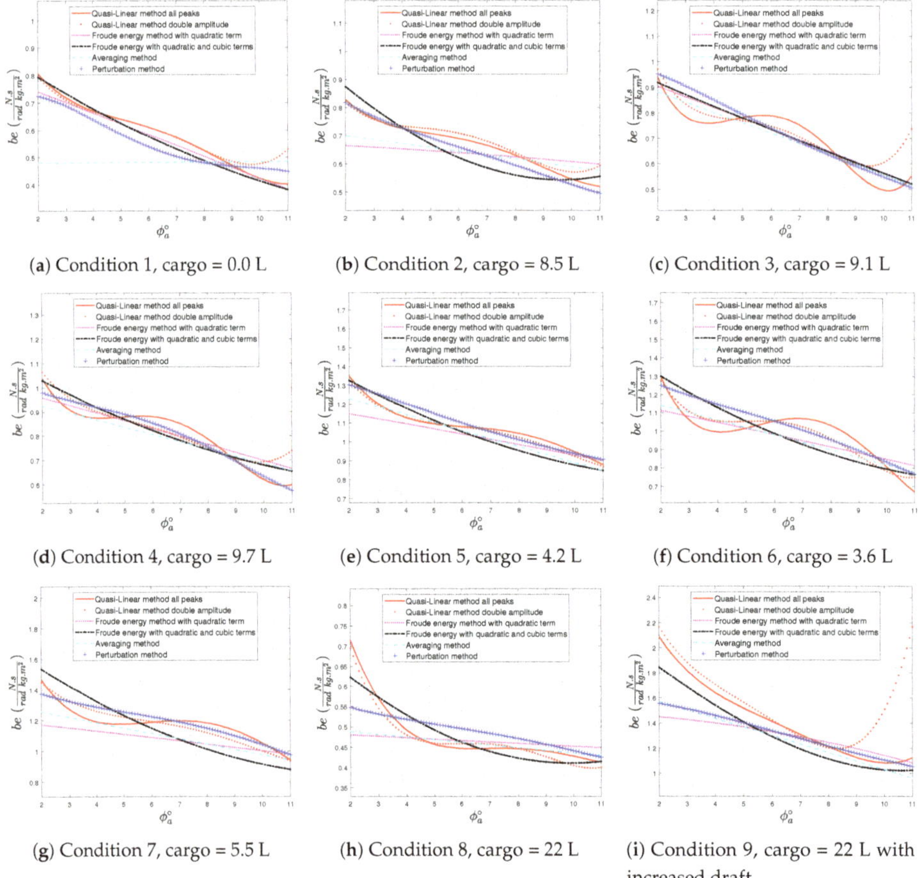

(**a**) Condition 1, cargo = 0.0 L (**b**) Condition 2, cargo = 8.5 L (**c**) Condition 3, cargo = 9.1 L

(**d**) Condition 4, cargo = 9.7 L (**e**) Condition 5, cargo = 4.2 L (**f**) Condition 6, cargo = 3.6 L

(**g**) Condition 7, cargo = 5.5 L (**h**) Condition 8, cargo = 22 L (**i**) Condition 9, cargo = 22 L with increased draft

Figure 11. Equivalent linear damping coefficient determined using different analysis methods as a function of roll angle amplitude for different liquid loading conditions.

4.2. Effect of Liquid Cargo

Looking at results of the Quasi-linear method using all peaks and double amplitudes, a relation between the total cargo volume and the slope of the equivalent linear damping coefficient is observed at higher roll angle amplitudes. For low loading conditions [5, 6 and 7] (Figure 11e–g) the the equivalent linear damping coefficient has a negative slope at higher roll amplitude angles. On the other hand for the high loading conditions [2, 3, 4, 8 and 9] (Figure 11b–d,h,i) the damping coefficient show an inflection point nearly at the roll angle amplitude 10° followed by an increase. It should be noted that the change in dampening at higher filling levels [2, 3, 4, 8], is solely due to the increase in liquid cargo and thus liquid cargo motion, since the volume displacement is only changed in condition 9.

To evaluate the effect of the amount of liquid cargo on the vessel damping, a comparison is done between the mean linear equivalent damping coefficient for the whole range of rolling amplitudes and the volume of liquid cargo loading for the 9 loading conditions, see Figure 12. The results show that the vessel is highly dampened for cases of low filling volumes (conditions [6, 5 and 7]) where the liquid cargo sloshing forces' effect on the tank walls does not affect the vessel motion. At condition 9, maximum condition with 22 L, and with increased draft, the vessel is observed to be very stable with the highest average equivalent damping coefficient due to the additional damping resistance applied by the increased draft. On the other hand and for high filling conditions [2, 3, 4 and 8], the cargo sloshing forces on tanks' walls affected the dynamic behaviour of vessel rolling, where the equivalent linear damping coefficient decreased to about 50 percent compared to the very low filling conditions.

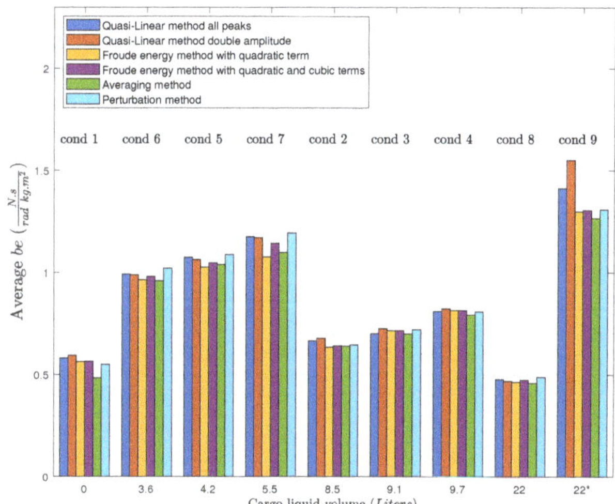

Figure 12. Average equivalent damping coefficient b_e and as a function of the liquid cargo volume for the different test conditions. Note that conditions 8 and 9 have the same cargo volume but different draft, see Tables 3 and A2.

4.3. Effect of Volume Displacement

Increasing the draft from 7.8 cm (condition 8) to 11.7 cm (condition 9), while keeping the liquid cargo volume constant, a dramatic increase in the damping is seen (Figure 13a,b). Figure 13a,b show the decay roll angle response for the two conditions with initial heel angle $\phi_0 = -15°$ where condition 8 require twice as many oscillations compared with condition 9 to reach steady state. A comparison of the equivalent linear damping coefficient between conditions 8 and 9, is shown in Figure 14 where a dramatic increase of the damping is observed when the draft increased. Analyzing condition 8 using methods that considers both quadratic and cubic terms such as (Quasi-linear and Froude energy method with quadratic and cubic terms), it becomes clear that the damping coefficient has nonlinear behaviour as it slightly decreases with increasing roll amplitude, and beyond roll angle $\phi_a = 5^\circ$ the damping is almost constant. However, using methods that only considers the quadratic damping terms such as (Averaging, Perturbation and Froude energy with quadratic term) the damping coefficient is observed to be almost constant for the whole range of roll angle amplitudes as shown in Figure 14.

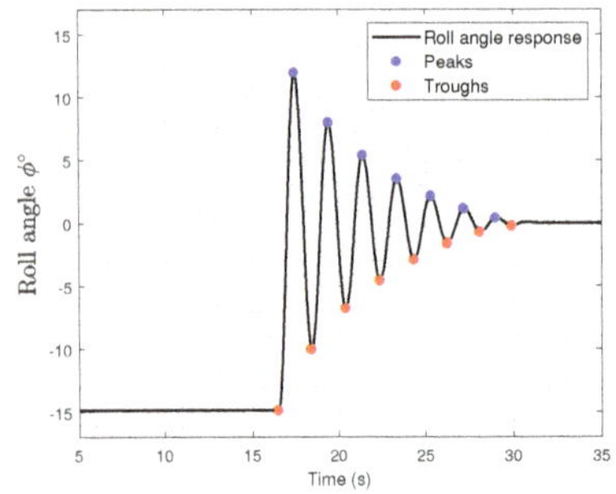

(**a**) Decay test roll angle response for condition 8 with draft = 7.6 cm and cargo = 22 L

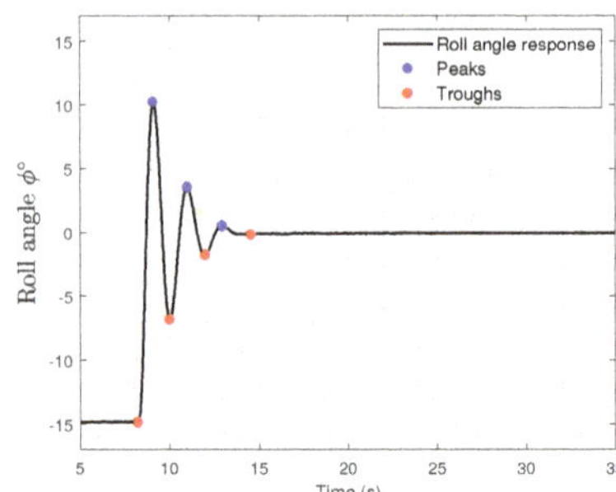

(**b**) Decay test roll angle response for condition 9 with draft = 11.7 cm and cargo = 22 L

Figure 13. Effect of vessel draft on the dynamic response of the FPSO model during roll decay test.

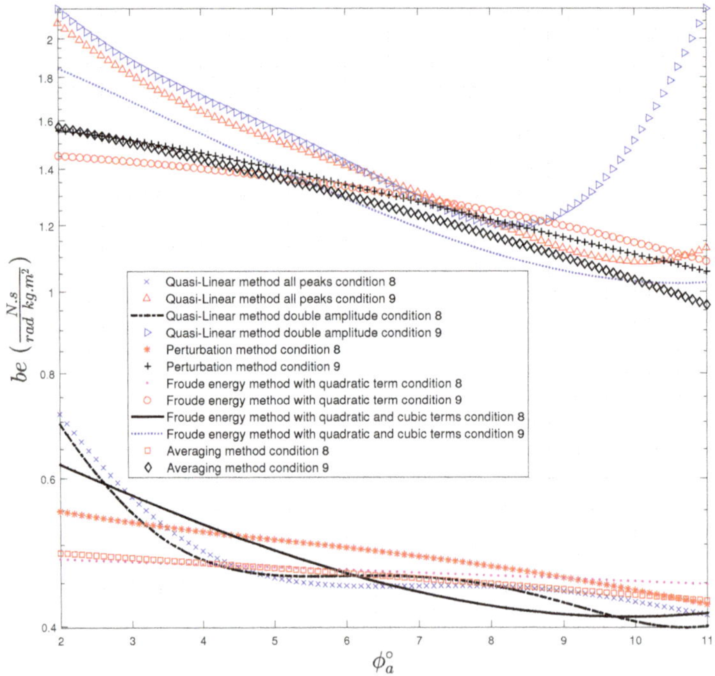

Figure 14. The effect of the vessel draft on the equivalent linear damping coefficient b_e using different decay analysis methods. Condition 8 has a shallow draft equivalent to an unloaded vessel, condition 9 has a draft equivalent to a fully loaded vessel. In both conditions all tanks have a 95± filling level.

5. Conclusions

In this paper an experimental study was conducted to evaluate the equivalent linear damping coefficient of an FPSO vessel with liquid cargo in two two-row tanks at nine different filling levels. The equivalent linear damping was obtained using four different methods. To evaluate the effect of liquid cargo, the draft is kept constant for all but one loading condition. The model is highly damped, therefore multiple roll decay tests at different initial disturbance angle were needed to get a sufficient number of oscillations to obtain the equivalent linear damping.

The effect of liquid cargo volume in the two-row tanks on the vessel roll damping was investigated at constant draft. The results showed that the vessel damping is inversely proportional to the loaded cargo volume, an increase in the liquid cargo results in more liquid sloshing on the tanks walls which produces additional forcing moments, thus increasing the rolling motion and decreasing the vessel damping.

For most of the tested cases, the linear equivalent damping coefficient is observed to be inversely proportional to the roll angle amplitude. As expected, the relationship between the roll angle and equivalent linear damping coefficient depends on the analyzing method; methods that only accounts for the linear and quadratic damping terms such as quadratic Froude energy method, Averaging method and Perturbation method, show a linear relationship. However a nonlinear relation is seen when using methods that accounts for the cubic damping term such as, Quasi-linear method and quadratic-cubic Froude energy method. This implies that the cubic damping term is necessary for this vessel type.

There are no dramatic differences between different roll decay test analysis methods for estimation the equivalent linear damping coefficient except the Averaging method that underestimates the coefficient for this highly dampened system.

The perturbation method is accurate in determining the linear and non-linear damping coefficients but it requires three unknown exponential curve fitting for the decay envelope. Hence, it is recommended to be used for roll decay tests that have large number of oscillations to ensure robust and stable curve fitting or the the curve fitting algorithm should be constrained for an expected range of the unknown values to be estimated.

Therefore, the Froude energy and the Quasi-linear methods are recommended, as they are able to compute the damping coefficient with no restrictions with regards to the system dampening (high or low), nor do they require a large number of peaks (minimum three). The Quasi-linear method has an advance over the Froude energy method; no curve fitting is required and the equivalent damping coefficient is computed directly with minimum.

Author Contributions: Conceptualization, J.-F.I. and M.F.; methodology, J.-F.I., O.S., M.F. and R.F.; software, O.S., M.F. and R.F.; validation, O.S. and M.F.; formal analysis, O.S.; investigation, J.-F.I., O.S., M.F. and R.F.; resources, M.F.; data curation, J.-F.I., O.S., M.F. and R.F.; writing—original draft preparation, J.-F.I., O.S., M.F. and R.F.; writing—review and editing, J.-F.I., O.S., M.F. and R.F.; visualization, J.-F.I., O.S., M.F. and R.F.; supervision, M.F.; project administration, M.F.; funding acquisition, J.-F.I. All authors have read and agreed to the published version of the manuscript.

Funding: This research was funded by Petroleum Technology Development Fund (PTDF) Nigeria.

Acknowledgments: Appreciation goes to the Petroleum Technology Development Fund (PTDF) Nigeria, Davidson Laboratory research group and members of Furth Lab, Stevens Institute of Technology.

Conflicts of Interest: The authors declare no conflict of interest.

Appendix A. Test Equipment

The locations of the ballast and measuring equipment used for the roll decay test are shown in Figure A1 and Table A1. All locations are measured from the aft longitudinally and from the keel for vertical distances. The model was fixed in sway, surge, yaw, pitch and heave, while free to roll. For condition 1 to 8, the model was locked in heave, fixing the draft at the unloaded ballasted condition.

For condition 9, sinkage was allowed by temporarily releasing the heave restrain, after the model reached the draft required to support the load heave was again clamped.

The model was free floating (not attached to the have post) for the inclination test, at the ballasted loading condition. Pivot box and drag balance (10), attachment bar (11) were not loaded in inclination test. The ballast weight distribution and incline weight (12) during the test is shown in Figure A1 and Table A1.

Table A1. Devices and Locations.

Unit Name and Figure Number	LCG [mm] from Aft	VCG [mm] from Keel	Mass [Kg]
FPSO Hull (1) with Ballast (2 and 3) and Empty Cargo Tanks (4 and 5)	1136.0	91.6	24.10
Forward Ballasts (2)			
(Roll Decay Test)			
Port Ballast (TCG = 44.45 mm)	1771.65	117.81	4.72
Starboard Ballast (TCG = −44.45 mm)	1771.65	117.81	4.72
(Inclination Test)			
Port Ballast (TCG = 44.45 mm)	1771.65	110.30	4.15
Starboard Ballast (TCG = −44.45 mm)	1771.65	110.30	4.15
Aft Ballast (3)			
(Roll Decay Test)	400.05	129.81	5.67
(Inclination Test)	400.05	121.80	4.77
Forward Cargo Tank (4)	770.64	Varies	Varies
Aft Cargo Tank (5) with Water Level Gauges	1529.08	Varies	Varies
Heave Post (6) with Pulleys and clamps (7 and 8)	412.50	685.80	3.20
Pivot Box and Drag Balance (10)	412.50	190.55	3.88
Attachment Bar (11)	1677.92	304.8	0.71
Incline Weight (12)	1328.42	280.80	2.04
Strings used to induce roll (9)			

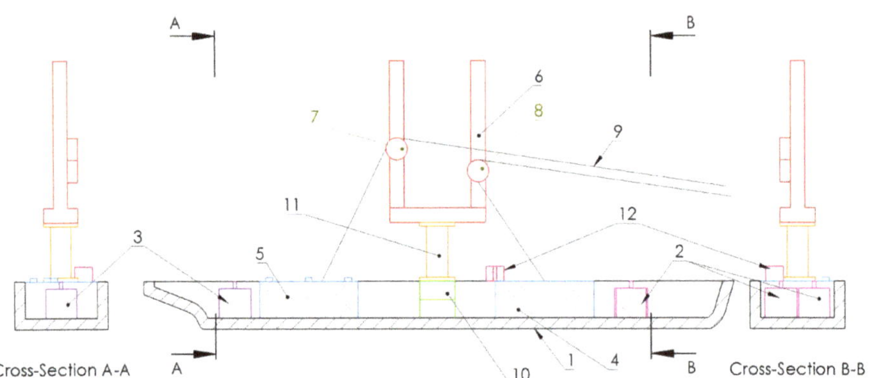

Figure A1. Position of sensors and weights in the model.

Table A2. Draft and draft measuring locations from Aft.

Incline Test (Port)	Location (mm)		612.77	1108.07	1743.07
	Draft (mm)		66	66	66
Roll Decay Test (Starboard)	Location (mm)		615.95	1098.55	1746.25
	Draft (mm)	Condition 1–8	70	78	81
		Condition 9	105	117	125

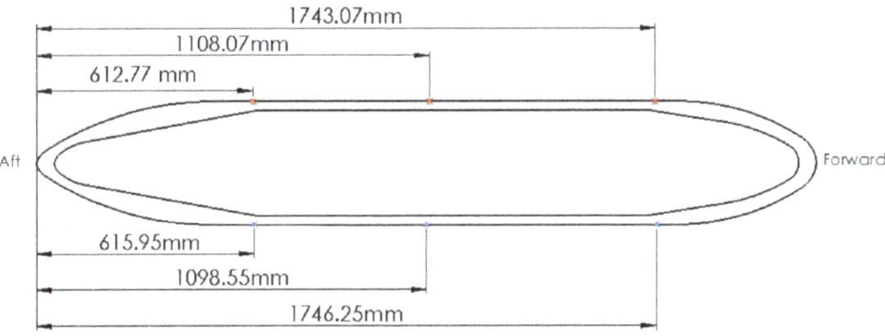

Figure A2. Measuring Locations.

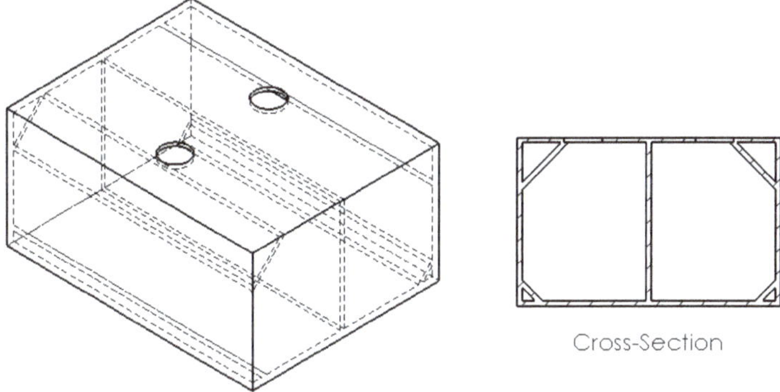

Figure A3. Schematic of Cargo Tank.

Appendix A.1. Ballast Weight and Weight Posts

The ballast weight were secured on three weight post, two in the forward and one in the aft as shown in Figure A1. The weight posts were fixed on the bottom of the model to minimize the motion induced by the ballast.

Appendix A.2. Cargo Tank

Two two-row tanks were installed in the forward and aft of the model as shown in Figure A1. The tanks consisted of two separated compartments, shown in Figure A3, one on the port side and one on the starboard side. The tank was machined with angled edges as shown in Figure A3 to reduce corner sloshing effect. Four water level gauges were installed at the center line of four walls in the port compartment of the aft cargo tank.

Appendix A.3. Towing Carriage and Heaving Post

Towing carriage was served as a testing platform in the roll decay test. Sensors, camera, heave post, pivot box and the model itself were attached to the carriage. The primary use of the carriage is to record the roll angle and to fix the movement of the model in surge, sway, yaw and heave. The heave post and the attachment bar are the component connecting the model to the towing carriage, see Figure A1. A pulley system was setup on the model for roll decay test as shown in Figure A1. Two pulleys were fixed on the heave post and two strings were secured on the starboard side of the model. To ensure a quick release during the test, the string selected has low flexibility and a weight realise system was implemented to avoid slack.

Appendix A.4. Pivot Box

The Pivot Box is attached to the base plate of the model and orientated so that only roll was allowed. The included rotational potentiometer is the primary method to measure roll in the roll decay test. The potentiometer is calibrated to statically to within an accuracy of $\pm$ 0.18 degrees.

References

1. Balthazar, J.M.; Gonçalves, P.B.; Lenci, S.; Mikhlin, Y.V. Models, Methods, and Applications of Dynamics and Control in Engineering Sciences: State of the Art. *Math. Probl. Eng.* **2010**, *2010*. [CrossRef]
2. Guze, S.; Wawrzynski, W.; Wilczynski, P. Determination of Parameters Describing the Risk Areas of Ships Chaotic Rolling on the Example of LNG Carrier and OSV Vessel. *J. Mar. Sci. Eng.* **2020**, *8*, 91. [CrossRef]
3. Kim, Y.; Park, M.J. Identification of the nonlinear roll damping and restoring moment of a FPSO using Hilbert transform. *Ocean Eng.* **2015**, *109*, 381–388. [CrossRef]
4. Arai, M.; Cheng, L.; Inoue, Y.; Sasaki, H.; Yamagishi, N. Numerical analysis of liquid sloshing in tanks of FPSO. In Proceedings of the Second International Offshore and Polar Engineering Conference, San Francisco, CA, USA, 14–19 June 1992.
5. Zhao, W.; Efthymiou, M.; McPhail, F.; Wille, S. Nonlinear roll damping of a barge with and without liquid cargo in spherical tanks. *J. Ocean Eng. Sci.* **2016**, *1*, 84–91. [CrossRef]
6. Chakrabarti, S. Empirical calculation of roll damping for ships and barges. *Ocean Eng.* **2001**, *28*, 915–932. [CrossRef]
7. ITTC. Numerical estimation of roll damping. In Proceedings of the 26th International Towing Tanks Conference, Rio de Janeiro, Brazil, 28 August–3 September 2011; Number 7.5-02-07-04.5.
8. Salvesen, N.; Tuck, E.; Faltinsen, O. *Ship Motions and Sea Loads*; The Society of Naval Architects and Marine Engineers: New York, NY, USA, 1970.
9. Prpić-Oršić, J.; Čorić, V. *Seakeeping*; University of Rijeka: Rijeka, Croatia, 2006.
10. Guo, Z.; Ma, Q.; Yu, S.; Qin, H. A body-nonlinear Green's function method with viscous dissipation effects for large-amplitude roll of floating bodies. *Appl. Sci.* **2018**, *8*, 517. [CrossRef]
11. Froude, W.; Abell, S.W.; Gawn, R.; Duckworth, A. *The Papers of William Froude, 1810–1879: With a Memoir by Sir Westcott Abell and an Evaluation of William Froude's Work by RWL Gawn: Collected Into One Volume*; Institution of Naval Architects: London, UK, 1955.
12. Ikeda, Y.; Himeno, Y.; Norio, T. A prediction method for ship roll damping. In *Report No. 00405 of the Department of Naval Architecture*; University of Osaka Prefecture: Osaka, Japan, 1978.
13. Kianejad, S.; Lee, J.; Liu, Y.; Enshaei, H. Numerical assessment of roll motion characteristics and damping coefficient of a ship. *J. Mar. Sci. Eng.* **2018**, *6*, 101. [CrossRef]
14. Remola, A.O. On Ship Roll Damping: Analysis and Contributions on Experimental Techniques. Ph.D. Thesis, Universidad Politécnica de Madrid, Madrid, Spain, 2018.
15. Handschel, S.; Feder, D.F.; Abdel-Maksoud, M. Estimation of ship roll damping-A comparison of the decay and the harmonic excited roll motion technique for a post panamax container ship. In Proceedings of the 12th International Conference on the Stability of Ships and Ocean Vehicles, Glasgow, UK, 14–19 June 2015; pp. 475–488.
16. Bačkalov, I.; Bulian, G.; Cichowicz, J.; Eliopoulou, E.; Konovessis, D.; Leguen, J.F.; Rosén, A.; Themelis, N. Ship stability, dynamics and safety: Status and perspectives from a review of recent STAB conferences and ISSW events. *Ocean Eng.* **2016**, *116*, 312–349. [CrossRef]
17. Kinnas, S.A.; Yi-Hsiang, Y.; Vinayan, V. Prediction of flows around FPSO hull sections in roll using an unsteady Navier-Stokes solver. In Proceedings of the Sixteenth International Offshore and Polar Engineering Conference, San Francisco, CA, USA, 28 May–2 June 2006.
18. Thiagarajan, K.P.; Braddock, E.C. Influence of bilge keel width on the roll damping of FPSO. *J. Offshore Mech. Arct. Eng.* **2010**, *132*, 011303-1–011303-7. [CrossRef]
19. Blume, P. Experimentell bestimmung von koeffizienten der wirksamen rolldämpfung und ihre anwedung zur abschätzung extremer rollwinkel. In *Institut für Schiffbau, Universität Hamburg, Bericht Nr. 1511, Zeitschrifft "Schiffstechnik', Band 26, 1979*; Universität Hamburg: Hamburg, Germany, 1979.

20. Handschel, S.; Abdel-Maksoud, M. Improvement of the harmonic excited roll motion technique for estimating roll damping. *Ship Technol. Res.* **2014**, *61*, 116–130. [CrossRef]
21. Wassermann, S.; Feder, D.F.; Abdel-Maksoud, M. Estimation of ship roll damping—A comparison of the decay and the harmonic excited roll motion technique for a post panamax container ship. *Ocean Eng.* **2016**, *120*, 371–382. [CrossRef]
22. Guarniz Avalos, G.O.; Wanderley, J.B. A Two-Dimensional Numerical Simulation of Roll Damping Decay of a FPSO Using the Upwind TVD Scheme of Roe-Sweby. In *International Conference on Offshore Mechanics and Arctic Engineering*; American Society of Mechanical Engineers: New York, NY, USA, 2012; Volume 44915, pp. 395–402.
23. Avalos, G.O.; Wanderley, J.B.; Fernandes, A.C.; Oliveira, A.C. Roll damping decay of a FPSO with bilge keel. *Ocean Eng.* **2014**, *87*, 111–120. [CrossRef]
24. Zhou, Y.H.; Ma, N.; Shi, X.; Zhang, C. Direct calculation method of roll damping based on three-dimensional CFD approach. *J. Hydrodyn.* **2015**, *27*, 176–186. [CrossRef]
25. Gu, Y.; Boulougouris, E.; Day, A. A Study on the Effects of Bilge Keels on Roll Damping Coefficient. In Proceedings of the 12th International Conference on the Stability of Ships and Ocean Vehicles, Glasgow, UK, 14–19 June 2015.
26. Rodríguez, C.A.; Esperança, P.T.; Oliveira, M.C. Estimation of Roll Damping Coefficients Based on Model Tests Responses of a FPSO in Waves. In *International Conference on Offshore Mechanics and Arctic Engineering*; American Society of Mechanical Engineers: New York, NY, USA, 2019; Volume 58851, p. V07BT06A021.
27. Zeraatgar, H.; Asghari, M.; Bakhtiari-Nejad, F. A study of the roll motion by means of a free decay test. *J. Offshore Mech. Arct. Eng.* **2010**, *132*, 031303. [CrossRef]
28. Oliva-Remola, A.; Bulian, G.; Pérez-Rojas, L. Estimation of damping through internally excited roll tests. *Ocean Eng.* **2018**, *160*, 490–506. [CrossRef]
29. Spouge, J. Non-linear analysis of large-amplitude rolling experiments. *Int. Shipbuild. Prog.* **1988**, *35*, 271–324.
30. Bulian, G. Estimation of nonlinear roll decay parameters using an analytical approximate solution of the decay time history. *Int. Shipbuild. Prog.* **2004**, *51*, 5–32.
31. ESilva, S.R.; Soares, C.G. Prediction of parametric rolling in waves with a time domain non-linear strip theory model. *Ocean Eng.* **2013**, *72*, 453–469.
32. Uzunoglu, E.; Soares, C.G. Automated processing of free roll decay experimental data. *Ocean Eng.* **2015**, *102*, 17–26. [CrossRef]
33. Sun, J.; Shao, M. Estimation of Nonlinear Roll Damping by Analytical Approximation of Experimental Free-Decay Amplitudes. *J. Ocean Univ. China* **2019**, *18*, 812–822. [CrossRef]
34. Roberts, J.B. *Estimation of Non-Linear Ship Roll Damping from Free-Decay Data*; Technical Report; Journal of Ship Research, Vol. 29, No. 2; The Society of Naval Architects and Marine Engineers; University of Sussex: Brighton, UK, 1985; pp. 127–138.
35. Flower, J.; Sabti Aljaff, W. Kryloff-Bogoliuboff's solution to decaying nonlinear oscillations in marine systems. *Int. Shipbuild. Prog.* **1980**, *27*, 225–230. [CrossRef]
36. Himeno, Y. *Prediction of Ship Roll Damping—A State of the Art*; Technical Report; University of Michigan: Ann Arbor, MI, USA, 1981.
37. Dalzell, J. *A Note on the Form of Ship Roll Damping*; Technical Report; Stevens Inst of Tech Hoboken N Davidson Lab.: Hoboken, NJ, USA, 1976.
38. Zubaly, R.B. *Applied Naval Architecture*; Cornell Maritime Press, Schiffer Publishing, Ltd.: Atglen, PA, USA, 1996.
39. Solutions, D.D. *ORCA3D User Manual*; DRS Technology: Largo, FL, USA, 2015.
40. Iqbal, K.S.; Bulian, G.; Hasegawa, K.; Karim, M.; Mashud, A. Interim guidelines for alternative assessment of the weather criterion Interim guidelines for alternative assessment of the weather criterion, 2006. *J. Mar. Sci. Technol.* **2008**, *13*, 282–290. [CrossRef]

Journal of
Marine Science and Engineering

Article

Ship Towed by Kite: Investigation of the Dynamic Coupling

Nedeleg Bigi [1], Kostia Roncin [2,*], Jean-Baptiste Leroux [3] and Yves Parlier [4]

1 Benjamin Muyl Design, 56400 Auray, France; nedelegbigi@gmail.com
2 French Air Force Academy, CREA, Base aérienne 701, F-13661 Salon Air, France
3 ENSTA Bretagne, FRE CNRS 3744, IRDL, F-29200 Brest, France; jean-baptiste.leroux@ensta-bretagne.fr
4 Beyond the Sea, 1010 avenue de l'Europe, 33260 La Teste de Buch, France; yves.parlier@beyond-the-sea.com
* Correspondence: kostia.roncin@ecole-air.fr

Received: 27 May 2020; Accepted: 28 June 2020; Published: 1 July 2020

Abstract: This paper presents a series of dynamic simulations for a ship towed by kite. To ensure time efficient computations, seakeeping analysis with forward speed correction factors is carried out in the frequency domain and then transformed in the time domain by convolution. The seakeeping modeling is coupled with a zero-mass kite modeling assuming linear dependence of aerodynamic characteristics with respect to turning rate. Decoupled (segregated) and coupled (monolithic) approaches are assessed and compared in different environmental conditions. Results show that in regular beam waves, strong interactions between the kite and the ship motions are captured by the monolithic approach. Around the wave frequency, especially for the lower one tested (0.4 rad/s), a kite lock-in phenomenon is revealed. It is concluded that the mean kite towing force can be increased whereas the ship roll amplitude can even be decreased compared to a non-kite assisted ship propulsion configuration.

Keywords: ship dynamics; kite assisted propulsion; time-domain seakeeping simulations; dynamic coupling lock in effects

---

1. Introduction

This work is part of the beyond the sea® research program launched by Yves Parlier and managed in partnership with French Ministry of Defence with the support of French Environment and Energy Management Agency (ADEME). The project attempts to develop a tethered kite system as an auxiliary device for the propulsion of merchant ships. The existing knowledge on ships towed by kite has demonstrated great prospects for this technology in terms of CO_2 emissions and fuel savings. However, studies on the limitations of this concept to guarantee the safety and the integrity of the ship are very limited.

Considering the mean kite towing force, Leloup et al. [1] and Naaijen et al. [2] solved the horizontal balance equations of a ship towed by kite to determine the fuel savings. Ran et al. [3] studied the contribution of a kite to the mean ship thrust, drift angle and rudder angle. All these previous studies neglected the interactions between the kite and the ship. The kite force is imposed as a predefined external force to the ship. The motions of such a system are highly dynamic since a kite experiences a periodic dynamic flight. In Bigi et al. [4], the influence of the kite attachment point on the deck is investigated on a fishing vessel equipped with a kite. This study is limited to horizontal ship motions. A maneuvering modeling is implemented with a monolithic coupling approach between the ship and the kite. The water is supposed to be calm and Bigi et al. [4] did not take into account the effect of the radiated waves on the ship motions. Thus, the influence of the kite excitation frequency on the added mass and damping of the ship is neglected. Since the hydrodynamic added mass and damping may depend strongly on the frequency of the motion [5–8], this assumption is questionable.

A ship sailing in waves is commonly studied through seakeeping codes based on the potential flow theory under the assumption of linear response of the ship to a given perturbation on a mean path. These studies are usually carried out in the frequency domain [9,10]. However, since the kite and the ship may be strongly coupled, their interactions cannot be directly computed through a spectral description of the kite excitation. Such computations are therefore limited to weak coupling. To perform a strong coupling between the kite and the ship, a time domain formulation is required. The aim of this paper is to assess the importance of taking into account the coupling between the kite and the ship motions by a time domain method.

As highlighted by Skejic [11], fast time-domain methods able to compute the 6 dof combining horizontal and vertical motions of a ship are based on the concept of linear convolution [12–14]. Such methods are first introduced by Cummins [15] in 1962. Böttcher [12] developed unified theories that couple the linear manoeuvring and seakeeping. In specific, he extended the method to simulate large amplitude motions taking into account nonlinear effects with Froude–Krilov effort and nonlinear roll damping for instance. To enable a fast computation of the linear convolution product, the use of state-space systems is introduced [13,16,17]. The identification of the state-space systems is detailed in [18] for the zero forward speed case. In this paper, this method is extended to the forward speed case.

Kite modeling is formulated under the zero mass assumption. Behrel et al. [19] presents experimental results comparing zero mass and point mass modeling. Differences are about few percent. Actually, both modelings give the same results when their coefficients identification is consistent. If taking into account the mass can be important for control issue, this is clearly not the case for performance assessment. Studies on kite performance assessment as ship propulsion device are based on zero-mass assumption (e.g., Wellicome and Wilkinson [20], Naaijen et al. [2], Dadd et al. [21] and Leloup et al. [1]). In addition, a linear evolution of the kite aerodynamics with the radius of trajectory curvature is proposed in this paper.

Section 2 presents the considered modeling of a ship towed by kite. Section 3 presents the results. First, the kite characteristics identification is detailed, then Section 3.2 defines the case study on the surface vessel combatant DTMB 5512. Sections 3.3 and 3.4 investigate respectively through a calm water case and a regular beam wave case, the coupling between ship and kite. A general discussion on the methods and results is provided in Section 4 before to conclude.

2. Ship Towed by Kite Modeling

2.1. Reference Frame and Parametrization

The Earth fixed frame **n**, the ship fixed frame **s** and the hydrodynamic frame **h** are sketched in Figure 1. The earth fixed frame is centered on O_n with $\underline{z}_n$ pointing downward. The ship fixed frame is composed of $\underline{x}_s$ pointing forward in the ship symmetry plane, $\underline{z}_s$ pointing downward and $\underline{y}_s$ completing the direct orthogonal basis. $\underline{z}_s$ is normal to free surface when the ship is at the equilibrium. The origin of **s** denoted by O_s is in the ship symmetry plane at a longitudinal distance l_x from the aft perpendicular and at a vertical distance l_z from the baseline. For a point P, the expression of its coordinates in a frame **f** is $\underline{P}^{(\mathbf{f})} = \left[p_x^{(\mathbf{f})},\, p_y^{(\mathbf{f})},\, p_z^{(\mathbf{f})}\right]^T$ or $\left[p_x,\, p_y,\, p_z\right]_{\mathbf{f}}^T$. The vectors of a frame **f** are $\underline{x}_f$, $\underline{y}_f$ and $\underline{z}_f$ and the origin **f** is denoted O_f. The generalized position vector of the ship denoted by $\underline{S} = \left[s_x^{(\mathbf{n})},\, s_y^{(\mathbf{n})},\, s_z^{(\mathbf{n})},\, \phi_s^{(\mathbf{n})},\, \theta_s^{(\mathbf{n})},\, \psi_s^{(\mathbf{n})}\right]^T$ is the assembly of the position of O_s and the ship Euler's angles with respect to the **n** frame. The generalized ship velocity at O_s expressed in **s** with respect to the **n** frame is denoted by $\underline{V}_s = \left[u_s,\, v_s,\, w_s,\, p_s,\, q_s,\, r_s\right]_{\mathbf{s}}^T$, where the first three components are the linear velocities and the last three components are the turning rates. The transformation of a vector expressed in the **s** frame denoted by $\underline{n}^{(\mathbf{s})}$ can be expressed in the **n** frame with $\underline{n}^{(\mathbf{n})} = \underline{\underline{T}}_{\mathbf{s}}^{\mathbf{n}}\underline{n}^{(\mathbf{s})}$, where $\underline{\underline{T}}_{\mathbf{s}}^{\mathbf{n}}$ is the direct cosine matrix expressed as:

$$\underline{\underline{T}}_{\mathbf{s}}^{\mathbf{n}} = \begin{bmatrix} c\psi_s c\theta_s & -s\psi_s c\phi_s + c\psi_s s\theta_s s\phi_s & s\psi_s s\phi_s + c\psi_s c\phi_s s\theta_s \\ s\psi_s c\theta_s & c\psi_s c\phi_s + s\phi_s s\theta_s s\psi_s & -c\psi_s s\phi_s + s\theta_s s\psi_s c\phi_s \\ -s\theta_s & c\theta_s s\phi_s & c\theta_s c\phi_s \end{bmatrix} \quad (1)$$

where, c and s denote respectively the cosine and the sinus functions. The turning rates and the time derivatives of the ship Euler's angles satisfy the following relationship:

$$\begin{bmatrix} p_s \\ q_s \\ r_s \end{bmatrix} = \underline{\underline{R}}_{\mathbf{n}}^{\mathbf{s}} \begin{bmatrix} \dot{\phi}_s \\ \dot{\theta}_s \\ \dot{\psi}_s \end{bmatrix}, \quad (2)$$

where,

$$\underline{\underline{R}}_{\mathbf{n}}^{\mathbf{s}} = \begin{bmatrix} 1 & 0 & -s\theta_s \\ 0 & c\phi_s & c\theta_s s\phi_s \\ 0 & -s\phi_s & c\phi_s c\theta_s \end{bmatrix}. \quad (3)$$

The hydrodynamic frame is used by the seakeeping theory. It can be considered as Galilean since **h** moves at the constant mean ship forward speed U_h on a straight path with respect to the earth fixed frame. The generalized position vector of the ship expressed in the **h** frame is denoted by $\underline{\xi} = [\xi_1, \xi_2, \xi_3, \xi_4, \xi_5, \xi_6]_{\mathbf{h}}^T$, where the first three components are the distance between O_h and a point H fixed to the ship. The three last components are the Euler's angle of the ship with respect to the **h** frame. It is assumed that H is in the symmetry plane of the ship and $\underline{O_s H} = [d_x, 0, d_z]_{\mathbf{s}}^T$ with respect to the **s** frame.

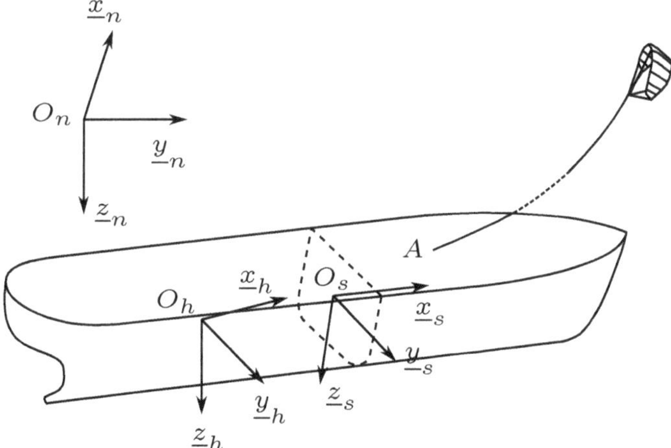

Figure 1. Earth fixed reference frame; hydrodynamic reference frame; ship fixed reference frame.

Figure 2 illustrates the notations used for the kite modeling. The relative wind coordinates system moves at the tether attachment point velocity $\underline{U}_A$ with respect to the Earth fixed frame. The relative wind velocity defined in Equation (4) is the difference between the true wind velocity and the velocity of the tether attachment point.

$$\underline{U}_{rw} = \underline{U}_{tw} - \underline{U}_A \quad (4)$$

$\underline{x}_{rw} = \underline{U}_{rw} / \|\underline{U}_{rw}\|$ and $\underline{y}_{rw}$ is defined as: $\underline{y}_{rw} = \underline{z}_n \times \underline{x}_{rw} / \|\underline{z}_n \times \underline{x}_{rw}\|$. In order to define a direct orthogonal coordinates system, $\underline{z}_{rw}$ is defined as follows: $\underline{z}_{rw} = \underline{x}_{rw} \times \underline{y}_{rw}$. It should be noticed that the relative wind frame depends on the considered altitude due to the wind gradient. Consequently, the relative wind frame is defined with respect to a given point P and is thus denoted $\mathbf{rw}(P)$. For instance, the relative wind frame drawn in Figure 2 is $\mathbf{rw}(K)$. At the kite location, the relative wind frame is obtained from:

$$\underline{U}_{rw(K)} = \underline{U}_{tw(K)} - \underline{U}_A \tag{5}$$

The apparent kite wind velocity is denoted by $\underline{U}_{aw}$ and can be expressed as follows:

$$\underline{U}_{aw} = \underline{U}_{\mathbf{rw}(K)} - \underline{U}_k \tag{6}$$

where $\underline{U}_k$ denotes the kite velocity with respect to the $\mathbf{rw}(K)$ frame; the kite reference frame centered on K is defined as follows: $\underline{z}_k = \underline{AK}/\ \|\underline{AK}\|$, $\underline{y}_k = \underline{z}_k \times \underline{z}_{rw}/\ \|\underline{z}_k \times \underline{z}_{rw}\|$ and $\underline{x}_k = \underline{y}_k \times \underline{z}_k$; the kite velocity is directed by $\underline{x}_{vk}$. The kite elevation angle is given by:

$$\theta_k = \frac{\pi}{2} - \arccos\left(\underline{z}_k \cdot \underline{z}_{rw}\right) \tag{7}$$

The kite azimuth angle is defined as follows:

$$\phi_k = \arccos\left(\underline{y}_k \cdot \underline{y}_{rw}\right) \tag{8}$$

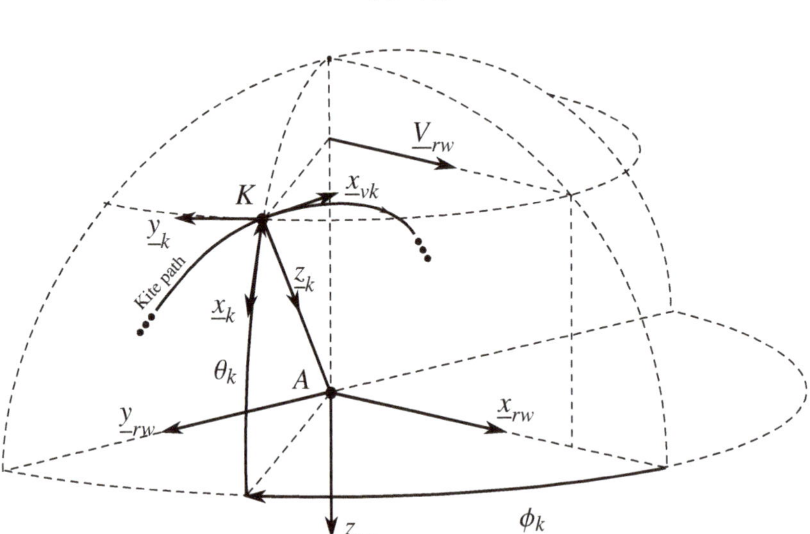

Figure 2. Kite reference frames.

2.2. Ship Modeling

As introduced in Section 1, a time domain formulation is a priori mandatory to represent the complete interaction between kite and ship motions. For the ship motion's modeling part, we implemented the very classical and efficient method described by Fossen [22]. Appendix A gives details on the state space technique used. Appendix A.1 describes the equations of motion for a ship in a seaway. Appendix A.2 shows how hydrodynamics are computed using strip method theory and how is carried out the computation of linear convolution terms. Appendix A.3 provides details on the calculation of wave forces.

2.2.1. Ship Motions

The equations describing the motion of the system can be transformed into a system of first order differential equations as in Equation (9). This system is obtained with the transformation of the generalized ship velocity vector from the **s** frame to the **n** frame and Equations (A9) and (A12).

$$\begin{cases} \dot{\underline{S}} & = \begin{bmatrix} \underline{\underline{T}}^{\mathbf{n}}_{\mathbf{s}} & \underline{\underline{0}} \\ \underline{\underline{0}} & \underline{\underline{R}}^{\mathbf{n}}_{\mathbf{s}} \end{bmatrix} \underline{V}_s \\ \dot{\underline{V}} & = \left[\underline{\underline{M}}_S + \underline{\underline{\tilde{A}}} \right]^{-1} \left[\underline{F} - \left[\underline{\underline{\tilde{B}}} + \underline{\underline{B}}_{\phi} + \underline{\underline{D}} \right] \underline{V}_s - \underline{\mu} - \underline{\underline{C}}\, \underline{S} \right] \\ \dot{\underline{y}}_{ij} & = \underline{\underline{A}}'_{ij} \underline{y}_{ij} + \underline{B}'_{ij} \delta V_{s,j}, \ \forall i,j \in [\![1;6]\!] \end{cases} \quad (9)$$

Equation (9) represents 12 scalar equations for the ship and 75 scalars equations for the state-space systems. Assuming, for instance, that the order of each state space system is 5 and taking into account the ship symmetry. Thus, with the present modeling, the ship is described by 87 scalar equations depending on the state space modeling orders. This system of differential equations is numerically integrated with the 4th order Runge–Kutta scheme with a fixed time step.

2.2.2. Ship Modeling Validation

The present ship modeling was compared to Experimental Fluid Dynamic (EFD) data and to the STF strip theory results on the David Taylor Model Basin (DTMB) 5512. The model tested is a 1:46.6 scale model. The hull form and its characteristics are respectively plotted in Figure 3 and summarized in Table 1. The experimental data are provided by the University of Iowa [23] and are presented in Irvine et al. [24]. The EFD data concern the heave and pitch motions in regular head waves, with and without forward speed.

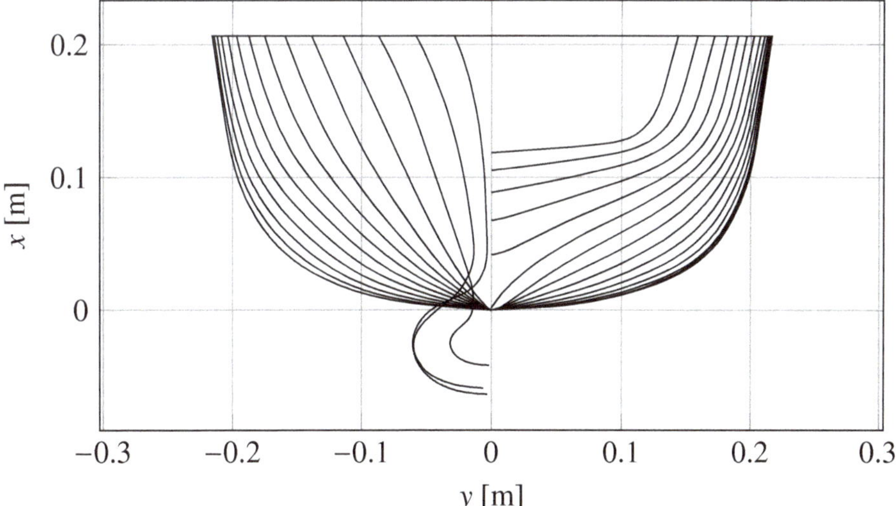

Figure 3. DTMB 5512 hull sections at the scale of 1:46.6.

Table 1. DTMB 5512 hull and full scale characteristics.

Parameter	Units	5512	Full Scale
Scale ratio	-	46.6	1
Length, L_{pp}	m	3.048	142.04
Beam, B	m	0.405	18.87
Draft, T	m	0.132	6.15
Weight	Kg - t	86.6	8763.5
LCG	m	1.536	71.58
VCG	m	0.162	7.55
Pitch radius of gyration, k_5	m	0.764	35.6

The computed ship motions were performed at zero forward speed and at a Froude number of 0.28, which corresponds to $U = 1.53$ m.s^{-1} and with frequency head waves from 1 rad.s^{-1} to 7.5 rad.s^{-1}. O_s was defined at the ship center of gravity, hence $l_x = LCG$ and $l_z = VCG$. Figures 4 and 5 plot the heave and pitch transfer function obtained with the experimental data, with the STF strip theory and with the present modeling. The experimental data are obtained for different wave steepness $s_w = \{0.025, 0.05, 0.075\}$. The amplitude of the transfer function for heave motion (a) is directly the ratio of the heave amplitude motion to the wave amplitude. The amplitude of the transfer function for the pitch motion (c) is given by the ratio of the motion amplitude (in radians) to the wave amplitude multiplied by the wave number k. The phase angle of the present modeling is obtained by cross correlation between the free surface elevation and the ship motion time series.

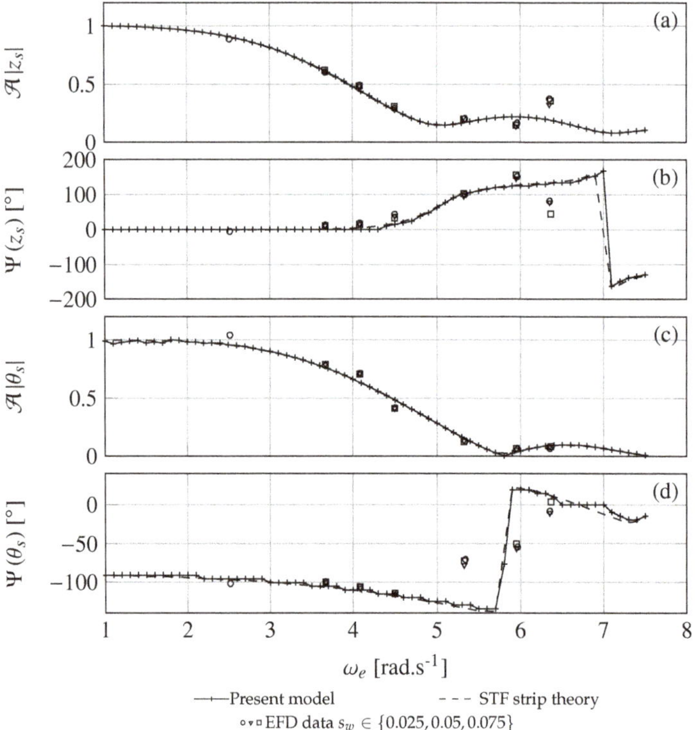

Figure 4. Heave and pitch transfer function at function of the frequency of encounter ω_e. Amplitudes are plotted for heave (**a**) and pitch (**c**). Phases are plotted for heave (**b**) and pitch (**d**). The results are obtained with the frequency domain and time domain approaches, experimental data for different wave steepness s_w and with the STF strip theory.

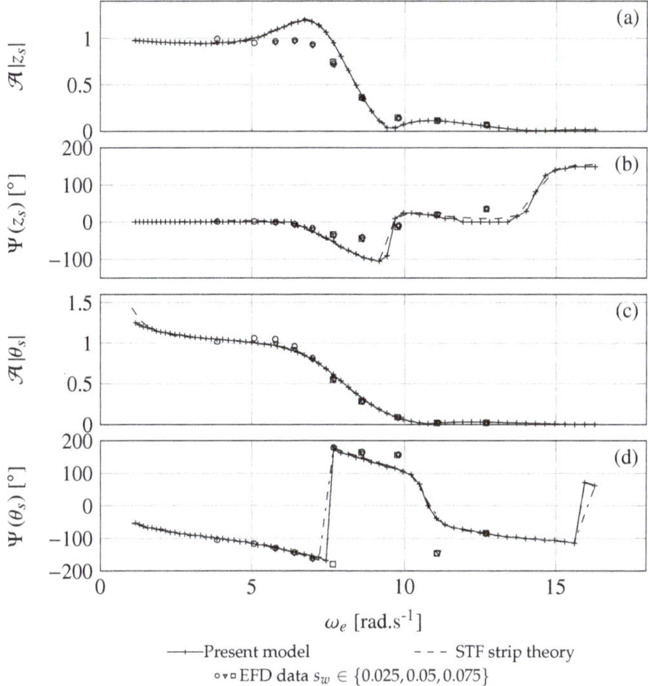

Figure 5. Heave and pitch transfer function at $U = 1.53$ m.s^{-1} according to the frequency of encounter ω_e. Amplitudes are plotted for heave (**a**) and pitch (**c**). Phases are plotted for heave (**b**) and pitch (**d**). The results are obtained with the frequency domain and time domain approaches, experimental data for different wave steepness s_w and with the STF strip theory.

Concerning the amplitude, an overall good agreement is found with (Figure 5) and without (Figure 4) forward speed between the EFD data, the STF strip theory, the time domain and the frequency domain modeling. For the considered waves, the influence of the wave steepness on the EFD data is not significant. As shown in Figures 4 and 5, the STF strip theory and the time domain approach match with a very good accuracy for the amplitude. Very small differences can be observed in terms of phase angle, but these differences are probably caused by the post-processing method (Figures 4 and 5). The very small differences with the STF strip theory are due to the approximations performed with the identification method of the transfer functions H_{ij}. As a conclusion for the heave and pitch motions, the transformation of the equation of motion into the ship reference frame and the state-space modeling identification method are consistent and accurate enough.

The roll motion is much more difficult to predict with a linear modeling. Full-scale results are presented in Figure 6. Results by Ikeda et al. [25] are considered as a reference. In the present method, the extra roll damping was set to $1.53 \cdot 10^8$ kg.m.s^{-1} in order to obtain the same roll amplitude at the natural frequency. As shown in Figure 6, for frequencies greater than 1 rad.s^{-1}, the roll motion amplitude is well predicted despite a slight difference in phase. Nevertheless predictions appear to be less accurate for the lower frequencies, except for the natural roll frequency, $\omega_{roll} = 0.56$ rad.s^{-1}, where, by identification, both modelings match.

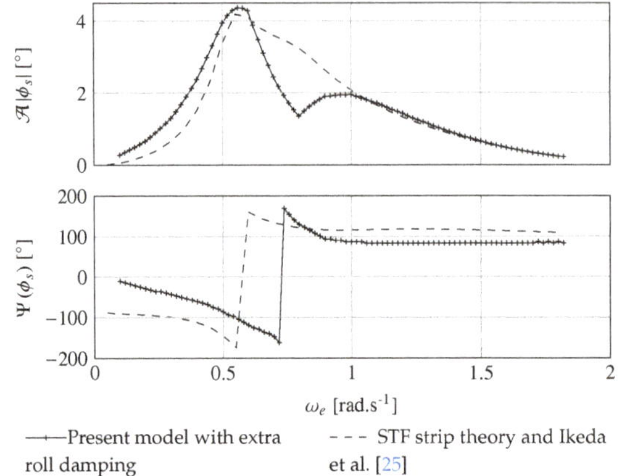

Figure 6. Roll response amplitude operator and phase at $U = 7.716$ m.s^{-1} as function of the frequency of encounter w_e with DTMB 5512 at full scale. The results are obtained with the present model and the STF strip theory with the roll damping modeled with the method proposed by Ikeda et al. [25].

2.3. Kite Modeling

2.3.1. Short Literature Survey

Kites are mostly studied as a power generation system since 1980 [26]. Comparing the weight of the kite on the one hand and the traction provided on the other, Loyd considers that it is relevant to neglect the mass of the kite. Thus, he developed the so-called zero-mass modeling. Taking the skysails system, the most advanced industrial example, gives us a good illustration of the relevance of this assumption. Paulig et al. [27] computes a pulling force to mass ratio of 50daN/kg in the case of a 320 m^2 kite with 360 m tether. Wellicome extended the zero-mass modeling defining a figure-8 trajectory for the dynamic mode [20,28]. Argatov et al. [29] proposed more recently a simpler trajectory definition. In addition, Argatov definition allows a much faster calculation and its formulation avoids discontinuities in kite acceleration. The size of the trajectory can easily be modified thanks only to azimuth and elevation amplitude parameters. In 2006, Naaijen et al. [2] computes the fuel savings on a full range of true wind angles for the British Bombardier, a 50000 DWT tanker. In 2011, Dadd et al. [21] improved the calculations with more realistic trajectories using the Wellicome formalism. In 2014, Leloup et al. [1,30] proposed a fully analytical solution for the zero-mass modeling that allows faster and more reliable computations avoiding optimization process failures. Leloup takes into account adequately the wind gradient effect and takes advantage of the Argatov trajectory definition for even faster computations.

2.3.2. Zero-Mass Kite Model

According to the zero-mass kite modeling, the masses of the tether and the kite are neglected and the tether is assumed to be straight and of constant length L_t. Consequently, the kite velocity with respect to the **rw** (K) frame is normal to $\underline{z}_{k0}$ and for any configuration the tether tension is opposed to the aerodynamic kite force. Assuming that the kite flies with a constant lift to drag ratio and that the apparent wind velocity is in its symmetry plane, Leloup et al. [1] expressed the kite velocity according to a given kite velocity direction $\underline{x}_{vk}$ in Equation (10):

$$\underline{U}_k = U_{rw}\left[(\underline{x}_{vk}\cdot\underline{x}_{rw}) + \sqrt{(\underline{x}_{vk}\cdot\underline{x}_{rw})^2 + \left(\frac{\underline{z}_{k0}\cdot\underline{x}_{rw}}{\sin\epsilon_k}\right)^2 - 1}\right]\underline{x}_{vk} \tag{10}$$

The kite velocity is a real number if the following condition is satisfied:

$$(\underline{x}_{vk} \cdot \underline{x}_{rw})^2 + \left(\frac{\underline{z}_{k0} \cdot \underline{x}_{rw}}{\sin \epsilon_k} \right)^2 - 1 \geqslant 0. \tag{11}$$

Then, the tether tension is given by the following formula:

$$\underline{T}_k = -\frac{C_{lk} \rho_a A_k U_{aw}^2}{2 \cos \epsilon_k} \underline{z}_k \tag{12}$$

The generalized tether force vector acting on the ship at O_s with respect to **s** frame is expressed as follows:

$$\underline{F}_k = \begin{bmatrix} \underline{T}_k^{(\mathbf{s})} & \underline{O_s A}^{(\mathbf{s})} \times \underline{T}_k^{(\mathbf{s})} \end{bmatrix}^T \tag{13}$$

In order to represent the wind friction with the sea, the true wind velocity $\underline{U}_{tw}$ is function of the altitude according to the wind gradient law recommended by the ITTC [31].

$$\underline{U}_{tw}(k_z^{(\mathbf{n})}) = \underline{U}_{10} \left(\frac{k_z^{(\mathbf{n})}}{10} \right)^{1/7} \tag{14}$$

2.3.3. Kite Control According to a Trajectory

The kite velocity direction $\underline{x}_{vk}$ is controlled in order to follow a trajectory denoted by $\mathscr{C}$ (Figure 7). The kite velocity direction is defined by the target point $\tilde{K}$ expressed as follows:

$$\tilde{K} = \mathscr{C} \left(\lambda + \| \underline{U}_k \| \, dt \right), \tag{15}$$

where λ is the curvilinear abscissa of C_λ the closest point of the trajectory from the current kite position K. Hence, the kite velocity direction $\underline{x}_{vk}$ is defined as follows:

$$\underline{x}_{vk} = \frac{\left(\underline{K\tilde{K}} \cdot \underline{x}_k \right) \underline{x}_k + \left(\underline{K\tilde{K}} \cdot \underline{y}_k \right) \underline{y}_k}{\left\| \left(\underline{K\tilde{K}} \cdot \underline{x}_k \right) \underline{x}_k + \left(\underline{K\tilde{K}} \cdot \underline{y}_k \right) \underline{y}_k \right\|} \tag{16}$$

The point C_λ is determined according to the following equation:

$$\underline{KC_\lambda} \cdot \underline{t}_\lambda = 0, \tag{17}$$

where $\underline{t}_\lambda$ is the tangent vector to the trajectory at curvilinear abscissa λ. Equation (17) is solved numerically with a Newton–Raphson algorithm.

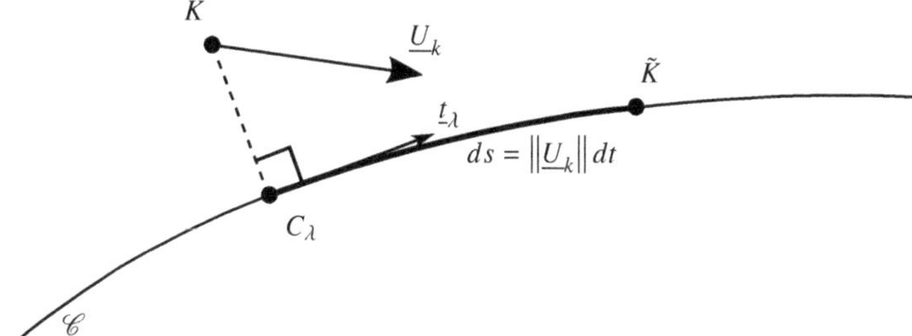

Figure 7. Schema describing the target point to control the kite velocity direction $\underline{x}_{vk}$.

2.3.4. Theoretical Lemniscate Trajectory

To perform a dynamic flight, the considered theoretical trajectory is the eight shaped Lissajous curve as used by Argatov et al. [29] and Leloup [32]. This type of trajectory has the advantage to avoid twists of the tether system. The Lissajous trajectories are defined in terms of elevation θ_C and azimuth ϕ_C:

$$\begin{cases} \theta_C &= \Delta\theta_8 \sin(2\alpha) + \theta_8 \\ \phi_C &= \Delta\phi_8 \sin(\alpha) + \phi_8 \end{cases}, \tag{18}$$

where, $\alpha \in [0; 2\pi]$. θ_8 and ϕ_8 are the elevation and the azimuth of the center of the trajectory denoted by C_8. The trajectory is defined with respect to $\mathbf{rw}\left(z_{ref}\right)$, the relative wind frame with respect to the wind measurement altitude. On a sphere of radius L_t, a point C of the trajectory is defined in terms of elevation θ_C and azimuth ϕ_C with respect to $\mathbf{rw}\left(z_{ref}\right)$ frame as follows:

$$\underline{C} = \begin{bmatrix} L_t \cos\theta_C \cos\phi_C \\ L_t \cos\theta_C \sin\phi_C \\ -L_t \sin\theta_C \end{bmatrix}_{\mathbf{rw}\left(z_{ref}\right)}, \tag{19}$$

As shown in Figure 8, the lemniscate trajectories can be rotated by an angle χ_8 around $\underline{C_8 A}$. Any point of the trajectory must satisfy Equation (11) insuring the realness of the kite velocity. The benefit to defining the trajectory with respect to the relative wind frame is that the realness of the kite velocity along the trajectory does not depend on the ship motions. Nevertheless, the ship motions may modify significantly the shape of the trajectory with respect to the Earth fixed frame $\mathbf{n}$. Vertical motions are especially known to cause unwanted overloads and flight instabilities [27]. Consequently, two trajectory definitions are investigated further in the paper. The first definition takes into account all components of the tether attachment point velocity. The second definition takes only the horizontal components of the tether attachment point velocity with respect to the earth fixed frame $\mathbf{n}$ to compute the relative wind frame. This modified relative wind frame of trajectory definition is denoted by $\tilde{\mathbf{rw}}\left(z_{ref}\right)$ and is defined from:

$$\underline{U}_{rw}^{(\mathbf{n})} = \underline{U}_{tw}^{(\mathbf{n})}\left(z_{ref}\right) - \left[u_a^{(\mathbf{n})}, v_a^{(\mathbf{n})}, 0\right]^T \tag{20}$$

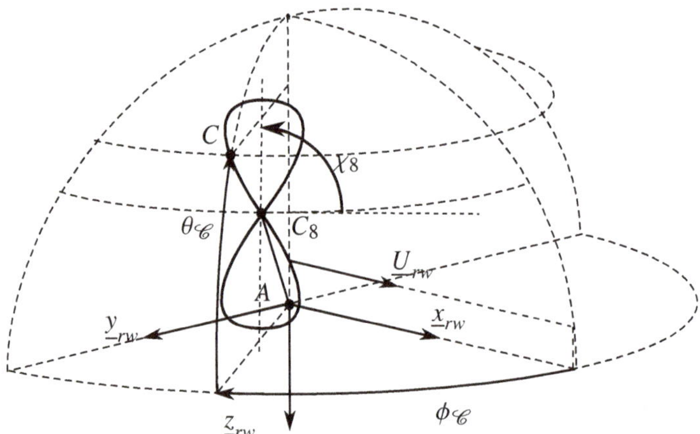

Figure 8. Lissajous trajectory parametrization.

2.4. Two Coupling Methods

The tether tension induces motions to the ship. The expression of generalized kite force acting on the ship is expressed as follows:

$$\underline{F}_k = \begin{bmatrix} \underline{T}_k^{(\mathbf{s})} & \underline{O_S A^{(\mathbf{s})}} \times \underline{T}_k^{(\mathbf{s})} \end{bmatrix}^T \tag{21}$$

In addition, according to the zero-mass kite modeling (cf. Section 2.3.2 and Equation (5)), the ship motions can modify the kite flight and the tether tension through the relative wind speed at the kite altitude with respect to the tether attachment point A:

$$\underline{U}_{rw}\left(k_z^{(\mathbf{n})}\right) = \underline{U}_{tw}\left(k_z^{(\mathbf{n})}\right) - \underline{U}_A \tag{22}$$

Two coupling approaches are investigated in this paper: a monolithic and a segregated approaches. The monolithic approach takes into account all the coupling terms between the kite and the ship modeling. Here, the considered segregated approach assumes a predefined kite force and then solves the ship equations of motion separately.

2.4.1. A Monolithic Approach

The monolithic equations of motion of a ship towed by kite are obtained with Equations (9) and (10) as follows:

$$\begin{cases} \underline{\dot{S}} &= \begin{bmatrix} \underline{\underline{T}}_{\mathbf{s}}^{\mathbf{n}} & \underline{\underline{0}} \\ \underline{\underline{0}} & \underline{\underline{R}}_{\mathbf{s}}^{\mathbf{n}} \end{bmatrix} \underline{V}_s \\ \underline{\dot{V}}_s &= \left[\underline{\underline{M}}_S + \underline{\underline{\tilde{A}}}\right]^{-1} \left[\underline{F} - \left[\underline{\underline{\tilde{B}}} + \underline{\underline{B}}_\phi + \underline{\underline{D}}\right] \underline{V}_s - \underline{\mu} - \underline{\underline{C}}\,\underline{S}\right] \\ \underline{\dot{y}}_{ij} &= \underline{\underline{A}}'_{ij} \underline{y}_{ij} + \underline{B}'_{ij} \delta V_{s,j},\ \forall i,j \in [\![1;6]\!] \\ \underline{\dot{K}}^{(\mathbf{n})} &= \underline{U}_A^{(\mathbf{n})} + U_{rw} \sqrt{\left(\underline{x}_{vk} \cdot \underline{x}_{rw}\right)^2 + \left(\frac{\underline{z}_{k0} \cdot \underline{x}_{rw}}{\sin \epsilon_k}\right)^2 - 1}\; \underline{x}_{vk}^{(\mathbf{n})} \end{cases} \tag{23}$$

Equation (23) adds 3 scalar equations for the kite. With the monolithic approach, the fully coupled system between the ship motions and the kite motions is solved. This monolithic system of differential equations is numerically integrated with the 4th order Runge–Kutta scheme with fixed time step.

2.4.2. A Segregated Approach

By contrast to the monolithic approach, the segregated approach considers only the mean tether attachment point velocity on the ship. The ship motions are computed applying the time series of the kite towing force as an external forces. Thus, the ship equation of motion can be expressed as follows:

$$\begin{cases} \underline{\dot{S}} = \begin{bmatrix} \underline{\underline{T}}_{\mathbf{s}}^{\mathbf{c}} & \underline{\underline{0}} \\ \underline{\underline{0}} & \underline{\underline{R}}_{\mathbf{s}}^{\mathbf{c}} \end{bmatrix} \underline{V}_s \\ \underline{\dot{V}}_s = \left(\underline{\underline{M}}_S + \underline{\underline{\tilde{A}}}\right)^{-1} \left[\underline{F}_k(t) + \underline{F}' - \left(\underline{\underline{\tilde{B}}} + \underline{\underline{B}}_\phi + \underline{\underline{D}}\right) \underline{V}_s - \underline{\mu} - \underline{\underline{C}}\,\underline{S}\right] \\ \underline{\dot{y}}_{ij} = \underline{\underline{A}}'_{ij} \underline{y}_{ij} + \underline{B}'_{ij} \delta V_{s,j},\ \forall i,j \in [\![1;6]\!] \end{cases} \tag{24}$$

where $\underline{F}'$ denotes the external forces such as rudder, propeller and windage forces excluding the kite forces applied as a time series $\underline{F}_k(t)$. This segregated approach could be very practical to study the motions of a ship towed by kite. Even if here, this approach is performed into the time domain, the segregated approach can be performed into the frequency domain applying the kite excitation spectrum directly in Equation (A1).

3. Results

3.1. Validation and Comments on the Kite Aerodynamic Characteristics

On the basis of the experimental data obtained by Behrel et al. [19,33], the zero-mass modeling is compared to the experimental data. The zero mass kite modeling is dependent of two parameters, the kite lift coefficient C_{lk} and the glide angle ϵ_k. These coefficients must be adapted in order to fit the data. The onshore full scale trials [19,33] were performed with a classical kite Cabrinha® Switchblade2016 of 5 m^2 designed for kite-surfing. The tether length was 80 m long. During the run, the kite performed eight shaped trajectory controlled by an autopilot based on the algorithm proposed in [34,35]. The experimental kite position is determined with a 3D load cell assuming that the tethers are straight. The evolution of the wind velocity with the altitude was identified thanks to a SOnic Detection And Ranging (SODAR). The experimental results presented here correspond to a phase averaging post-processing of a 5-minute kite flight run. For the presented case, the wind velocity was interpolated from the SODAR data with the following linear function:

$$U_{tw} = 3.16 - 0.035 k_z^{(\mathbf{n})} \tag{25}$$

The wind gradient function in Equation (25) is used for the following presented simulation results.

The zero-mass modeling presented in [1] assumes a constant glide angle and a constant lift coefficient. The kite velocity depends only on its direction and position in the wind window and on the glide angle ϵ_k. Consequently, the present zero-mass modeling results are obtained with a glide angle enabling that the simulated kite trajectory has the same period than the experimental data, which is 5.86 s. The kite towing force is dependent of the kite lift coefficient. Hence, the kite lift coefficient is adjusted in order to obtain the same mean towing force. The kite trajectory used for the simulation is the same than the kite trajectory of the experimental measurements. The position of the kite is integrated with the 4th order Runge–Kutta scheme with a time step of 0.1 s.

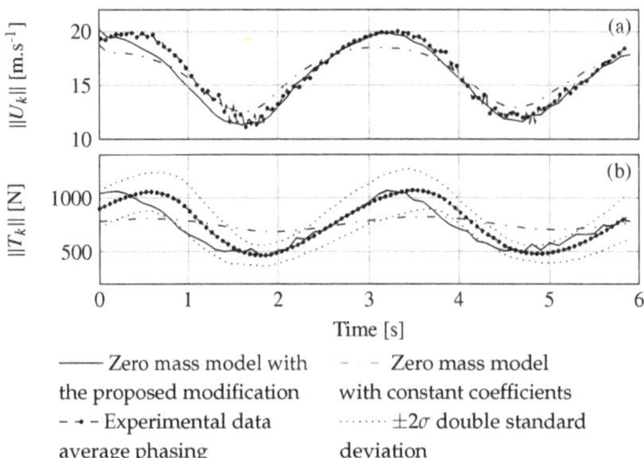

Figure 9. Evolution of the kite velocity (**a**) and of the tether tension at A, along the average trajectory (**b**); comparison between the zero mass model with constant aerodynamics and the model with a linear modification of the kite aerodynamic characteristics using phase averaging of experimental data from Behrel et al. [33].

Figure 9 shows the evolution of the magnitude of the kite velocity (a) and the evolution of the magnitude of the tether tension at A on the average trajectory (b). The experimental velocity is obtained by finite differentiation, which induces some noise in the signal sampled at 50 Hz. This noise in terms of speed and tension for the simulated results is still present since the simulation uses the average kite trajectory of the experimental data. However, this noise is barely visible since the sampling frequency of the simulation is only 10 Hz. The kite glide angle and the lift coefficient used for the zero-mass modeling are respectively 12.45° and 0.855. The main difference between the experimental data and the results concerns the tether tension and the kite velocity amplitude. The simulated amplitude is slightly underestimated for the velocity and is particularly underestimated for the tension. The tether tension given by Equation (12) is a quadratic function of the apparent wind velocity. Therefore, for a relative kite velocity error, the relative tension error is twice. In order to perform an eight trajectory, the back tethers are steered. The difference between the back tethers lengths leads to important modifications of the kite flying shape. Hence, it can be expected a modification of the kite lift coefficient and of the glide angle leading to the significant differences observed between the zero-mass modeling with constant aerodynamic characteristics and the measurements. Leloup [36] and Duport [37] demonstrated the variation of the aerodynamic characteristics with respect to the turning rate. Even without considering tip vortices, the yaw rate induces an evolution of the local inflow velocity along the kite span. According to Equation (10), the kite velocity is independent of its area. Consequently, for a given radius of curvature of the trajectory, the evolution of the local inflow along the span and the effect of the flying shape deformation are more consequent with larger kites. Here it is assumed that the evolution of the kite glide angle and lift coefficient are proportional to the ratio of a characteristic length of the kite with the radius of curvature of the trajectory. For a given aspect ratio, the characteristic length of the kite can be $\sqrt{A_k}$. Denoting the trajectory radius of curvature $R_{\mathscr{C}}$, the glide angle and the lift coefficient could written as follows:

$$\begin{cases} \epsilon_k &= \epsilon_0 + \kappa_\epsilon' \frac{\sqrt{A_k}}{R_{\mathscr{C}}} \\ C_{lk} &= C_{l0} + \kappa_l' \frac{\sqrt{A_k}}{R_{\mathscr{C}}} \end{cases} \tag{26}$$

where, κ_ϵ' and κ_l' are two coefficients. Equation (26) can be rewritten in terms of heading rate according to the relationship between the radius of curvature of the trajectory and the kite velocity. Furthermore, according to Equation (10), the kite velocity is proportional to the relative wind speed, thus the correction proposed can be rewritten as functions of the time derivative of the heading and the relative wind speed as follows:

$$\begin{cases} \epsilon_k &= \epsilon_0 + \frac{\sqrt{A_k}}{U_{rw}} \kappa_\epsilon \left|\dot{\gamma}\right| \\ C_{lk} &= C_{l0} + \frac{\sqrt{A_k}}{U_{rw}} \kappa_l \left|\dot{\gamma}\right| \end{cases} \tag{27}$$

The determination of the coefficient ϵ_0, κ_ϵ, C_{l0} and κ_l can be evaluated by comparison with the experimental data. First, ϵ_0 and κ_ϵ are identified in order to obtain respectively the same maximum and minimum speed than the experiments. Then, C_{l0} and κ_l are identified in order to obtain respectively the same maximum and minimum tether tension than the experiments. According to this method, the following coefficient values are identified:

$$\begin{cases} \epsilon_0 &= 0.2013 \text{ rad} \\ \kappa_\epsilon &= 0.0422 \text{ s} \\ C_{l0} &= 0.9856 \\ \kappa_l &= -0.3718 \text{ s.rad}^{-1} \end{cases} \tag{28}$$

These values are consistent with the ones by Behrel et al. [19], who did the identification with a quite different and more straightforward method. The results in terms of kite velocity and tether tension are plotted in Figure 9. With the modification of the glide angle and the lift coefficient, the noise in the kite velocity and the tether tension time series is more significant. Indeed, the computation of the kite yaw rate requires two finite differentiations of the kite position. As expected with the identification method of ϵ_0, κ_ϵ, C_{l0} and κ_l, the amplitude of the kite velocity and the amplitude of the tether tension are respected.

An overall good agreement is found in terms of velocity but a slight phase difference is observable between the simulation results and the experimental data. The linear modifications of the glide angle and of the lift coefficient with the kite yaw rate lead to a better agreement with the experimental results than the zero mass modeling. However, these modifications should be confirmed with more experimental cases. Since at the first order the ship motions are proportional to the amplitude of the excitation forces, these modifications enhancing the prediction of the kite excitation force amplitude are crucial.

Moreover, the minimum allowable radius of curvature of the trajectory can be estimated according to Equation (26). Indeed, during a kite flight C_{lk} must be positive. This means that the trajectory radius of curvature must satisfy the following condition:

$$R_C \geqslant -\frac{\kappa_l^{'}}{C_{l0}}\sqrt{A_k} \tag{29}$$

As a numerical application, the minimum trajectory radius of curvature of the 5 m^2 Cabrinha® Switchblade is $2.23\sqrt{A_k} \approx 5.0$ m, Which is a little less than twice the projected wingspan. Given that the lift coefficient must have a strictly positive value in order to fly, we are here outside the validity domain of the proposed modeling. However, the order of magnitude is confirmed by Fagiano who evaluated the minimum turning radius at 2.5 times the wingspan [38,39]. The coefficients ϵ_0, κ_ϵ, C_{l0} and κ_l are dimensionless quantities. Consequently, the presented modifications in Equations (27) and (28) are retained as formulated for the rest of the paper.

3.2. *Study Case*

In order to simplify the analysis, only vertical ship motions (heave, roll and pitch) of the DTMB 5512 at full scale were considered here. The analysis was focused on the roll motion. Thus, in the scope to observe significant roll motion, a true wind angle $\beta_{tw} = 90°$ was chosen. A true wind speed of reference U_{10} =10 $m.s^{-1}$ corresponding to the high range of a fresh breeze from the Beaufort scale was considered. The ship speed was set to U_h =7.5 $m.s^{-1}$ since it corresponds to a common sailing speed condition of the world merchant ship fleet. A kite with an area of A_k =500 m^2 and with the aerodynamic characteristics determined in Equation (28) was used. The tether attachment point was in the center plane, 7.9 m above the water line and 25 m in front of the center of gravity.

The kite flight trajectory corresponded to a Lissajous trajectory as defined in Section 2.3.4. The amplitudes of the trajectory were arbitrarily set to $\Delta\phi_8 = 20°$ and $\Delta\theta_8 = 8°$. According to the formulation of the lift coefficient in Equation (26), the tether length condition to obtain a positive lift coefficient was $L_t \geqslant 360$ m. The kite excitation was studied by varying the tether length. Tether length between 360 m and 1000 m were investigated. The center of the trajectory $[\phi_8, \theta_8]$ and the angle of the trajectory χ_8 around the axis $\underline{C_8A}$ were determined optimizing of the longitudinal kite towing force with the same code used by Leloup et al. [1]. In this configuration, the kite provided about 16% of the propulsive power, provided the data from [40].

Figure 10a–c show respectively the evolution of the ϕ_8, θ_8 and χ_8 with the tether length. A calm water case and three regular beam wave cases are considered. A regular beam wave of 2.5 m high consistent with a fresh breeze. Three wave frequencies were investigated $\{0.4, 0.56, 0.8\}$ $rad.s^{-1}$. The 0.56 $rad.s^{-1}$ wave frequency corresponded to the natural roll ship frequency. For all following results, the simulation time was 1640 s with a time step of 0.3 s.

3.3. *Calm Water Case*

3.3.1. Kite Excitation Spectrum

Figure 11a,b show respectively the time series and the spectrum of the kite roll excitation moment obtained with the segregated approach for a tether length $L_t = 500$ m. Only the varying part of the kite excitation is taken into account to compute the kite excitation spectrum. With the segregated approach, the kite flight is not modified by the ship motions. It can be noticed in Figure 11 that the roll excitation moment is mainly composed of several harmonics. For convenience, the harmonics are denoted

ω_{ki} where i is a natural number. Only the first, the second and the fourth harmonics are significant. The whole spectrum is not represented but the other harmonics that occur at higher frequencies are not significant. The second and the fourth harmonics are the most powerful. The second harmonic should appear to be the most critical for the ship motions due to its proximity with the natural roll ship frequency.

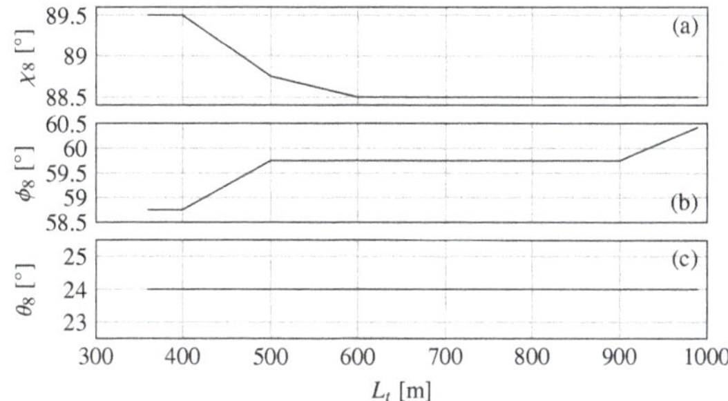

Figure 10. Kite flight trajectory parameter versus tether length. (**a**) Trajectory angle χ_8; (**b**) Azimuth of the center of the trajectory; (**c**) elevation of the center of the trajectory.

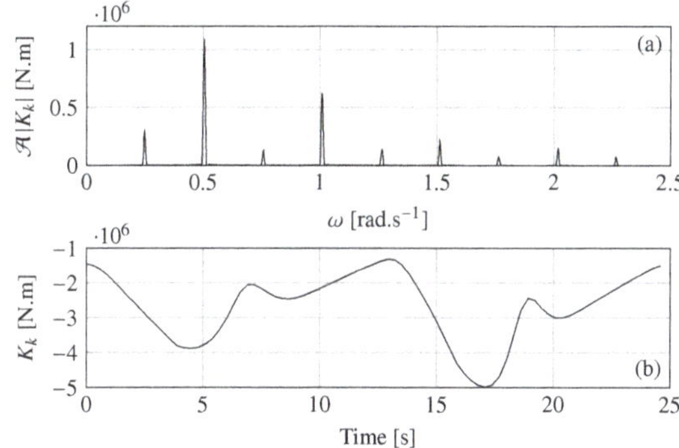

Figure 11. With a tether length $L_t = 500$ m; (**a**) spectrum of the kite excitation moment around the longitudinal ship axis $\underline{x}_s$; (**b**) time history of the kite excitation moment around the longitudinal ship axis $\underline{x}_s$ over the last loop.

3.3.2. Comparison of the Segregated Approach with the Monolithic Approach

Figure 12 shows the evolution of the roll amplitude, first kite excitation harmonic frequency and amplitude kite moment with respect to the tether length. Three methods are compared: the segregated approach in dashed-dotted line and the two monolithic approaches with the kite trajectories defined in $\mathbf{r\tilde{w}}$ (A) and in $\mathbf{rw}$ (A) respectively in solid and dashed lines.

As expected, the first kite excitation harmonic ω_{k1} decreases with the tether length since the angular amplitude of the Lissajous trajectories is remained constant. No major difference can be noticed between the three approaches in terms of harmonics frequencies. The roll amplitude $\Delta\phi_s$ and the amplitude of the kite roll moment ΔK_k predicted by the segregated are higher than those predicted by the two monolithic approaches. The two trajectory definitions in $\mathbf{r\tilde{w}}$ and $\mathbf{rw}$ give almost the same results both in terms of roll amplitude and kite roll moment amplitude. For all approaches presented, the roll amplitude evolution is similar. The kite moment amplitude increase quasi linearly with the tether length. Four ruptures (mainly three) can be observed in the evolution of the kite moment amplitude. These ruptures corresponds to the evolution of the trajectory parameters with the tether length as shown in Figure 10. Due to the wind gradient, the longer the tether is, the larger the kite roll moment is. This increase in kite roll moment explains the continuous increase in roll amplitude. However, two important increases can be noticed between $L_t = 360$ m and $L_t = 500$ m and from $L_t = 790$ m, which are not explained by the kite roll moment. In fact, these two increases are due to the proximity between the two most powerful kite roll excitation harmonics and the vessel natural roll frequency. Indeed, for $L_t = 440$ m the second harmonic frequency is almost equal to the natural roll frequency of the ship. From $L_t = 790$ m, the fourth harmonic frequency approaches the natural roll ship frequency. In case of a slower true wind speed, the increase of the kite roll moment with the tether length would be less significant. Thus, it could be observed a maximum of roll amplitude for tether length corresponding to a match between a kite harmonic frequency and the natural roll frequency of the ship. According to these results in calm water, the segregated approach is slightly conservative with respect to both monolithic approaches.

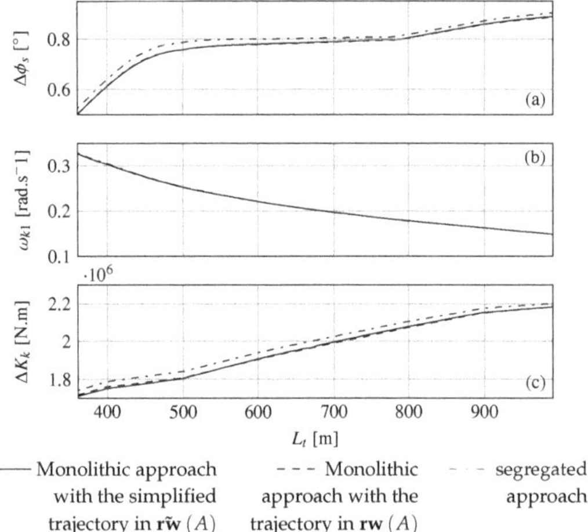

Figure 12. (**a**) Amplitude of the ship roll motion, (**b**) first kite harmonic frequency and (**c**) amplitude of the kite moment of excitation for different tether lengths from 360 m to 990 m by step length of 10 m in calm water.

3.4. Regular Beam Wave Case

To investigate a case closer to a real ocean environment, the influence of regular beam waves is studied. The wave considered is a 2.5 m high at the frequencies ω_w of 0.4 rad.s^{-1}, 0.56 rad.s^{-1} and 0.8 rad.s^{-1}. As for the calm water case, the frequency domain of kite excitation is scanned with different tether lengths from $L_t = 360$ m to $L_t = 990$ m with a tether length step of 10 m.

3.4.1. Ship Vertical Motion Influence on Kite Trajectory

This section focuses on the influence of ship vertical motion on kite flight trajectory. Paulig et al. [27] underlined the major disturbances caused by this motion in terms of flight control and overload. Although this is in line with the rare feedback from experiences at sea, this has never been simulated before, all published studies being carried out so far in cases of calm water. Figure 13 shows the trajectory defined neglecting the ship vertical motion in solid line and the trajectory defined in the actual relative wind frame in dashed line. The latter has considerably reduced turning radius. It is not in line with the relationship on the minimum kite flight radius highlighted by Fagiano nor with the experiments by Behrel presented in Section 3.1. Such a sharp trajectory could not be achieved in the real world. As a consequence, in the rest of the paper, all the trajectories will be defined in $\tilde{\mathbf{rw}}$ (A) and will be called.

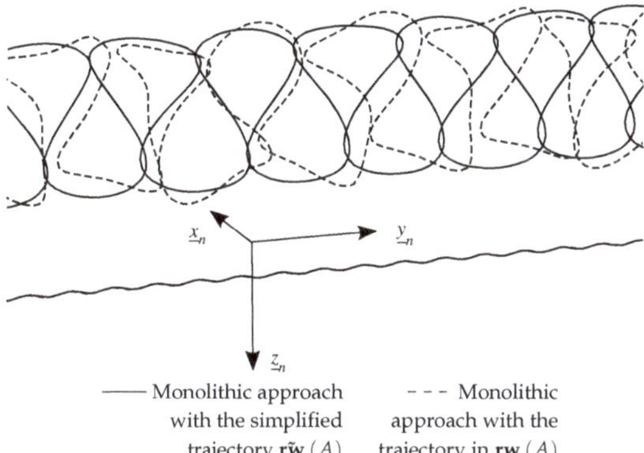

Figure 13. Kite and ship path with respect to **n** for $L_t = 390$ m with a wave of 2.5 m high at the frequency $\omega_w = 0.8$ rad.s^{-1}.

3.4.2. Interactions with Regular Beam Waves

Figures 14–16 show respectively for three wave frequencies, the roll amplitude of the ship, the first kite harmonic frequency and roll moment. For each wave cases, the monolithic approach is compared to the segregated approach. The dashed-dotted line corresponds to the roll amplitude of the ship due to the wave excitation without kite. For all wave cases, the roll amplitude and the amplitude of the kite roll moment increase globally with the tether length as for the calm water case. In contrary to the calm water case, the amplitude of the kite roll moment obtained with the monolithic approach is globally larger than the one obtained with the segregated approach. However, it should be noticed some drops in the evolutions of the roll amplitude and amplitude of the kite roll moment. For instance, in Figure 15a, for the tether length $L_t = 440$ m, a significant drop of the roll amplitude can be noticed in the segregated case. This phenomenon of interaction between the kite and the ship can explained the fact that the monolithic approach predicts a smaller roll amplitude despite a larger kite roll moment amplitude compared to the segregated approach. Indeed, the secondary kite harmonic at the wave frequency has a relative phase with the ship motion. As an example, for the case plotted in Figure 17, the difference in phase angle between the ship motion and the kite excitation at $\omega = 0.4$ rad.s^{-1} is 60.7°. For the case with the frequency wave $\omega_w = 0.8$ rad.s^{-1}, the interaction between the kite and the ship is less significant since the wave frequency is far from the most powerful kite harmonic frequencies. Moreover, in contrary to the calm water case, the segregated approach is not necessarily conservative with respect to the monolithic approach as shown by Figure 16a.

Concerning the evolution of the first harmonic frequency, differences can be noted between the two approaches. Indeed, for certain ranges of tether length, the harmonic frequency obtained with the monolithic approach remains constant. For instance, on Figure 15b, the first harmonic frequency remains constant to half the wave frequency for tether lengths within the range [390; 470] m. A sharp evolution of the roll amplitude concerns both approaches. In the monolithic case, this phenomenon does not occur only when a harmonic frequency of the kite corresponds to the wave frequency. With a wave frequency of 0.4 rad.s^{-1} and a tether length $L_t = 840$ m, a drop of the ship roll amplitude can be noticed. Figure 17 shows that no principal kite harmonic frequency corresponds to the wave frequency. However, due to coupling between kite and ship motions, secondary harmonics appear. The secondary harmonic frequencies are denoted ω'_{ki}. This phenomenon occurs when the frequency gap between the closest principal harmonic frequency and the wave frequency is a submultiple of the first kite harmonic frequency. For the case presented in Figure 17, $\omega_{k1} = 3\omega'_{k1}$. Furthermore, it should be noted that these drops in roll amplitude are due to the interactions between the kite and the ship. With the monolithic approach, these drops are independent from initial conditions. The most important drops in roll amplitude occur when $\omega_{k1} = \omega'_{k1}$ or $\omega_{k1} = 2\omega'_{k1}$. The importance of the phenomenon occurring at $\omega_{k1} = n\omega'_{k1}$ decreases with the increasing value of the integer n. Moreover, the importance of the phenomenon decreases when the interaction with the wave concerns high harmonic orders.

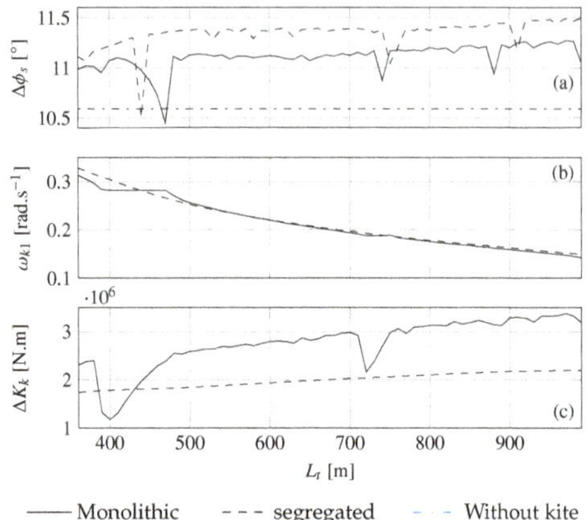

Figure 14. (**a**) Amplitude of the ship roll motion, (**b**) first kite harmonic frequency and (**c**) amplitude of the kite moment of excitation for different tether lengths from 360 m to 990 m by step length of 10 m with a beam regular wave of 2.5 m high at a frequency of $\omega_w = 0.4$ rad.s^{-1}.

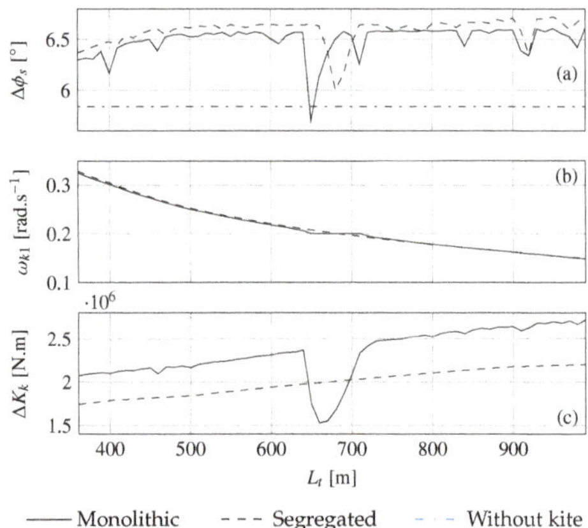

Figure 15. (**a**) Amplitude of the ship roll motion, (**b**) first kite harmonic frequency and (**c**) amplitude of the kite moment of excitation for different tether lengths from 360 m to 990 m by step length of 10 m with a beam regular wave of 2.5 m high at a frequency of $\omega_w = 0.56$ rad.s^{-1}.

3.4.3. Kite Lock-in Phenomenon

As shown in the previous section, an important interaction phenomenon between the kite and the ship occured. We observed the appearance of plateaus which were all the more pronounced as they were located near the first harmonics of the moment of excitation produced by the kite. It can be observed in Figures 14a and 15a respectively at $L_t = 650$ m and $L_t = 470$ m, that the drop in roll amplitude predicted by the monolithic approach is important enough to lead to smaller roll amplitude than the case without kite. A particular attention is paid to the case where the wave frequency $\omega_w = 0.4$ rad.s^{-1} and $L_t = 470$ m, because the interaction between the kite and the ship is win-win. Indeed, for this case the mean kite towing force predicted by the monolithic approach is 8.0%

more important than the mean kite towing force predicted with the segregated approach. In addition, the ship roll amplitude is 1.4% weaker than without kite.

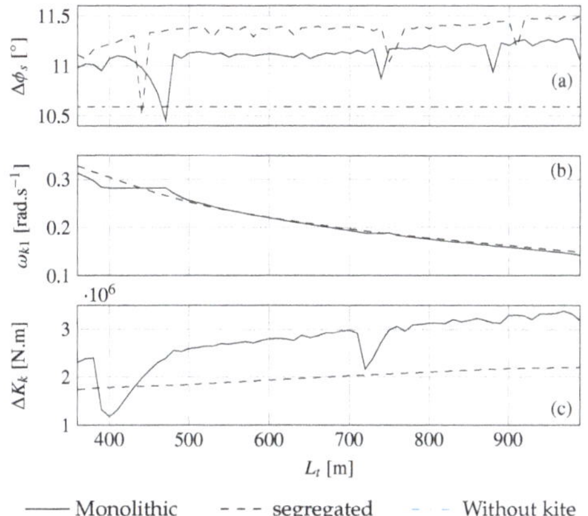

Figure 16. (**a**) Amplitude of the ship roll motion, (**b**) first kite harmonic frequency and (**c**) amplitude of the kite moment of excitation for different tether lengths from 360 m to 990 m by step length of 10 m with a beam regular wave of 2.5 m high at a frequency of $\omega_w = 0.8$ rad.s^{-1}.

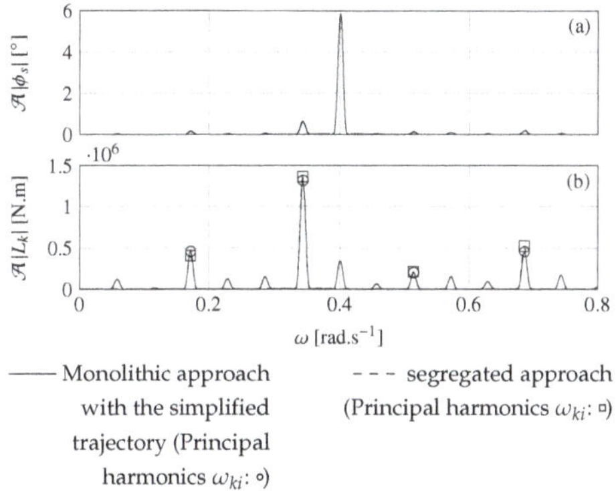

Figure 17. With a tether length $L_t = 840$ m at the wave frequency $\omega_w = 0.4$ rad.s^{-1}; (**a**) Spectrum of the roll motion of the ship; (**b**) Spectrum of the kite excitation moment around the longitudinal ship axis.

Figure 18 shows the spectrum of the roll motion (a) and the spectrum of the kite excitation (b). It can be noticed that the second harmonic of the kite roll moment is attracted towards the wave frequency. Since the kite harmonic frequency is increased towards the wave frequency, the kite performed the whole trajectory faster. The time to perform the trajectory is decreased by 5.7%. This leads to a higher apparent wind speed and therefore to a higher average kite towing force. Moreover, the difference in phase angle between the roll motion and the kite roll moment is 18.2°, leading to a reduction of the roll amplitude. This attraction towards the wave frequency can be noticed also on the secondary harmonic corresponding to the case $\omega_{k1} = 2\omega'_{k1}$. However, the effects are less significant. This phenomenon is similar to the lock-in phenomenon happening with the vortex-induced vibrations [41,42]. Henceforth, this phenomenon is called the kite lock-in.

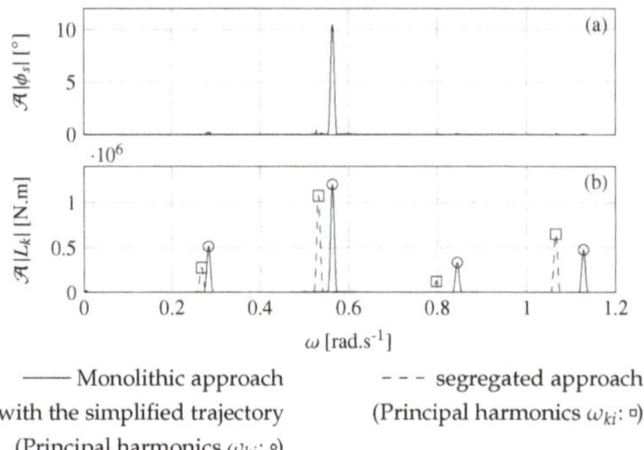

Figure 18. With a tether length $L_t = 470$ m at the wave frequency $\omega_w = 0.56$ rad.s^{-1}; (**a**) Spectrum of the roll motion of the ship; (**b**) Spectrum of the kite excitation moment around the longitudinal ship axis $\underline{x}_s$.

4. Discussion

A method coupling a time domain seakeeping and a zero-mass kite modeling was developed to investigate the dynamic motion of ship towed by kite. The zero-mass kite modeling neglects the inertial forces, its deformation and the dynamic behavior of the tether. The ship modeling was based on the linear seakeeping STF strip theory [9]. A monolithic approach coupling the kite and the ship modeling was compared to a segregated approach, which neglected the coupling terms. Both methods allowed fast computation. Implemented in Python, the monolithic approach computed faster than the real-time. The computation of the considered cases were five time faster than the real time on a processor Intel(R) Xeon(R) CPU E3-1220 v3 with a CPU frequency of 3.10 GHz.

The ship modeling provides satisfactory results in terms of heave and pitch motions. Since the roll motion is highly nonlinear, the linear prediction of the roll motion is less accurate. However, a similar evolution of the predicted roll motion was noticed between the proposed roll motion modeling and the modeling by Ikeda et al. [25] used as reference here. Consequently, a good confidence in the kite influence on the roll motion could be attributed in terms of evolution. Nevertheless, the value of the results in terms of roll motion should be considered with caution. Moreover, the ship motions were restricted to heave, pitch and roll motions and existing couplings, such as roll and sway coupling, were neglected.

A linear dependency of the kite aerodynamic characteristics was proposed to correct the kite velocity amplitude and the tether tension amplitude in order to get a better agreement between the experiments and the proposed kite modeling. The identification of the kite modeling was carried out with experimental data on a single run of 5 minutes. A more extensive validation should investigate the influence of the true wind speed and the influence of the trajectory position and orientation. Moreover, the determination of the kite aerodynamic coefficients was entirely based on experimental data. It could be interesting to be able to estimate by simulation the kite aerodynamic characteristics for a deeper understanding.

In case of a dynamic flight, the high tensions developed lead to small tether sags, justifying the assumption that the lines are straight and inelastic. However, the tether acts as an interface between the kite and the ship. The dynamic tether effect may have a significant influence on the interaction between the kite and the ship. In particular, additional theoretical research should be carried out to study the effect of the tether on the kite lock-in phenomenon.

Moreover, in order to validate the modeling approach proposed in this paper, several strategies could be implemented. Full scale experiments on a 6-meter long dedicated experimental boat towed by kite were carried out by Behrel et al. [33,43]. However, measurement of the environment is challenging at full scale. With towing tank tests, similarity issues are also difficult to solve, as Martin et al. [44]

emphasize this in the case of offshore floating wind turbines. In order to get around the problem of similitude and to take benefit from the towing tank test conditions, the kite could be modeled with a hardware in the loop method such as proposed by Giberti and Ferrari [45] for classical sailing yacht.

5. Conclusions and Future Work

A dynamic modeling of a ship towed by a kite was developed by combining two modelings. One for the ship and the other for the kite. Each modeling was validated independently. Two coupling approaches between the kite and the ship were compared: a segregated approach where the kite force was applied as a predefined time series and a monolithic approach where all the coupling terms were taken into account.

The computation of the linear convolution term for the ship motion modeling was performed with a state-space approach in order to provide time efficient computations. The identification of the state-space modeling was revisited in order to fit the forward speed case requirements. Kite modeling was modified using linear correction terms for aerodynamic characteristics. These modifications improved significantly the predicted amplitude of kite velocity and tether tension.

In the calm water case, the coupling between the ship and the kite decreases the kite forces and consequently, the segregated approach was conservative compared to the monolithic one. Different tether lengths were tested with a constant angular trajectory size leading to an evolution of the kite excitation frequencies. For the considered case, the effect of the wind gradient led to a continuous increase in roll amplitude according to the tether length and excitation frequency. The vicinity between the kite harmonics and the ship natural roll frequency led to further local significant increases.

In case of regular beam waves, the interaction between the kite and the ship was more significant. The definition of the kite trajectory into the relative wind frame taking into account the vertical ship velocity led to unrealistic trajectories with very short radius of curvature. This problem was avoided by neglecting the vertical component of the vessel motion when defining the kite's trajectory. Moreover, a kite lock-in phenomenon to the frequency of the waves was highlighted, which can be win-win. Indeed, for some configurations, the mean kite towing force was increased while the ship roll amplitude remained smaller than the case without kite.

Author Contributions: Conceptualization, K.R. and N.B.; methodology, K.R., J.-B.L.; Software, N.B.; formal analysis, J.-B.L.; Validation, J.-B.L., K.R. and N.B.; Writing—original draft preparation, N.B.; Writing—review and editing, K.R.; Supervision, K.R., J.-B.L.; Project administration, Y.P.; Funding acquisition, Y.P. All authors have read and agreed to the published version of the manuscript.

Funding: This research was funded by the French Environment and Energy Management Agency. GRANT NUMBER: 1482C0090

Conflicts of Interest: The authors declare no conflict of interest. The funders had no role in the design of the study; in the collection, analyses, or interpretation of data; In the writing of the manuscript, or in the decision to publish the results.

Nomenclature

dof	Degree(s) of freedom
EFD	Experimental Fluid Dynamic
Representations	
n	Earth fixed reference frame (inertial) $\left(O_n, \underline{x}_n, \underline{y}_n, \underline{z}_n\right)$
s	Ship fixed frame $\left(O_s, \underline{x}_s, \underline{y}_s, \underline{z}_s\right)$
h	Hydrodynamic reference frame $\left(O_h, \underline{x}_h, \underline{y}_h, \underline{z}_h\right)$
rw	Relative wind frame $\left(\underline{x}_{rw}, \underline{y}_{rw}, \underline{z}_{rw}\right)$
k	Kite reference frame $\left(K, \underline{x}_k, \underline{y}_k, \underline{z}_k\right)$

Representations

l_z	Vertical position of O_s with respect to ship the baseline [m]
l_x	Longitudinal position of O_s with respect to the oft perpendicular of the ship [m]
Variables	
$\underline{V}_s$	Generalized velocity vector of the ship at O_s with respect to the **n** frame expressed in the **s** frame [m.s^{-1}, rad.s^{-1}]
$\underline{\xi}$	Generalized position vector of the ship with respect to the **h** frame expressed in the **h** frame[m, rad]
$\underline{S}$	Generalized position vector of the ship with respect to the **n** frame expressed in the **n** frame[m, rad]
$\underline{U}_h$	Mean ship forward speed [m.s^{-1}]
Parameters	
K	Kite position (k_x, k_y, k_z)
A	Tether attachment point
O_s	Origin of the Ship reference frame
O_h	Origin of the hydrodynamic reference
O_n	Origin of the Earth reference frame
O_k	Origin of the kite reference frame
$\underline{U}_{rw}$	Relative wind velocity [m.s^{-1}]
$\underline{U}_{tw}$	True wind velocity [m.s^{-1}]
$\underline{U}_{10}$	Reference wind speed at 10m altitude [m.s^{-1}]
$\underline{U}_A$	Kite attachment point velocity [m.s^{-1}]
$\underline{U}_{aw}$	Apparent wind velocity to the kite [m.s^{-1}]
$\underline{U}_k$	Kite velocity with respect to the **rw** frame [m.s^{-1}]
Parameters	
ω	Angular frequency of the motion [rad.s^{-1}]
ω_w	Angular frequency of the wave [rad.s^{-1}]
ω_e	Angular frequency of encounter [rad.s^{-1}]
ω_e	Angular frequency of encounter [rad.s^{-1}]
ω_{ki}	Angular frequency of the ith principal kite roll moment harmonic [rad.s $^{-1}$]
ω'_{ki}	Angular frequency of the ith secondary kite roll moment harmonic [rad.s $^{-1}$]
k	Wave number [m^{-1}]
L_t	Tether length [m]
$\underline{T}_k$	Tether tension [N]
ϵ_k	Glide angle of the kite [rad]
C_{lk}	Kite lift coefficient
A_k	Kite surface area [m^2]
z_{ref}	Measurement altitude of the wind [m]
SODAR	Sonic Detection And Ranging
FFT	Fast Fourier Transform
g	Gravitational constant (9.81) [m.s^{-2}]
S_w	Wave spectrum
ψ_w	Wave direction with respect to the **c** frame [rad]
Parameters	
$\underline{\underline{\mathscr{K}}}$	Impulse response function of the retardation matrix
$\underline{\underline{K}}$	Laplace transform of the retardation matrix
$\mathscr{C}$	Kite trajectory
$\underline{\underline{R}}_{\mathbf{s}}^{\mathbf{n}}$	Transformation matrix of the time derivatives of the Euler's angle **s** with respect to **n** to the turning rates in **s**
$\underline{\underline{A}}$	Generalized added mass matrix with respect to **s** for a given frequency of motion and $\underline{\underline{\tilde{A}}}$ at infinite frequency; $\underline{\underline{A}}^*$ with respect to **h**
$\underline{\underline{B}}$	Generalized damping matrix with respect to **s** for a given frequency of motion and $\underline{\underline{\tilde{B}}}$ at infinite frequency; $\underline{\underline{B}}^*$ with respect to **h**
$\underline{\underline{C}}$	Generalized restoring matrix with respect to **s**; $\underline{\underline{C}}^*$ with respect to **h**
ϕ_8	Azimuth of the centre of the kite trajectory [rad]

θ_8	Elevation of the centre of the trajectory [rad]
χ_8	Rotation angle of the trajectory around the axis defined by its centre and the tether attachment point A
ϕ_s	Heeling angle of the ship [rad]
ϕ_w	Phase angle of the Froude-Krilov force with respect to the free surface elevation [rad]
$\underline{U}_{ref}$	True wind velocity at altitude of measurement [m.s^{-1}]
$R_{\mathscr{C}}$	Trajectory radius of curvature [m]
Δ_{K_k}	Amplitude of the kite roll moment [N.m]
Δ_{ϕ_s}	Amplitude of the ship heeling angle [rad]
$\mathcal{A}$	Amplitude of the Fourier transform
Ψ	Phase of the Fourier transform

Appendix A. Ship Modeling

Appendix A.1. Time Domain Equations of Motion

This section is an overview of the work of Fossen [22]. The starting point of the mathematical modeling is the linearized frequency domain equation of motion, Equation (A1), used by the STF strip theory [9]:

$$\left[\underline{\underline{M}}^*_S + \underline{\underline{A}}^*\right]\ddot{\underline{\xi}} + \left[\underline{\underline{B}}^* + \underline{\underline{B}}^*_{\phi}\right]\dot{\underline{\xi}} + \underline{\underline{C}}^*\underline{\xi} = \underline{F}^* - \bar{\underline{F}}^*, \tag{A1}$$

where, $\underline{\underline{M}}^*_S$, $\underline{\underline{A}}^*$, $\underline{\underline{B}}^*$ and $\underline{\underline{C}}^*$ denote respectively the generalized mass matrix, added mass matrix, damping matrix and the restoring matrix with respect to the **h** frame. $\underline{\underline{B}}^*_{\phi}$ is an extra generalized damping matrix accounting only for the roll motion as proposed in [9]. $\underline{F}^*$ denotes the sum of the generalized external forces (forces and moments) applied to the ship expressed in the **h** frame. $\bar{\underline{F}}^*$ is the mean value of $\underline{F}^*$. It should be noticed that Equation (A1) holds only for a given frequency, ω, with small amplitude sinusoidal motions. Since, $\underline{\underline{A}}^*$and $\underline{\underline{B}}^*$ are frequency dependent. This assumption leads to the relationship:

$$\ddot{\underline{\xi}} = -\omega^2\underline{\xi} \tag{A2}$$

In addition, the direct cosine matrix between the **h** frame and the **s** frame can be simplified considering small angles:

$$\underline{\underline{T}}^{\mathbf{h}}_{\mathbf{s}} = \begin{bmatrix} 1 & -\xi_6 & \xi_5 \\ \xi_6 & 1 & -\xi_4 \\ -\xi_5 & \xi_4 & 1 \end{bmatrix} \tag{A3}$$

Defining $\delta\underline{V}_s = [u_s - U_h, v_s, w_s, p_s, q_s, r_s]^T_{\mathbf{s}}$, $\underline{\xi}$ can be expressed from:

$$\begin{cases} \dot{\underline{\xi}} = & \underline{\underline{J}}\delta\underline{V}_s - \frac{U}{\omega_e^2}\underline{\underline{L}}\delta\dot{\underline{V}}_s \\ \ddot{\underline{\xi}} = & \underline{\underline{J}}\delta\dot{\underline{V}}_s + U\underline{\underline{L}}\delta\underline{V}_s \end{cases}, \tag{A4}$$

where,

$$\underline{\underline{J}} = \begin{bmatrix} 1 & 0 & 0 & 0 & z_h & 0 \\ 0 & 1 & 0 & -z_h & 0 & x_h \\ 0 & 0 & 1 & 0 & -x_h & 0 \\ 0 & 0 & 0 & 1 & 0 & 0 \\ 0 & 0 & 0 & 0 & 1 & 0 \\ 0 & 0 & 0 & 0 & 0 & 1 \end{bmatrix}, \tag{A5}$$

and,

$$\underline{\underline{L}} = \begin{bmatrix} 0 & 0 & 0 & 0 & 0 & 0 \\ 0 & 0 & 0 & 0 & 0 & 1 \\ 0 & 0 & 0 & 0 & -1 & 0 \\ 0 & 0 & 0 & 0 & 0 & 0 \\ 0 & 0 & 0 & 0 & 0 & 0 \\ 0 & 0 & 0 & 0 & 0 & 0 \end{bmatrix}. \tag{A6}$$

The detailed of this transformation is presented in [22]. Using Equation (A4), the equation of motion in Equation (A1) can be expressed in terms of $\delta \underline{V}_s$ as:

$$\left[\underline{\underline{M}}_S + \underline{\underline{A}}\right] \delta \dot{\underline{V}}_s + \left[\underline{\underline{B}} + \underline{\underline{B}}_\phi + \underline{\underline{D}}\right] \delta \underline{V}_s + \underline{\underline{C}}\,\underline{\xi} = \underline{F} - \underline{\bar{F}}, \tag{A7}$$

where,

$$\begin{cases} \underline{\underline{M}}_S & = \underline{\underline{J}}^T \underline{\underline{M}}_S^* \underline{\underline{J}} \\ \underline{\underline{A}} & = \underline{\underline{J}}^T \underline{\underline{A}}^* \underline{\underline{J}} \\ \underline{\underline{D}} & = \underline{\underline{J}}^T \underline{\underline{M}}_S^* \underline{\underline{L}} \\ \underline{\underline{B}} & = \underline{\underline{J}}^T \left[\underline{\underline{B}}^* + U_h \underline{\underline{A}}^* \underline{\underline{L}}\right] \underline{\underline{J}} \\ \underline{\underline{B}}_\phi & = \underline{\underline{J}}^T \underline{\underline{B}}_\phi^* \underline{\underline{J}} \\ \underline{\underline{C}} & = \underline{\underline{J}}^T \underline{\underline{C}}^* \\ \underline{F} & = \underline{\underline{J}}^T \underline{F}^* \end{cases} \tag{A8}$$

Equation (A7) can be solved directly for a single frequency excitation. Nevertheless, since a kite and a ship may have strongly coupled motions, it is preferable to transform Equation (A7) into the time domain using the impulse response function as proposed by Cummins [15,17,46]. The steady state corresponds to $u_s = U_h$ and $\delta \underline{V}_s = \underline{0}$. Due to the special structure of $\underline{\underline{C}}$, it can be noticed that $\underline{\underline{C}}\,\underline{\xi} = \underline{\underline{C}}\,\underline{S}$. Consequently, the ship equation of motion for arbitrary motions and using the parametrization in $\underline{V}_s$ is:

$$\left[\underline{\underline{M}}_S + \underline{\underline{\tilde{A}}}\right] \dot{\underline{V}}_s + \left[\underline{\underline{\tilde{B}}} + \underline{\underline{D}}\right] \underline{V}_s + \underline{\mu} + \underline{\underline{C}}\,\underline{S} = \underline{F}, \tag{A9}$$

where, $\underline{\underline{\tilde{A}}} = \lim\limits_{\omega \to +\infty} \underline{\underline{A}}(\omega)$ and $\underline{\underline{\tilde{B}}} = \lim\limits_{\omega \to +\infty} \underline{\underline{B}}(\omega)$. $\underline{\mu}$ is defined as follows:

$$\underline{\mu} = \int_0^t \underline{\underline{\mathcal{K}}}(t-\tau)\, \delta \underline{V_s}(\tau)\, d\tau, \tag{A10}$$

where $\underline{\underline{\mathcal{K}}}$ denotes the impulse response of the retardation matrix. Strictly speaking, the integral boundaries of the convolution term should be $+\infty$ and $-\infty$. However, assuming that $\delta \underline{V_s} = \underline{0}$ for $t < 0$, the left boundary can be replaced by zero. In addition, assuming that the ship is a causal system, the right boundary can be truncated to t. $\underline{\underline{K}}$, the Laplace transform of the retardation matrix, can be identified with Equation (A1) assuming unit sinusoidal motions in Equation (A9).

$$\underline{\underline{K}}(j\omega) = \underline{\underline{B}}(\omega) - \underline{\underline{\tilde{B}}} + j\omega \left[\underline{\underline{A}}(\omega) - \underline{\underline{\tilde{A}}}\right], \tag{A11}$$

where $j^2 = -1$.

Appendix A.2. Hydrodynamic Data and Linear Convolution Term

Each convolution component $\mu_{i\in[\![1,6]\!]}$ can be approximated by state space systems in Equation (A12), as introduced by Kristiansen and Egeland in [16].

$$\begin{cases} \mu_i & = \sum_{j=1}^{6} \mu_{ij} \\ \dot{\underline{y}}_{ij} & = \underline{\underline{A}}'_{ij}\underline{y}_{ij} + \underline{B}'_{ij}\delta V_{s,j} \\ \mu_{ij} & = \underline{\underline{C}}'_{ij}\underline{y}_{ij} \end{cases}, \tag{A12}$$

where, $\left\{\underline{\underline{A}}'_{ij}, \underline{B}'_{ij}, \underline{\underline{C}}'_{ij}\right\}$ represents the state-space modeling corresponding to a transfer function denoted by H_{ij} fitting $K_{ij}(j\omega)$, for $i, j \in [\![1,6]\!]$. At zero forward speed, the properties of the retardation functions as described in [18] impose the form of the transfer function as follows:

$$H_{ij} = \frac{\mathscr{K}_{ij}(t=0)\, p^{n-1} + \ldots + a_1 p}{p^n + b_{n-1}p^{n-1} + \ldots + b_0} \tag{A13}$$

In order to get finite results, according to the Riemann–Lebesgue Lemma, the transfer functions must be stable [47]. The denominator should respect the Routh–Hurwitz criterion [48,49]. With forward speed, the retardation function may not tend towards zero at zero frequency. Indeed $B_{ij}(\omega = 0)$ can be different from $B_{ij}(\omega = \infty)$ [5]. Consequently, on the contrary to the zero speed case, the limit towards zero of K_{ij} can be non zero. This condition is not satisfied with the form given in Equation (A13). Thus, in case of forward speed, H_{ij} should have the form:

$$H_{ij} = \frac{\mathscr{K}_{ij}(t=0)\, p^{n-1} + \ldots + a_1 p + a_0}{p^n + b_{n-1}p^{n-1} + \ldots + b_0}, \tag{A14}$$

where the coefficient a_0 and b_0 should respect the following condition:

$$\frac{a_0}{b_0} = B_{ij}(\omega = 0) - B_{ij}(\omega = \infty) \tag{A15}$$

The data $K_{ij}(j\omega)$ are obtained with the added mass and damping obtained according to the STF strip theory [9]. The 3D added mass and damping are expressed in terms of sectional added mass and damping. For instance, the STF strip theory expressed $A^*_{33}$and B^*_{33} in terms of sectional added mass a_{33} and damping b_{33} as follows:

$$\begin{cases} A^*_{33}(\omega) & = \int_{L_{pp}} a_{33}(\omega, x)\, dx - \frac{U_h}{\omega^2} b_{33}(\omega, x_a) \\ B^*_{33}(\omega) & = \int_{L_{pp}} b_{33}(\omega, x)\, dx + U_h a_{33}(\omega, x_a) \end{cases}, \tag{A16}$$

where x_a is the longitudinal position of the aft perpendicular section. The sectional added mass and damping are obtained with the Shipmo seakeeping software developed by the Marin® assuming an infinite depth. The frequency range of the data depends on the ship size, but for a commercial ship, the low frequency limit is generally about 0.1 rad.s^{-1} and the high frequency limit does not generally exceed 3 rad.s^{-1}. To improve the quality of the identification method an extrapolation of the hydrodynamic data toward the asymptotic value is necessary. As shown by Newman [5], assuming a potential flow, at zero and infinite frequency, the sectional damping is zero. Hence, each sectional damping are then extrapolated at high frequency with the function in Equation (A17), proposed by Greenhow [50]:

$$f_e(\omega) = \frac{\beta_1}{\omega^4} + \frac{\beta_2}{\omega^2}, \tag{A17}$$

where β_1 and β_2 are two constants chosen in order to provide continuity and differentiability. At zero frequency, the sectional damping is linearly extrapolated to zero. The sectional added mass are linearly extrapolated. At high frequency, the sectional added mass remains almost constant, consequently, a constant extrapolation is performed. The identification of H_{ij} can be identified either into the frequency domain or into the time domain, see [18]. Here, a time domain identification method is used to initialize the frequency domain identification method. The time domain identification is performed with the singular value decomposition method proposed by Kung [51]. This step is performed with a modified Matlab® function "imp2ss" to control the order. This method is efficient but the identified transfer function has the following form:

$$H_{ij} = \frac{a_n p^n + \ldots + a_1 p + a_0}{p^n + b_{n-1} p^{n-1} + \ldots + b_0}. \tag{A18}$$

Consequently, to comply with the form imposed with Equation (A14), a_n is set to zero, a_{n-1} is set to $\mathscr{K}_{ij}(t=0)$ and a_0 is set to $a_0 = b_0\left[B_{ij}(\omega=0) - B_{ij}(\omega=\infty)\right]$. This first estimate of the transfer function is used as initial solution of the frequency domain identification method. The frequency identification step is performed with the "oe" function of the Matlab® system identification toolbox. This function use local optimization under constraints algorithm based on gradient methods. The structure of the transfer function, as proposed in Equation (A14), can be imposed to the frequency domain optimization algorithm. These two steps are repeated for several transfer function orders, for instance from 2 to 10. Then, the best transfer function order is selected according to a criterion based on the normalized quadratic error e_{tot} from:

$$\begin{cases} e_{tot} & = \frac{1}{2}\left(e_\omega + e_t\right) \\ e_\omega & = \frac{\sum_k \left[\left|H_{ij}(j\omega_k) - K_{ij}(j\omega_k)\right|\right]^2}{\sum_k \left|K_{ij}(j\omega_k)\right|^2} \\ e_t & = \frac{\sum_k \left[H_{ij}(t_k) - K_{ij}(t_k)\right]^2}{\sum_k K_{ij}(t_k)^2} \end{cases} \tag{A19}$$

Appendix A.3. Wave Forces

The Froude-Krilov and diffraction forces are obtained with the STF 2D strip theory [9]. Assuming an infinite depth, the dispersion relationship is $kg = \omega_w^2$, where k is the wave-number and g the gravity. Thus the frequency of encounter denoted by ω_e can be approximated by the following relationship:

$$\omega_e = \omega_w - \frac{u_s}{g}\omega_w^2 \cos\beta_w \tag{A20}$$

where ω_w and β_w denote respectively the wave angular frequency in rad.s^{-1} and the angle of the waves with respect to the ship heading. β_w is given by $\beta_w = \psi_s - \psi_w$, where ψ_w denotes the wave angle with respect to $\underline{x}_n$. With $i \in [\![1;6]\!]$, each component f_{wi} of the Froude–Krilov and diffraction forces generated by a single unit wave can be expressed by the following expression:

$$f_{wi}\left(u_s, \beta_w, \omega_w\right) = E_i\left(u_s, \beta_w, \omega_w\right) \cos\left(k \cos\left(\psi_w\right) s_x^{(\mathbf{n})} + k \sin\left(\psi_w\right) s_y^{(\mathbf{n})} - \omega t - \phi\left(u_s, \beta_w, \omega_w\right) + \epsilon_w\right) \tag{A21}$$

where E_i is the amplitude of the ith component of $\underline{f}_w$ and ϵ_w is a random phase.
For any wave spectrum $S_w\left(\omega_w, \psi_w, \epsilon_w\right)$, the Froude–Krilov and diffraction forces can be expressed as follows:

$$\underline{F}_w\left(u_s, \psi_s, S_w, \epsilon_w\right) = \sum_{j=1}^{N} \sqrt{2S_w\left(\omega_{wj}, \psi_{wj}\right)\Delta\psi_w\Delta\omega_w}\, \underline{f}_w\left(u_s, \psi_s - \psi_{wj}, \omega_j, \epsilon_{wj}\right) \tag{A22}$$

where ϵ_{wj} is a random phase equidistributed between 0 and 2π to obtain a Gaussian wave spectrum.

References

1. Leloup, R.; Roncin, K.; Behrel, M.; Bles, G.; Leroux, J.B.; Jochum, C.; Parlier, Y. A continuous and analytical modeling for kites as auxiliary propulsion devoted to merchant ships, including fuel saving estimation. *Renew. Energy* **2016**, *86*, 483–496. [CrossRef]
2. Naaijen, P.; Koster, V.; Dallinga, R. On the power savings by an auxiliary kite propulsion system. *Int. Shipbuild. Prog.* **2006**, *53*, 255–279.
3. Ran, H.; Janson, C.E.; Allenström, B. Auxiliary kite propulsion contribution to ship thrust. In Proceedings of the 32nd International Conference on Ocean, Offshore and Arctic Engineering, Nantes, France, 9–14 June 2013; pp. 1–9.
4. Bigi, N.; Behrel, M.; Roncin, K.; Leroux, J.B.; Nême, A.; Jochum, C.; Parlier, Y. Course Keeping of Ship Towed by Kite. In Proceedings of the 15th Journees de l'Hydrodynamique, Brest, France, 22–24 November 2016; pp. 1–14.
5. Newman, J.N. *Marine Hydrodynamics*; MIT Press: Cambridge, MA, USA, 1977.
6. Molin, B. *Hydrodynamique des Structures Offshore*; Editions Technip: Paris, France, 2002.
7. Faltinsen, O.M. *Hydrodynamics of High-Speed Marine Vehicles*; Cambridge University Press: Cambridge, UK, 2005; pp. 1–476. Available online: http://xxx.lanl.gov/abs/arXiv:1011.1669v3 (accessed on 5 June 2020).
8. Bertram, V. *Practical Ship Hydrodynamics*; Elsevier: Amsterdam, The Netherlands, 2012.
9. Salvesen, N.; Tuck, E.; Faltinsen, O. Ship motions and sea loads. *Trans. SNAME* **1970**, *78*, 250–287. [CrossRef]
10. Lee, C.H. *WAMIT Theory Manual*; Massachusetts Institute of Technology, Department of Ocean Engineering: Cambridge, MA, USA, 1995.
11. Skejic, R. Ships Maneuvering Simulations in a Seaway—How close are we to reality? In Proceedings of the International Workshop on Next Generation Nautical Traffic Models, Delft, The Netherlands, 6 November 2013; pp. 91–101.
12. Böttcher, H. *Simulation of Ship Motions in a Seaway*; Technical report; Institut für Schiffbau der Universität Hamburg: Hamburg, Germany, 1989.
13. Sutulo, S.; Guedes Soares, C. An implementation of the Method of Auxiliary State Variables for solving seakeeping problems. *Int. Shipbuild. Prog.* **2005**, *52*, 357–384.
14. Sutulo, S.; Soares, C. A generalized strip theory for curvilinear motion in waves. In Proceedings of the International Conference on Offshore Mechanics and Arctic Engineering—OMAE, Estoril, Portugal, 15–20 June 2008; Volume 6, pp. 359–368. [CrossRef]
15. Cummins, W. *The Impulse Response Function and Ship Motions*; Technical report; David Taylor Model Basin: Washington, DC, USA, 1962.
16. Kristiansen, E.; Hjulstad, A.; Egeland, O. State-space representation of radiation forces in time-domain vessel models. *Ocean Eng.* **2005**, *32*, 2195–2216. [CrossRef]
17. Fossen, T.I.; Smogeli, Ø.N. Nonlinear time-domain strip theory formulation for low-speed manoeuvring and station-keeping. *Model. Identif. Control* **2004**, *25*, 201–221. [CrossRef]
18. Pérez, T.; Fossen, T.I. Time-vs. frequency-domain Identification of parametric radiation force models for marine structures at zero speed. *Model. Identif. Control* **2008**, *29*, 1–19. [CrossRef]
19. Behrel, M.; Roncin, K.; Leroux, J.B.; Montel, F.; Hascoet, R.; Neme, A.; Jochum, C.; Parlier, Y. Application of phase averaging method for measuring kites performance: Onshore results. *J. Sail. Technol.* **2018**.
20. Wellicome, J.F.J.F.; Wilkinson, S. *Ship Propulsive Kites: An Initial Study*; Technical report; Department of Ship Science, University of Southampton: Southampton, UK, 1984.
21. Dadd, G.M.; Hudson, D.A.; Shenoi, R.A. Determination of kite forces using three-dimensional flight trajectories for ship propulsion. *Renew. Energy* **2011**, *36*, 2667–2678. [CrossRef]
22. Fossen, T. A Nonlinear Unified State-Space Model for Ship Maneuvering and Control in a Seaway. *Int. J. Bifurc. Chaos* **2005**, *15*, 2717–2746. [CrossRef]
23. The University of Iowa. EFD Data, The University of Iowa, Iowa City, USA, 2013. Available online: https://www.iihr.uiowa.edu/shiphydro/efd-data/ (accessed on 30 June 2020).
24. Irvine, M.; Longo, J.; Stern, F. Pitch and Heave Tests and Uncertainty Assessment for a Surface Combatant in Regular Head Waves. *J. Ship Res.* **2008**, *52*, 146–163.
25. Ikeda, Y.; Himeno, Y.; Tanaka, N. *A Prediction Method for Ship Roll Damping*; Technical report; Department of Naval Architecture, University of Osaka Prefecture: Osaka, Japan, 1978.

26. Loyd, M.L. Crosswind kite power (for large-scale wind power production). *J. Energy* **1980**, *4*, 106–111. [CrossRef]
27. Paulig, X.; Bungart, M.; Specht, B. Conceptual design of textile kites considering overall system performance. In *Airborne Wind Energy*; Springer: Berlin/Heidelberg, Germany, 2013; pp. 547–562.
28. Wellicome, J. Some comments on the relative merits of various wind propulsion devices. *J. Wind Eng. Ind. Aerodyn.* **1985**, *20*, 111–142. [CrossRef]
29. Argatov, I.; Rautakorpi, P.; Silvennoinen, R. Estimation of the mechanical energy output of the kite wind generator. *Renew. Energy* **2009**, *34*, 1525–1532. [CrossRef]
30. Leloup, R.; Roncin, K.; Bles, G.; Leroux, J.B.; Jochum, C.; Parlier, Y. Kite and classical rig sailing performance comparison on a one design keel boat. *Ocean Eng.* **2014**, *90*, 39–48. [CrossRef]
31. Group, I.Q.S. ITTC Symbols and Terminology List Version 2014; Technical report. In Proceedings of the International Towing Tank Conference, Copenhagen, Denmark, 31 August–5 September 2014.
32. Leloup, R. Modelling Approach and Numerical Tool Developments for Kite Performance Assessment and Mechanical Design; Application to Vessels Auxiliary Propulsion. Ph.D. Thesis, Université de Bretagne Occidentale, Brest, France, 2014.
33. Behrel, M.; Roncin, K.; Leroux, J.B.; Neme, A.; Jochum, C.; Parlier, Y. Experimental set up for measuring onshore and onboard performances of leading edge inflatable kites—Presentation of onshore results. In Proceedings of the 4th Innov'Sail Conference, Lorient, France, 28–30 June 2017.
34. Fagiano, L.; Zgraggen, A.U.; Khammash, M.; Morari, M. Automatic control of tethered wings for airborne wind energy: Design and experimental results. In Proceedings of the European Control Conference (ECC), Zurich, Switzerland, 17–19 July 2013; IEEE: Piscataway, NJ, USA, 2013; pp. 992–997.
35. Fagiano, L.; Zgraggen, A.U.; Morari, M. On modeling, filtering and automatic control of flexible tethered wings for airborne wind energy. In *Green Energy and Technology*; Springer: Berlin/Heidelberg, Germany, 2013; pp. 167–180.
36. Leloup, R.; Roncin, K.; Blès, G.; Leroux, J.B.; Jochum, C.; Parlier, Y. Estimation of the effect of rotation on the drag angle by using the lifting line method. In Proceedings of the 13th Journees de L'Hydrodynamique, Chatou, France, 21–23 November 2012.
37. Duport, C.; Leroux, J.B.; Roncin, K.; Jochum, C.; Parlier, Y. Comparison of 3D non-linear lifting line method calculations with 3D RANSE simulations and application to the prediction of the global loading on a cornering kite. In Proceedings of the 15th Journees de L'Hydrodynamique, Brest, France, 22–24 November 2015.
38. Fagiano, L. Control of Tethered Airfoils for High–Altitude Wind Energy Generation. Ph.D. Thesis, Politecnico di Torino, Torino, Italy, 2009.
39. Fagiano, L.; Milanese, M.; Piga, D. Optimization of airborne wind energy generators. *Int. J. Robust Nonlinear Control* **2012**, *22*, 2055–2083. [CrossRef]
40. Longo, J.; Stern, F. Resistance, sinkage and trim, wave profile, and nominal wake tests and uncertainty assessment for DTMB model 5512. In Proceedings of the 25th American Towing Tank Conference, Iowa City, IA, USA, 1998. Available online: https://pdfs.semanticscholar.org/3799/7310985bd531109d91e38c08e6508353f4c9.pdf (accessed on 30 June 2020).
41. Bearman, P.W. Vortex shedding from oscillating bluff bodies. *Annu. Rev. Fluid Mech.* **1984**, *16*, 195–222. [CrossRef]
42. Sarpkaya, T. A critical review of the intrinsic nature of vortex-induced vibrations. *J. Fluids Struct.* **2004**, *19*, 389–447. [CrossRef]
43. Behrel, M.; Roncin, K.; Iachkine, P.; Hascoet, R.; Leroux, J.B.; Montel, F.; Parlier, Y. Boat towed by kite: Methodolgy for sea trials. In Proceedings of the 16th Journées de l'Hydrodynamique, Marseille, France, 27–29 November 2018.
44. Martin, H.R.; Kimball, R.W.; Viselli, A.M.; Goupee, A.J. Methodology for Wind/Wave Basin Testing of Floating Offshore Wind Turbines. *J. Offshore Mech. Arct. Eng.* **2014**, *136*, 021902. [CrossRef]
45. Giberti, H.; Ferrari, D. A novel hardware-in-the-loop device for floating offshore wind turbines and sailing boats. *Mech. Mach. Theory* **2015**, *85*, 82–105. [CrossRef]
46. Ogilvie, F.T. Recent Progress Towards the Understanding and Prediction of Ship. In Proceedings of the 6th Symposium on Naval Hydrodynamics, Bergen, Norway, 10–12 September 1964; pp. 3–79.

47. Riemann, B. *Ueber die Darstellbarkeit Einer Function Durch eine Trigonometrische Reihe;* Dieterichschen Buchhandlung: Göttingen, Germany, 1867.
48. Routh, E.J. *A Treatise on the Stability of a Given State of Motion: Particularly Steady Motion*; Macmillan and Company: London, UK, 1877.
49. Hurwitz, A. Ueber die Bedingungen, unter welchen eine Gleichung nur Wurzeln mit negativen reellen Theilen besitzt. *Math. Ann.* **1895**, *46*, 273–284. [CrossRef]
50. Greenhow, M. High- and low-frequency asymptotic consequences of the Kramers-Kronig relations. *J. Eng. Math.* **1986**, *20*, 293–306. [CrossRef]
51. Kung, S.Y. A New Identification and Model Reduction Algorithm via Singular Value Decompositions. In Proceedings of the 12th Asilomar Conference on Circuits, Systems and Computers, Pacific Grove, CA, USA, 6–8 November 1978; pp. 705–714.

Journal of
Marine Science and Engineering

Article

Effect of Maneuvering on Ice-Induced Loading on Ship Hull: Dedicated Full-Scale Tests in the Baltic Sea

Mikko Suominen [1,*], Fang Li [1], Liangliang Lu [1], Pentti Kujala [1], Anriëtte Bekker [2] and Jonni Lehtiranta [3]

1 Department of Mechanical Engineering, Aalto University, 02150 Espoo, Finland; fang.li@aalto.fi (F.L.); liangliang.lu@aalto.fi (L.L.); pentti.kujala@aalto.fi (P.K.)
2 Department of Mechanical and Mechatronic Engineering, Stellenbosch University, Stellenbosch 7599, South Africa; annieb@sun.ac.za
3 Finnish Meteorological Institute, 00560 Helsinki, Finland; jonni.lehtiranta@fmi.fi
* Correspondence: mikko.suominen@aalto.fi; Tel.: +358-41-441-6435

Received: 20 August 2020; Accepted: 21 September 2020; Published: 28 September 2020

Abstract: Maneuvers in level ice are common operations for icebreakers and polar supply vessels. Maneuvering exposes the midship and stern area to ice interaction, influencing the magnitude and frequency of ice-induced loading in these areas. However, full-scale measurements do not typically cover the midship and stern areas, as measurements have commonly focused on the bow area. Controlled maneuvering tests were conducted during the ice trials of S.A. Agulhas II in the Baltic Sea. During these tests, ice-induced loading at different hull areas was measured simultaneously with ship control, navigation, and ice condition data. This work studied the effect of maneuvers on the characteristics and statistics of ice-induced loading at different hull areas and compared the impact to ahead operations. The study showed that the maneuvers had minor impact to the magnitude, frequency, and duration of loading at the bow and bow shoulder. On the other hand, maneuvers had a clear effect on the load magnitude and frequency at the stern shoulder. Additionally, a statistical analysis showed that the load magnitude increased as a function of load duration in all hull areas. Furthermore, the analyzed measurement data are presented and made available with the paper.

Keywords: full-scale; ice-induced load; load duration; load rise time; maneuvering; ice load frequency

---

1. Introduction

Merchant vessels are commonly designed to operate independently in light ice conditions or with icebreaker assistance where the navigation takes place in a broken ice field behind an icebreaker. Icebreakers and icebreaking supply vessels on the other hand are designed to operate independently in ice conditions. Their operations include transits between locations along the shortest possible or the most convenient route, mainly straight ahead. However, maneuvers are required from icebreaking vessels while assisting merchant vessels and performing ice management and operations in the near vicinity of destinations. These maneuvers include turning in an intact ice field and breaking out from an ice channel.

Maneuvering operations expose the midship and stern to ice interaction that increases the frequency and magnitude of ice-induced loading in these areas of the vessel at the side opposite to the turning direction [1]. For ships equipped with conventional shaft lines and rudders, the magnitude of increased loading has been shown to correlate with rudder forces [2,3]. Ships are commonly designed to operate ahead, i.e., the bow first. Thus, the bow is commonly designed to break ice with low frame angles favoring the bending failure of ice, whereas the frames at the midship and stern areas are often vertical or close to vertical. The vertical frames enhance breaking of ice through crushing that increases the load level on the hull in comparison to bending, and full-scale measurements have shown that the load magnitude at the stern area can match load levels at the bow [4,5]. The measured ice-induced loads have even exceeded the design load of ice class assigned to the vessel in the stern area [4,6].

A lot of effort has been afforded to determine the frequency of ice-induced loads for offshore structures as resonating frequency could cause severe damage to the structure (e.g., [7]). However, the speed of the ship affects the frequency of ice-induced loads. It is reasoned that a resonating situation for a ship can be considered rare as the vessel speed and ice conditions vary constantly. Consequently, the frequency and duration of ice-induced loads on a ship hull have not attracted as much attention as on offshore structures. However, the duration, rise time, and decay time related to ice-loading magnitudes are important to understand the load dynamics. If a dynamic numerical simulation with ice-induced loads is to be performed, knowledge on the load duration is required. Furthermore, the duration and frequency of loading are important when extreme ice-induced loads are considered, as these have a direct impact on exposure that affects the predicted extreme loads. In addition, the duration and frequency have an effect on the number of stress cycles that impacts fatigue life.

Commonly, the duration of ice-induced loading is reported briefly, and the accuracy is not clear. Typically, durations in an area of one frame spacing at the bow region range from 0.1 to 0.6 s, but durations of 0.007 and 1.5 s have also been reported [8–15]. Hänninen et al. [16] presented the distribution of ice-induced load duration on a frame at the bow, indicating the mode value of 2 s and durations ranging from 0.5 to 6 s. More thorough work on characterizing ice-induced loading events has been done by Lee et al. [17] and Ahn et al. [18]. They categorized ice-induced loading events based on the occurrence of intermediate load drops, and whether the possible load drop occurred before or after the peak load. They reported load durations between 0.1 and 14 s, with the mean being 1.2 s [18]. The load rise time varied between 0.05 and 7.2 s and the highest loads occurred between durations of 0.3 and 0.8 s. Furthermore, separate laboratory tests have been carried out to study the intermediate load drops applied in the categorization presented in these papers [19]. However, all these studies focused on the bow area of the vessel during ahead operations or ramming situations. The stern areas or maneuvers were not considered in these studies.

In order to gain insight into ice-induced loading at the stern area during different operations, dedicated maneuvering tests were conducted during the ice trial of the Polar Supply and Research Vessel (PSRV) S.A. Agulhas II in the Baltic Sea [1,20]. In these dedicated maneuvering tests, the ship controls and operations were predetermined, and the interference of the ship crew was minimized in order to reduce the possible effect from ship crew actions and experience. The maneuvering tests included breaking out from the channel, turning tests in an intact ice field, and straight-ahead operations for comparative purposes. This study focused on the characteristics of ice-induced loading events at different hull regions during the dedicated maneuvering tests. The load characteristics consisted of the peak load rise time, duration, and magnitude, and the trigger value load event duration and the highest load occurring during the event. The peak load and trigger value event are described in Section 2.2. Furthermore, the frequency of loading during different operations was studied.

2. Data Analysis Methods

2.1. Ice-Induced Loading on Hull

At the onset of ship–ice interaction, the ice crushes against the hull as the ship penetrates into the ice sheet. As the penetration continues and the total contact area and loading increases, the downward force increases in hull areas with small frame angles (Figure 1a). Ultimately, the ice fails through bending, and the broken ice piece is rotated and pushed aside or submerged, depending on the location at the hull. This process forms cusps in the ice field and is repeated when new contact is obtained. Thus, the ice-induced loading process is commonly described using a triangularly shaped loading history (Figure 2). In the case of vertical frame angles, the ice flakes from the top and bottom surfaces of the ice sheet, while the crushing process continues in the contact area between surfaces (Figure 1b).

This can be observed as intermediate load drops in the force signal before a larger failure and load drop (e.g., Daley [21]).

However, in many cases the contact point of the ice-induced load moves along the hull as the ice does not fail through bending or large fracturing, but the local crushing continues (Figure 3). Generally, the load movement can appear at the parallel midship region in all ice conditions. In level ice, contact can be initiated at the midship by turning or other movements of the vessel in relation to the surrounding level ice. In this case, local crushing and compaction might occur, but the surrounding level ice is strong enough to withstand higher loads and does not break through larger failure modes. As the ice is strong enough to prevent the movement of the vessel sideways, the load preventing the movement occurs at the contact point and moves with the contact point along the hull. The load travelling distance at the waterline is expected to be reduced in level ice conditions at the bow as sailing ahead into level ice requires ice to be broken and displaced in order to advance forward.

Besides the load movement, the width of the contact area can also vary due to local crushing and failures (Figure 3). Furthermore, intermediate failures occur, such as flaking, causing intermediate load drops. Thus, when a single frame is observed, it is not possible to confidently determine if the ice feature had a major failure on the observed frame or the ship–ice interaction point progressed onto the next frame along the hull. If several adjacent frames are instrumented, it is possible to observe how the load travels on the hull (Figure 3). Furthermore, observing several adjacent frames enables an estimation as to whether a major failure occurred at the instrumented area. If the loading clearly decreases when moving from one frame to another, it can be assumed that a major failure occurred [22].

In order to identify ice-induced loads from continuous measurement signals, two methods are applied—trigger values and Rayleigh separation. The trigger values are applied to determine the total duration of the ice-induced loading event, while Rayleigh separation is utilized to identify and separate load events into intermediate load events, referred to as peak load events, based on intermediate load drops, which are described in Section 2.2.

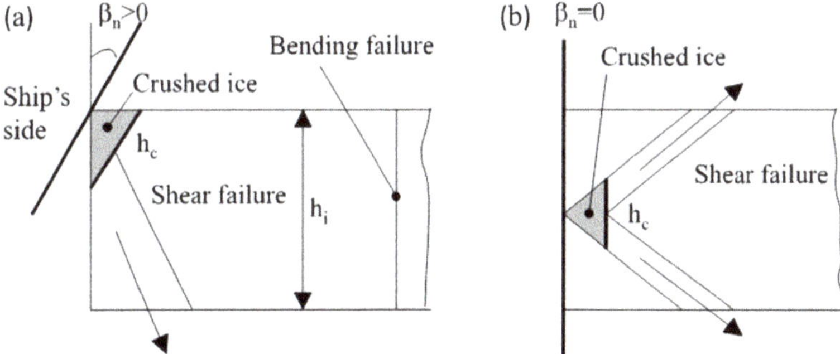

Figure 1. An idealization of the ice edge failure process in a case of (**a**) small to large frame angles and (**b**) vertical frames. Reproduced from [23], with permission from Finnish Academy of Technology, 1994. β_n, h_c, and h_i denote the frame angle, contact height, and ice thickness, respectively.

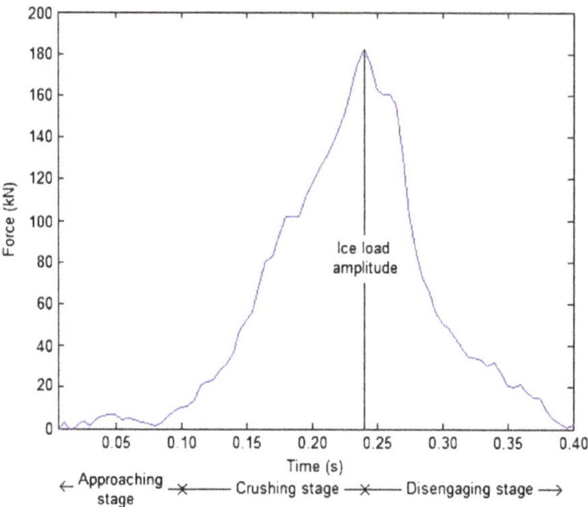

Figure 2. Ideal ice breaking process. Reproduced from [24], with permission from Elsevier, 2017.

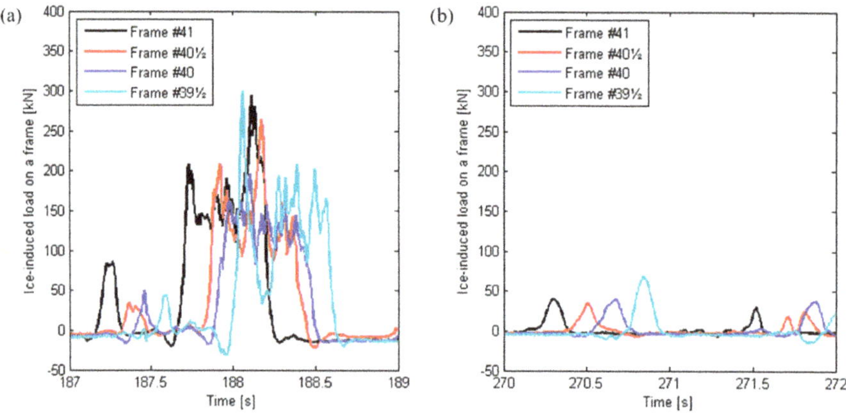

Figure 3. (**a**) Wide and (**b**) narrow ice-induced loads travelling from the foremost frame to the aftmost, from the frame #41 to the frame # $39\frac{1}{2}$, when the ship moves forward. Reproduced from [25], with permission from Elsevier, 2017.

2.2. Ice-Induced Load Duration and Rise Time

Ice-induced loading events are determined from the measured load signal utilizing two methods—the trigger value exceedance and Rayleigh separation. In the trigger value exceedance method, the load event starts and ends when the measured load crosses the preset load level. The events determined with this method shall be referred to as trigger load events in this paper. The maximum load value for a trigger load event is the highest measured load between the triggering points. This is demonstrated in Figure 4a.

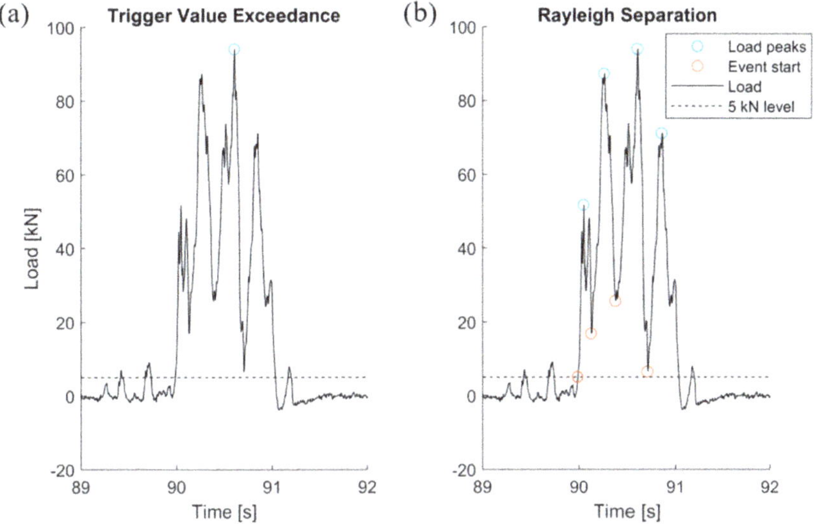

Figure 4. (**a**) Load event identified and separated with Rayleigh separation and (**b**) load identified with trigger value exceedance on the right.

In this work, the triggering value was set to 5 kN. However, the trigger load events having a maximum load not exceeding 10 kN were disregarded. As demonstrated in Figure 4a, the smaller load events before and after higher loading were not considered as ice-induced load events. Here, it was considered that setting the minimum accepted load equal to the trigger value would stress the number of short loads, as the load level repeatedly increased to 5 kN load level. Furthermore, the measurements showed that the highest load level in open-water condition was approximately 5 kN. Therefore, the threshold was set to 10 kN in order to be confident that all the loads identified from the signal were ice-induced loads. This was in line with the earlier analysis of the data [25,26].

As discussed in Section 2.1 and demonstrated in Figure 4, significant load level drops can occur without the load level decreasing below the threshold value. In this type of cases, it is considered that ice has experienced a significant failure, or the load has moved to the adjacent frame. Thus, the load event can be considered to consist of several separate peak load events. The peak load events can be separated utilizing the Rayleigh separation method. According to the method, two adjacent local maxima are compared to the minimum load level between the maxima. If the magnitude of the local minimum is below the level of smaller maximum multiplied by the separator value, the load event is separated into two peak load events at the location of the local minimum. The separator value was set to 0.5 in this study. There was no physical basis for setting the value to 0.5, but it was reasoned that a 50% decrease in the loading level constituted a significant load drop, and the value was in accordance with earlier studies [25,26]. Figure 4b presents an example of the Rayleigh separation, where blue circles indicate the separate peak loads. The load events determined with the Rayleigh separation will henceforth be referred to as peak load events.

In the case where the load event was separated with Rayleigh separation, as in Figure 4b, the starting point of the first peak load event was set to the point where the load exceeded the threshold and the starting points of the following events were set to the local minimum load between the local maxima. The end point of the peak load event was set to the starting point of the following event or when the load level decreased below the threshold. The peak load duration was the time difference between the end and starting point of an event. The load rise time was defined as the time interval from the starting point of a peak load event to the time of the peak load maximum. Thus, load drops smaller than the definition by Rayleigh separation were included in the peak load rise time.

3. Description of Vessel and Instrumentation

Full-scale measurements were conducted on-board the PSRV S.A. Agulhas II, which was built by STX Finland at Rauma Shipyard and was delivered in April 2012. The main dimensions of PSRV S.A. Agulhas II are presented in Table 1. She was built to the polar ice class PC5 and the hull was constructed in accordance with DNV ICE-10. The ship gains its power from a diesel electric propulsion system of four diesel generators each producing 3 MW. Two 4.5 MW electric motors provide power for two controllable pitch propellers [20].

Table 1. The main dimensions of the ship.

Length, between perpendiculars.	121.8 m
Breadth, moulded.	21.7 m
Draught, design	7.65 m
Deadweight at design displacement	13 687 t
Speed, service	14.0 kn

Three areas of the starboard side of the hull were instrumented with strain gauges when she was under construction in 2011/2012. The upper and lower parts of the frame were instrumented with V-shaped strain gauges, which measure the shear strains occurring in the frame. The instrumentation consists of two, three, and four adjacent frames at the bow, bow shoulder, and stern shoulder respectively (Figure 5). In addition, the hull plating was instrumented with strain gauges in these areas. The ice-induced loads on the frames were determined from the load–strain relation utilizing influence coefficient matrices. The matrices were determined using finite element models of the structure that were validated with calibration pulls. The method and models are described in detail by Suominen et al. [25]. In this study, the load–strain relation was determined by applying a point load on the midspan of the frame that was expected and realized at the waterline level during the Baltic Sea ice trial. Identifying the load–strain relation with more distributed loading would lead to smaller load–strain relation, i.e., higher load magnitudes would be assigned with smaller changes in the measured strain magnitudes.

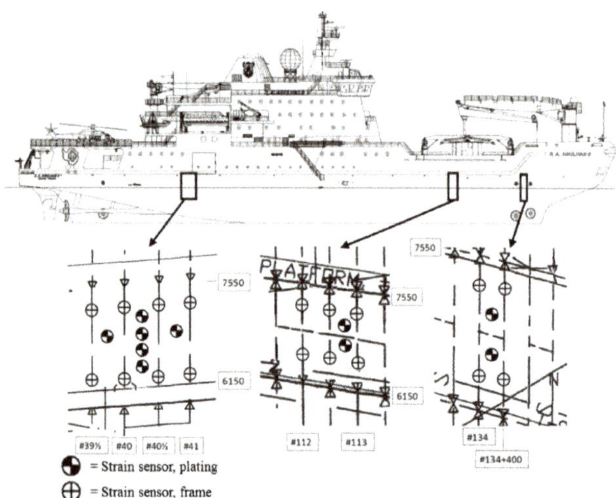

Figure 5. Instrumentation of S.A. Agulhas II. Reproduced from [27], with permission from the Committee of the Port and Ocean Engineering under Arctic Conditions, 2015.

It should be noted that the frames #111$\frac{1}{2}$, #112, and #112$\frac{1}{2}$ at the bow shoulder were connected to each other at the midspan of the frame structures with horizontal plates, which match the height of the frame. The load values recorded, especially on frame #112, sometimes indicated significant negative loads. This was expected to result from the load acting on the non-instrumented frames. As the frame #111$\frac{1}{2}$ was not instrumented, but connected to the instrumented frames, it was expected that the frame-to-frame interaction would be magnified, causing the measurement of negative values.

The navigational, machinery control, and ice-induced load data were collected with the central measurement unit, which is permanently installed onboard. The ice thickness was measured visually, with a stereo camera system and an electromagnetic (EM) device at the same time. Each of the thickness measurements utilized separate computers. The time synchronization of the computers was conducted with a wristwatch.

The navigational data (coordinates, speed over ground, course over ground, and heading) and machinery control data (motors' power, propellers' pitch, shafts' rotation speed, and rudders' angle) were measured with the ship's own systems and sampled at a frequency of approximately 1 Hz and 0.5 Hz, respectively. Ice-induced loads were measured and sampled at a frequency of 600 Hz and 200 Hz, respectively.

The EM measurements utilized a Geonics EM-31 instrument. The EM measurement frequency was typically set to 20 Hz and later down-sampled in data processing to approximately 1–5 Hz depending on the moving speed of the vessel. The device was located approximately at the front perpendicular. The stereo camera system was placed at the location of bow shoulder strain gauge instrumentation (frame #112) to capture images during the voyage. The images collected with the stereo camera system were manually analyzed after the voyage. Here, the thickness was measured at points where the broken ice pieces had turned upwards revealing the cross-section to the downward-looking camera system. As the measurement points were dependent on upward-turned ice pieces, no predefined measuring frequency could be applied. The measurement methods of the stereo camera system and EM device are described by Kulovesi and Lehtiranta [28] and Lensu et al. [29], respectively.

The visual observations contained ice thickness, ridge sail height, and ridge density. The thickness was estimated with the help of a measurement yardstick, which was placed over the side of the ship. The observations were collected in 10-min time intervals. A more detailed description of the ice condition observations is given by Suominen et al. [30]. In addition to the ice-thickness measurements, the bending and compressive strength of ice were measured during the ice trials between the tests. The methodology is reported in detail by Suominen et al. [20].

It should be noted that the EM measures the total thickness of ice, including the loose pieces under the solid ice sheet. The stereo camera system measures the thickness of an ice layer, i.e., the thickness of the broken level ice, or the thickness of the turning loose ice pieces. Thus, the ice thicknesses measured with the two systems deviate significantly in some occasions, especially at the location of ridges (e.g., Figure 6). The visual observations, by methodology, range between the values obtained from the other two systems, as the level ice thickness is estimated from the turning ice pieces, while the size of ridges can be calculated based on the observed sail height [31].

The controlled maneuvering tests were conducted in "Ice Mode". In this mode, the machinery automation seeks to maintain the shaft rotation speed at 140 revolutions per minute. In the case of propeller–ice interaction, the automated control may decrease the pitch of the propeller, thus reducing the thrust, if the rotational speed of the propeller is about to be inhibited.

The data from the tests described in this paper are collected and provided in separate files with this paper as a Supplementary Material. The content of the data files is described in Appendix B.

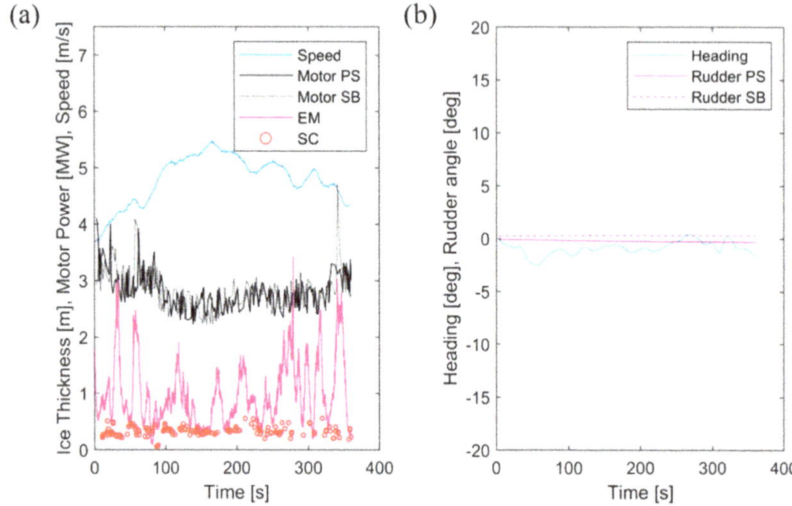

Figure 6. (**a**) Applied motor power, measured ice thickness with the electromagnetic device (EM) and stereo camera system (SC), and obtained speed, and (**b**) applied rudder angles with obtained heading during test L1.

4. Description of Conducted Tests and Measurements

4.1. Overview of Conducted Tests

Three types of tests were conducted: straight ahead in level ice, breaking out from the ice channel, and turning in level ice. The overview on the conducted tests are presented in Table 2 and detailed descriptions of different tests are given in the following chapters. In total, 16 controlled tests were conducted. In some cases, the following test was conducted directly after another test. These tests were separated in the data analysis, e.g., L2_1 and L2_2, but not in Table 2.

Table 2. Conducted maneuvering and straight-ahead tests. SB and PS refer to starboard and port side, respectively.

Test ID	Run	Ice Condition	Start time UTC	End time UTC	Notes
L1	Ahead	Level ice field	21.3. 6:59:00	21.3. 7:05:00	-
L2_1 and L2_2	Ahead	Level ice field	21.3. 10:13:00	21.3. 10:41:00	Two motor power levels
L3	Ahead	Level ice field	21.3. 14:12:30	21.3. 14:22:30	-
BC1	Break out (PS)	Channel	21.3. 9:32:40	21.3. 9:36:20	-
BC2	Break out (PS)	Channel	21.3. 9:52:45	21.3. 9:54:30	-
BC3	Break out (SB)	Channel	21.3. 11:17:00	21.3. 11:19:40	-
T1_1 and T1_2	Turning (SB + PS)	Level ice field	21.3. 10:46:0	21.3. 10:57:30	Turn to SB, turn to PS
T2_1 and T2_2	Turning (PS + PS)	Level ice field	21.3. 11:21:00	21.3. 11:31:00	Turn to PS, turn to PS
BC4 + T3	Break out (PS) + Turning (PS)	Channel + Level ice field	22.3. 9:03:25	22.3. 9:11:20	Break out to PS, turn to PS
BC5, T4_1, and T4_2	Break out (PS) + Turning (PS) + Turning (SB)	Channel + Level ice field	22.3. 9:31:00	22.3. 9:54:30	Break out to PS, turn to PS, turn to SB

The test locations were chosen based on visual observations. Level ice conditions and small deformation levels were preferred in the selection of test locations. However, ice conditions vary naturally, and rafting below the surface and conditions further away are not possible to observe. Thus, it is not possible to fully avoid a variation in conditions in longer tests. The visual observations at the time of the tests are presented in Table 3.

Table 3. Visual observations from the ice conditions during the tests. The time indicates the beginning of the period.

Date	Time	Concentration %		Ice Level cm (in Tenths)			Ridge Sail Height m (in Tenths)			Ridge Density
	(UTC)	Ice Field	Channel	<10	10–30	30–70	<0.5	0.5–1.0	1.0–1.5	in Ship Length
	21.3.2012									
L1	6:50	90				9				
	7:00	100				10	10			2
BC1	9:30	90	10		1	9		10		3
BC2	9:50	80	20			10	5	5		1
L2_1 & L2_2; T1_1 & T1_2	10:10	90	10	1		9	10			3
	10:20	90	10	1		9	5	5		3
	10:30	100				10	6	4		4
	10:40	80	20			10	9	1		1
	10:50	70	30		5	5	10			1
BC3	11:10	100				10	8	2		3
T2_1 & T2_2	11:20	90	10		1	9	3	6	1	3
	11:30	90	10			10	2	8		3
	22.3.2012									
BC4 & T3	9:00	50	50		2	8	10			1
	9:10	100			2	8	10			3
BC5 & T4_1 & T4_2	9:30	90	10			10	10			5
	9:40	100			4	6	10			3
	9:50	90	10		4	6	9	1		2

The measured bending strength of ice varied between 321 and 472 kPa and the average strength was approximately 404 kPa. The compressive strength varied between 0.65 and 1.98 MPa in the horizontal direction and between 1.41 and 2.79 MPa in the vertical direction, while the average compressive strengths were 1.28 and 2.02 MPa in the horizontal and vertical directions, respectively [31].

4.2. Straight-Ahead Tests

In the ahead test, the rudders were kept in zero or close to zero position, i.e., no turning was initiated, and the thrust request was kept constant. However, changes in the heading and course over ground were possible due to the ship–ice interaction in different parts of the hull. It was also observed that small rudder angle (approximately two degrees) was applied at the end of test L2 to obtain the same heading as in the beginning of the test. Figure 6 presents an example of the applied motor power and rudder angles, obtained speed and heading, and measured ice thickness during test L1. It should be noted that the heading indicates the heading with respect to the heading at the beginning of the test, not the actual heading.

The measured ice conditions are summarized in Table 4, and the ship navigational and machine control data are summarized in Table 5. Table 4 indicates similar ice conditions between the tests, although the level ice was slightly thinner during test L2_1, while the maximum ice thickness was the smallest in L3. Visual observations indicated that the ship encountered a heavy ridge at the end of L2_2 that caused a decrease in speed. As shown in Table 5, the rudder angles were kept close to zero, and the turning rate was negligible. Figure 7 gives the measured load at the bow, bow shoulder, and stern shoulder. Appendix A gives similar plots for tests L2_1, L2_2, and L3.

Table 4. Measured ice conditions during the straight-ahead tests.

Test ID	Ice Thickness [m]					
	Stereo Camera			Electromagnetic Device		
	Mean	Standard Deviation	Maximum	Mean	Standard Deviation	Maximum
L1	0.33	0.09	0.56	0.99	0.56	3.42
L2_1	0.29	0.09	0.55	0.74	0.46	2.86
L2_2	0.27	0.08	0.52	0.90	0.56	3.33
L3	0.33	0.10	0.80	0.89	0.15	1.59

Table 5. Ship navigation and machine control data during straight-ahead tests.

Test ID	Speed	Turning Rate	Engine Power		Rudder Angle		Propeller Pitch	
			PS	SB	PS	SB	PS	SB
	[m/s]	[deg/s]	[kW]	[kW]	[deg]	[deg]	[%]	[%]
L1	4.8	0.00	2804	2804	0	0	65	65
L2_1	4.5	−0.01	2189	2226	−1	0	58	57
L2_2	6.2	0.01	3723	3746	1	1	80	81
L3	2.9	−0.01	1420	1451	0	0	39	39

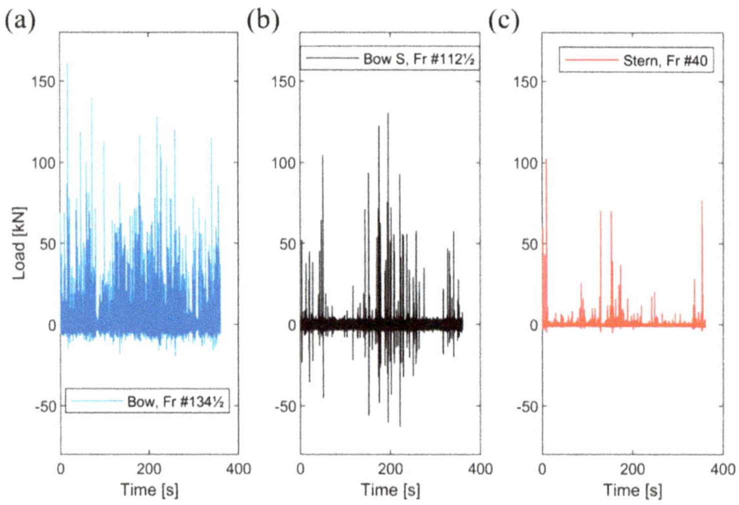

Figure 7. Measured ice-induced loads at (**a**) bow, (**b**) bow shoulder, and (**c**) stern shoulder during test L1.

4.3. Maneuvering Tests

The maneuvering tests consisted of breaking out from the ice channel and turning tests in level ice sheet. The breaking out from the channel tests were conducted in newly broken channels by PSRV S.A. Agulhas II. In the test, the ship was first accelerated to a speed achieved in the channel with a certain motor power. The power was kept constant during the tests. After the ship had gained speed, the rudders were turned to a certain predetermined angle that was retained until the ship had broken out from the channel. Figure 8 presents the applied motor power and rudder angles, obtained speed and heading, and measured ice thickness during tests BC5, T4_1, and T4_2, as an example of the breaking out from the channel test followed by two turning tests. The measured ice-induced loads during these tests are presented in Figure 9. The other tests are presented in Appendix A.

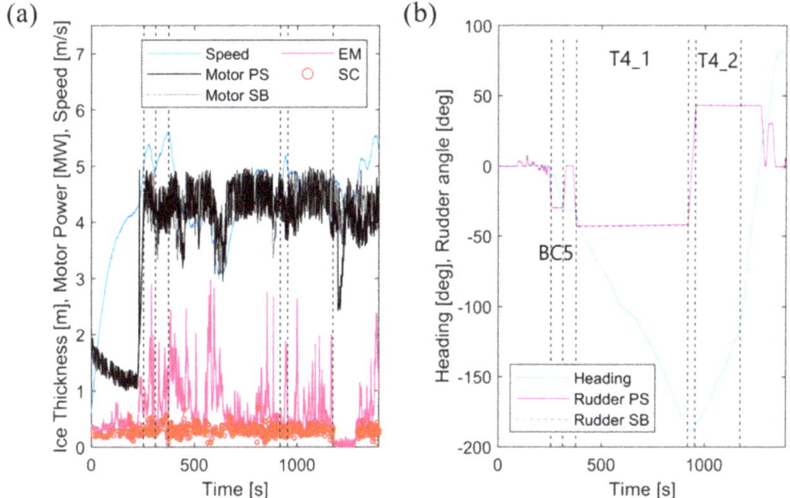

Figure 8. (**a**) Applied motor power, measured ice thickness with the electromagnetic device (EM) and stereo camera system (SC), and obtained speed, and (**b**) applied rudder angles with obtained heading during tests BC5, T4_1, and T4_2.

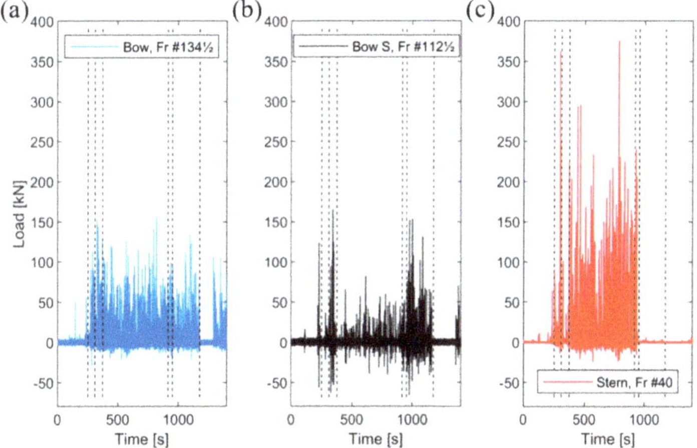

Figure 9. Measured ice-induced loads at (**a**) bow, (**b**) bow shoulder, and (**c**) stern shoulder during tests BC5, T4_1, and T4_2.

Turning tests were conducted in a similar manner in level ice sheets. The motor power and rudder angle were kept constant during the turn. The turning was continued until the heading had changed approximately 90 degrees. As in the straight-ahead tests, ice sheets with low deformation levels were preferred as the turning test locations, but it was not always possible to find such a field. Figures 8 and 9 present examples of turning tests with the applied control and measured ice condition and ice-induced loads. Positive rudder angles indicate turning to starboard and negative values turning to port side.

As only the starboard side of the vessel was instrumented, the maneuvering tests were conducted on the port side and starboard to study the loads on different sides of the vessel with respect to turning direction. As turning to port side exposes the instrumented starboard side of the stern shoulder to ice interaction, more turning tests to port side were conducted. In the data analysis, only the parts where the rudder angle was at the target angle were utilized (dashed lines in Figures 8 and 9). If a clear open-water section was encountered during a turning test, it was cut out from the data analysis (see Figure 8 for test T4_2). Tables 6 and 7 present the measured ice conditions, ship navigational data, and applied machinery control data for the maneuvering tests.

Table 6. Measured ice conditions during the maneuver tests.

Test ID	Ice Thickness [m]					
	Stereo Camera			Electromagnetic Device		
	Mean	Standard Deviation	Maximum	Mean	Standard Deviation	Maximum
BC1	0.26	0.07	0.45	0.46	0.21	1.11
BC2	0.23	0.03	0.29	0.30	0.11	0.78
BC3	0.22	0.09	0.45	0.89	0.48	2.32
BC4	0.27	0.08	0.41	0.61	0.23	1.11
BC5	0.33	0.09	0.53	1.02	0.50	2.91
T1_1	0.25	0.08	0.55	0.73	0.21	1.77
T1_2	0.25	0.05	0.39	0.72	0.34	2.5
T2_1	0.29	0.07	0.49	0.70	0.33	2.09
T2_2	0.34	0.12	0.65	1.60	0.68	3.94
T3	0.28	0.09	0.59	0.64	0.28	1.78
T4_1	0.31	0.09	0.71	0.83	0.51	2.99
T4_2	0.29	0.06	0.45	0.64	0.38	2.65

Table 7. Ship navigation and machine control data during the maneuver tests.

Test ID	Speed	Turning Rate	Engine Power		Rudder Angle		Propeller Pitch	
			PS	SB	PS	SB	PS	SB
	[m/s]	[deg/s]	[kW]	[kW]	[deg]	[deg]	[%]	[%]
BC1	4.1	−0.61	1966	1708	−36	−35	49	48
BC2	3.2	−0.80	3783	3798	−30	−30	75	76
BC3	5.0	0.82	1743	2361	29	30	57	57
BC4	5.4	−0.58	2726	2510	−29	−29	65	65
BC5	5.2	−0.43	4464	4280	−30	−29	81	81
T1_1	4.3	0.24	2132	2330	28	29	57	57
T1_2	4.3	−0.64	2446	2171	−28	−28	57	57
T2_1	4.2	−0.48	2367	2229	−29	−29	57	57
T2_2	3.0	−0.27	2726	2759	−30	−30	57	57
T3	4.1	−0.73	3755	3546	−38	−38	72	73
T4_1	4.3	−0.28	4300	4294	−42	−43	80	81
T4_2	4.7	0.32	4354	4461	43	43	80	81

5. Data Analysis

In this section, the straight-ahead tests and maneuver tests are analyzed from the intervals indicated in Section 4. The load events identified from the load signal with the triggering values and Rayleigh separation were referred to as trigger and peak load events, respectively (see Section 2.2). The load events duration, load magnitude, and rise time, were identified as described in Section 2.2.

The analysis showed that the number of peak and trigger load events was small at the bow shoulder area in tests BC1, BC3, and BC4. Similarly, only a few or no load events were measured at the stern shoulder area during the tests involving turning to starboard (tests T1_1, T4_2, and BC3). Due to the small number of load events, the statistical parameters calculated for these tests in this section had high uncertainty.

When the data were interpreted, it should be considered that the duration of maneuver tests varied significantly (Table 2). As the exposure to higher loads increased as a function of the duration or travelled distance, it was not considered valid to compare the maximum values between the tests. Thus, quantiles were utilized in the data analysis and comparison of load levels, as the consideration of number of loads was embedded in the quantile analysis.

5.1. Peak Load Frequency

The frequency of the peak load events during a test was determined for each frame by first identifying the peak load events from the measurement signal. Then, the number of peak loads was divided by the duration of the test. In order to determine a representative frequency for different hull areas, the frequency was determined for each frame separately and then the mean frequency was calculated from the frames in that area. The mean frequencies for different hull areas during different tests are presented in Figure 10 and Table 8 with the mean speed and turning rate during the tests. Note that the tests were organized based on type and increasing ship speed.

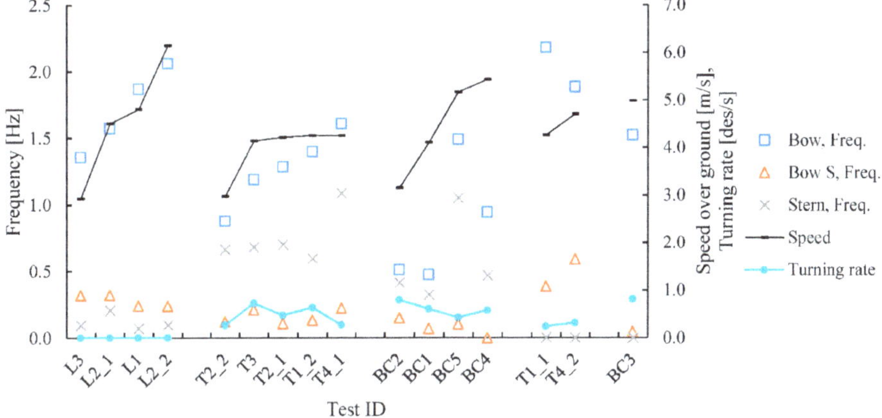

Figure 10. The frequency of measured peak loads, obtained speed, and turning rate in different tests. In the legend, Bow S, Stern, and Freq. refer to Bow Shoulder, Stern Shoulder, and Frequency, respectively.

Table 8. The frequency of measured peak loads in Hertz in different tests.

	L3	L2_1	L1	L2_2	T2_2	T3	T2_1	T1_2	T4_1	BC2	BC1	BC5	BC4	T1_1	T4_2	BC3
Bow	1.4	1.6	1.9	2.1	0.9	1.2	1.3	1.4	1.6	0.5	0.5	1.5	0.9	2.2	1.9	1.5
Bow Shoulder	0.3	0.3	0.2	0.2	0.1	0.2	0.1	0.1	0.2	0.2	0.1	0.1	0.0	0.4	0.6	0.0
Stern Shoulder	0.1	0.2	0.1	0.1	0.7	0.7	0.7	0.6	1.1	0.4	0.3	1.1	0.5	0.0	0.0	0.0

The frequency of peak loads at the bow showed an increasing trend as a function of the speed in the straight-ahead test. A comparison of the straight-ahead tests with the turning tests in an intact level ice field indicated that the frequency slightly decreased or remained the same on the bow side opposite to the turning direction, while the frequency seemed to increase on the bow side toward the turning direction. Likewise, breaking out from the channel indicated a similar trend. However, only one test was done toward the starboard. In the case of a port-side turn, the calculated frequency highly depended on when the bow broke into the ice. As an example, the first ice interaction at the instrumented bow frames was observed approximately at the midpoint of the test in BC2, while the first contact was reached after a short duration in BC5.

The peak load frequency levels at the bow shoulder area were clearly smaller than that at the bow, but the frequencies at the bow shoulder showed similar trends to those at the bow. The frequency level decreased or remained the same when the vessel was turned to the opposite direction with respect to the instrumentation and increased when the instrumented ship side matched the turning direction. The breaking out of the channel tests showed only a few interactions in tests BC1, BC3, and BC5.

Thus, no clear trend was determined. Similarly, no clear trend between the speed and the frequency of peak load events at the bow shoulder was observed.

At the stern shoulder area, the occurrence and frequency of ice-induced loads had a clear dependency on ship maneuvers. Some ice-induced loads were observed at the stern shoulder area in the straight-ahead operations due to ship motions caused by contact with the ice field. When the ship turned, the frequency of the loading increased significantly at the opposite side of the ship with respect to the turning direction, while, generally, no interaction on the hull was observed at the side of the ship toward the turning direction. The turning rates were relatively constant in similar test runs. However, this did not seem to have any effect on the frequency of the peak ice loading.

5.2. Peak Load Magnitude

For the ice-induced load magnitude analysis, the peak loads were identified and separated on each frame for each test. Due to the small number of peak loads in some tests, the peak loads measured on different frames in the same area were combined for each test, e.g., the measurements on frames #134$\frac{1}{2}$ and 134 were combined to represent the loading at the bow. This was considered to give a better statistical representation of the loads. Here, it was assumed that the minor differences in the frame angles and locations at the same hull area did not affect ice-induced loading. After the sets of peak loads for different hull areas for each test were identified, 25%, 50%, 75%, and 95% quantiles were determined (Figure 11, Table 9). Note that the tests were organized based on increasing maximum ice thickness measured with the stereo camera system in Figure 11.

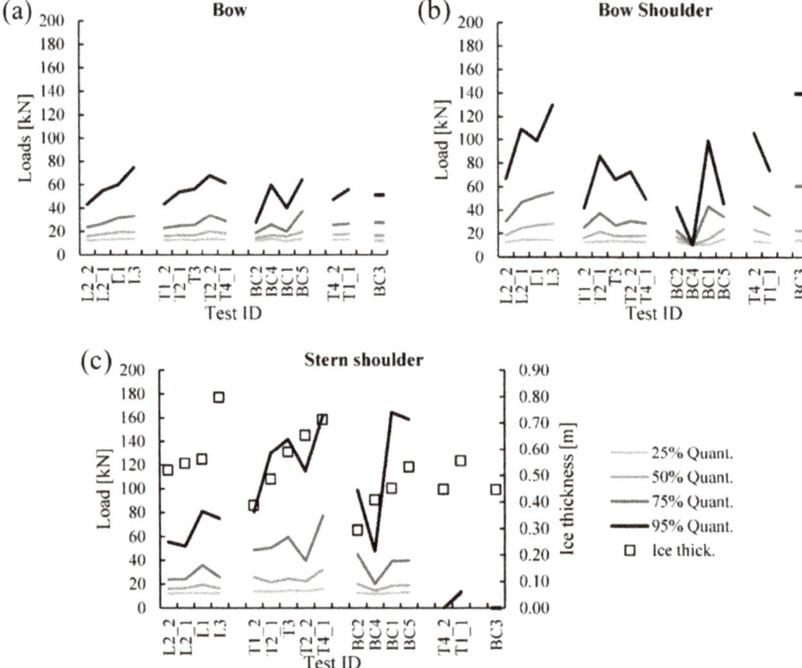

Figure 11. The 25%, 50%, 75%, and 95% quantiles of measured ice-induced load magnitudes at (**a**) bow, (**b**) bow shoulder, and (**c**) stern shoulder, with measured maximum ice thickness with the stereo camera system.

Table 9. The 25%, 50%, 75%, and 95% quantiles of measured ice-induced load magnitudes at bow, bow shoulder, and stern shoulder.

	Quantile	L2_2	L2_1	L1	L3	T1_2	T2_1	T3	T2_2	T4_1	BC2	BC4	BC1	BC5	T4_2	T1_1	BC3
Bow	25%	12.3	12.9	13.6	13.5	12.3	12.9	12.5	13.4	13.1	11.7	13.7	11.6	13.6	12.8	13.0	12.1
	50%	16.1	17.6	19.2	19.3	16.1	16.7	16.5	20.0	18.3	14.4	16.8	15.7	19.3	17.0	17.6	16.6
	75%	23.6	26.6	31.5	33.0	22.7	24.7	25.4	33.7	28.9	18.8	25.8	19.9	37.2	25.4	26.4	27.5
	95%	43.2	54.9	60.0	74.5	43.5	53.8	56.3	67.7	61.6	27.6	59.7	40.3	64.0	47.3	55.9	51.1
Bow Shoulder	25%	12.7	14.5	14.6	14.5	12.8	13.1	13.6	13.4	12.9	13.0	10.5	10.6	15.4	13.9	12.4	14.4
	50%	18.8	24.7	27.0	28.2	16.5	21.4	17.9	18.0	18.3	17.1	10.5	15.5	23.9	23.6	19.4	22.8
	75%	30.3	46.8	51.6	55.0	25.2	37.5	27.0	30.5	28.8	22.3	10.5	43.5	34.6	43.4	35.8	60.7
	95%	66.8	109.1	99.1	129.9	42.0	85.9	66.1	72.8	49.5	42.7	10.5	99.2	45.6	105.7	74.0	139.6
Stern Shoulder	25%	12.3	12.6	12.4	12.3	13.9	13.9	14.6	14.4	15.8	12.6	11.7	12.6	13.3	0.0	11.5	0.0
	50%	16.1	16.8	19.5	16.5	25.7	21.6	24.6	22.4	32.2	20.0	14.7	18.7	19.0	0.0	12.3	0.0
	75%	24.1	24.1	35.8	25.8	48.9	50.7	59.7	39.6	77.4	44.8	20.3	39.3	39.7	0.0	13.3	0.0
	95%	55.6	52.2	80.9	75.2	80.5	130.0	141.8	114.8	161.2	98.7	47.6	164.4	158.6	0.0	13.6	0.0

As shown in Figure 11, the highest loads in the straight-ahead operations were measured in the bow shoulder area, as the 75% quantile loads were approximately at the same level as 95% quantile loads at the bow. The magnitude of the highest measured loads in these areas was similar (approximately 300 kN), but the bow had a large number of peak ice-induced loads of a small magnitude. Only some ice-induced loads occurred at the stern shoulder, but the magnitudes of these were relatively large, which increased the quantile levels.

Generally, the maneuvers did not seem to affect the load levels measured at the bow and bow shoulder, as the load levels of the quantiles remained at the same level. However, a significant impact on the load magnitudes was seen at the stern shoulder area. The load level at the side of the stern opposite to the turning direction increased clearly above the load levels at the bow or bow shoulder areas, while on the other side of the stern the load diminished. Figure 11 shows increasing trend in ice-induced loading as a function of the measured maximum ice thickness in all hull regions.

5.3. Peak Load Rise Time and Duration and Trigger Load Duration

After the trigger load events and peak loads were identified from the measurement data, peak load rise time and durations, and the trigger load event durations were determined. The duration of all the peak loads were gathered from different tests for the bow, bow shoulder, and stern shoulder areas and the peak load magnitudes were plotted as a function of peak load duration (Figure 12). Similar plots were presented for peak load rise time and trigger load event duration in Figures 13 and 14. The mean, median, and maximum values for the peak load rise time and duration, and load event duration are presented in Table 10.

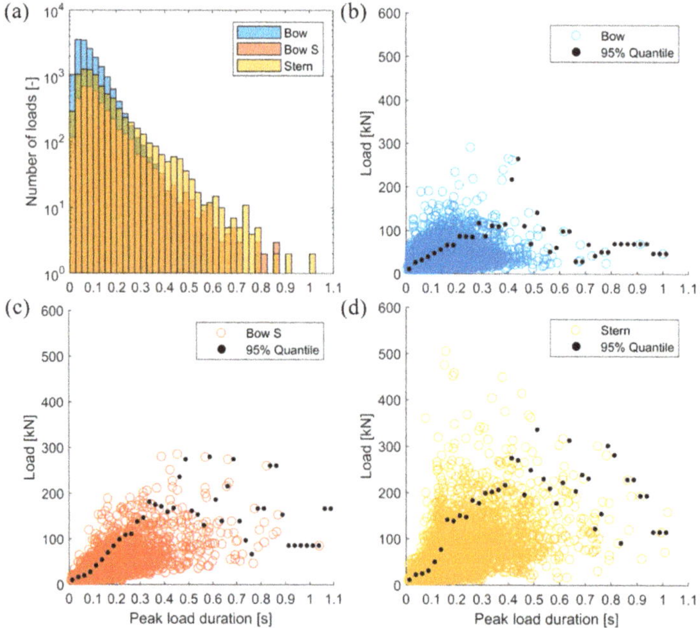

Figure 12. Duration of peak loads from all the tests at (**a**) all hull regions in histogram format, and individual loads with calculated 95% quantiles at (**b**) bow, (**c**) bow shoulder, and (**d**) stern shoulder.

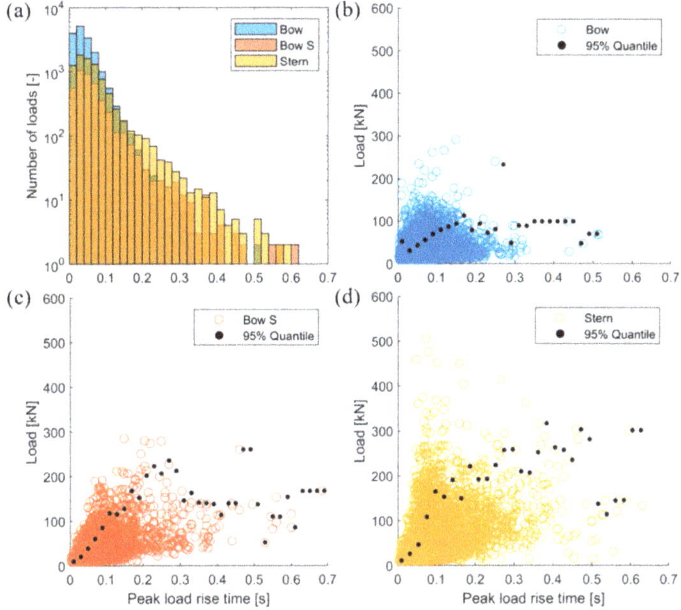

Figure 13. Peak load rise times from all the tests at (**a**) all hull regions in histogram format, and individual loads with calculated 95% quantiles at (**b**) bow, (**c**) bow shoulder, and (**d**) stern shoulder.

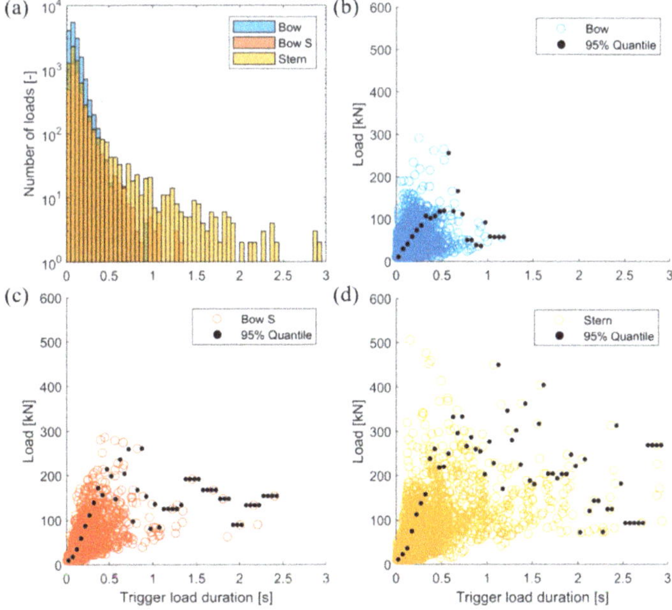

Figure 14. Duration of load events from all the tests at (**a**) all hull regions in histogram format, and individual loads with calculated 95% quantiles at (**b**) bow, (**c**) bow shoulder, and (**d**) stern shoulder.

Table 10. Statistical figures for the rise time and duration of peak loads and duration of load events.

	Number of Loads		Rise Time [s]			Peak Load Duration [s]			Load Event Duration [s]		
	Peak Loads	**Load Events**	**Mean**	**Median**	**Max**	**Mean**	**Median**	**Max**	**Mean**	**Median**	**Max**
Bow	16922	15731	0.045	0.035	0.516	0.094	0.075	1.001	0.101	0.080	1.157
Bow Shoulder	4249	3831	0.066	0.050	0.686	0.135	0.105	1.166	0.150	0.105	2.403
Stern Shoulder	8246	6722	0.067	0.050	0.631	0.137	0.105	1.021	0.168	0.095	2.939

The load times were the shortest at the bow (Figures 12–14, Table 10). The mean peak load and trigger load event durations were 0.094 and 0.101 s, respectively, while the maximum durations at bow were around 1.0 and 1.2 s. The mean and median peak load rise time and duration at the bow shoulder and stern shoulder were approximately the same, indicating very similar load durations in these areas. The trigger load event durations in these areas were significantly higher than at the bow as the maximum values were 2.4 and 2.9 s for the bow shoulder and stern shoulder, respectively. However, it should be noted that the maximum peak load duration in these areas was approximately 1.1 s. This indicated that the longest triggering load events consisted of several peak load events, as illustrated in Figure 4. The mean peak load rise time was quite accurately half of the mean peak load duration for each hull area (Table 10). This indicated that the mean peak load event would have an equal rise and decaying time. This suggested that, on average, the ice did not fail on the instrumented frame, but the contact point traveled over the instrumented area and the failure occurred outside the instrumented area. This assumed that, in the case of a failure, the loading dropped rapidly.

The shortest loads occurred at the bow (Figures 12–14). No clear trend between the peak load magnitude and peak load rise time or duration, or trigger load event duration was observed at the bow, bow shoulder, or stern shoulder when separate loadings were observed. The durations of the highest peak loads were less than 0.2 s at the stern shoulder area.

However, the 95% quantiles showed a clear trend that the load magnitude increased as a function of peak load duration and rise time, and trigger load event duration in all hull areas (Figures 12–14). The quantiles were calculated for the same bin ranges, as presented in the histograms in Figure 12a, Figure 13a, and Figure 14a. The 95% quantiles first increased in a linear fashion and then stabilized around a value when the amount of data in a bin decreased. As an example, the peak load magnitude as a function of peak load duration for the stern shoulder showed a linear trend up to a duration of 0.5 s, and then stabilized for longed durations (Figure 12). As the number of data points was small in the bin ranges where the 95% quantiles stabilized, it cannot be concluded if this was due to a physical process or lack of data points.

As the 95% quantiles for the load magnitude as a function of peak load rise time increased linearly, the slope of the increase gave the loading rate for the 95% quantiles. The obtained loading rates, based on Figure 13, were 600 kN/s, 1000 kN/s, and 1000 kN/s, for the bow, bow shoulder, and stern shoulder, respectively. It is noted that the frame angle at the bow favors bending failure of ice while the vertical frames at the bow shoulder and stern shoulder areas favor ice failure through crushing. Thus, it is possible that the difference in the loading rates is related to the failure mode of ice. However, it is not possible to draw solid conclusions from this data. Dedicated tests are required for this matter.

Figure 15 presents the 25%, 50%, 75%, and 95% quantiles of the peak load durations in each test for all the hull areas. The tests were organized based on increasing speed. The quantiles for the bow and bow shoulder showed decreasing trends for all the test types, which indicated that the load duration in these areas was affected by the speed of the vessel. Similar trends were not observed for the stern shoulder area. Figure 15 suggests that turning does not affect the peak load duration at the bow or bow shoulder areas, but has a significant impact on the load duration at the stern shoulder.

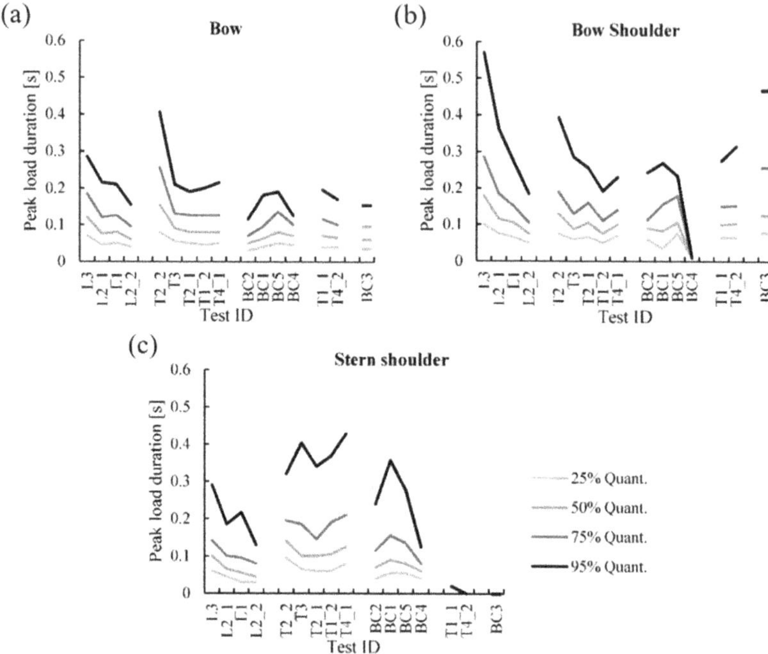

Figure 15. The 25%, 50%, 75%, and 95% quantiles of measured peak ice-induced load duration at (**a**) bow, (**b**) bow shoulder, and (**c**) stern shoulder.

6. Discussion

The applied threshold in ice-induced load identification had a significant impact on the number of identified loads as the number of loads decreased exponentially as a function of the load magnitude. However, as the system was designed to measure loads over 1 MN, the measurement accuracy at smaller load levels hindered the load identification from the signal due to noise. Thus, a threshold was applied in the load identification in order not to account for signal noise as ice-induced loads. Thus, the threshold was set to 10 kN as it was considered that this level would not contain noise or open-water-related loads. Although there was no solid basis to set the threshold to this level and the selection of a threshold was subjective, the impact on the results was considered negligible, as the same threshold was applied invariably. Thus, the threshold did not affect the comparison in this study, but should be accounted for if compared to other studies. Future work is needed to neglect the threshold or determine the physical basis for it.

In addition to the applied threshold, the determined frequency was affected by the applied load identification method (triggering value vs. Rayleigh separation) and the Rayleigh separation value, in a case this method was applied. Despite the effect, it was considered that these did not affect the results where the loading frequency was compared between different speeds and maneuvering operations as the same methods and limits were applied through the study. However, these should be accounted for when comparing with other studies. In general, a lower threshold and applying smaller Rayleigh separation values increased the frequency of ice-induced loads. Applying triggering values had a two-fold effect as smaller triggering values increased the number of small loads, whereas higher triggering values separated some loading events into a larger number of loading events. Future research on the applicable Rayleigh separator and triggering value is needed to unify the applied methods and to make the different studies more comparable. In the future studies, these should be linked to physical processes.

In the maneuvering tests, the frequency of the peak loads was determined over the duration the rudders were kept at the predetermined position. If the frequencies had been determined from the point of first contact, the determined frequencies would have been higher. However, this would have hindered the comparison of different hull areas as the duration of the test would have been different.

Due to the complex failure process, the loading on the hull varies significantly. Thus, it is difficult to give a clear, but not too simplified characterization. Lee et al. [17] and Ahn et al. [18] categorized the ice-induced load events based on the possible occurrence of the intermediate load drops before and after the peak load. However, especially at the midship and stern areas, where the structure does not penetrate straight ahead to ice, the loading may have constant phases and great drops before and after the peak loads that are related to the relative movement between the ice field and ship. Thus, these are not related to the failure processes. It is considered that criteria for the intermediate load drops and separation of load events should be set for the characterization, see e.g., Ahn et al. [19]. When criteria for the ice-induced load events based on failure mode can be established, the load characterization and load event separation can be improved. This is left for the future studies.

7. Conclusions

The study focused on the analysis of ice-induced loads at different hull areas during controlled maneuvering tests on-board PSRV S.A. Agulhas II in the Baltic Sea during the ice trial. According to the analyses, maneuvers may have a minor impact on the frequency of the loading at the bow and bow shoulder area and a major impact on the frequency at the stern shoulder area. The measured frequencies at the bow and bow shoulder were 0.9–2.2 Hz and 0.1–0.6 Hz, respectively, in an intact ice field. At stern shoulder, the frequencies were 0.1 Hz in straight-ahead operations, but can increase up to 1.1 Hz in turnings. Similarly, maneuvers did not have a clear impact on the load magnitude at the bow and bow shoulder areas, but had a significant effect at stern shoulder. The ice thickness had an increasing impact on the load magnitudes in all hull regions.

The ice-induced load duration analysis showed that the mean duration of the peak loads was 0.09 s at the bow and 0.14 s at the bow shoulder and stern shoulder. Furthermore, the peak load duration did not exceed 1.2 s. When the trigger load event duration was observed, the durations increased to 0.10, 0.15, and 0.17 s for the bow, bow shoulder, and stern shoulder, respectively; the corresponding longest durations were 1.2, 2.4, and 2.9 s. The maneuvers did not seem to affect the durations of ice-induced loading on the bow and bow shoulder, but had a clear impact on the durations at the stern shoulder. Furthermore, increasing speed decreased the load duration at bow and bow shoulder. The calculated 95% quantiles indicated a clear increasing trend of load magnitudes as a function of load durations. The loading rates calculated from the 95% quantiles indicated a loading rate of 600 kN/s for the bow and 1000 kN/s for the bow shoulder and stern shoulder.

The load duration and frequency of loading are important parameters when the probability of different magnitude ice-induced loads is determined, as the duration and frequency affect the exposure that has a direct impact on the predicted extreme loads. However, the criteria for identification and characterization of separate ice-induced loads from the continuous measurements are not solidly based on physical interpretation, but have a direct impact on the obtained load durations and frequencies. Thus, future research on these are required.

As noted, the analyzed and presented measurement data are made available with the paper as Supplementary Materials. This is considered valuable for other researchers for several topics on ice-induced loads, and the validation of their simulation models as ship performance can be estimated from the data.

Supplementary Materials: The following are available online at http://www.mdpi.com/2077-1312/8/10/759/s1. The measurement data as described in Appendix B.

Author Contributions: Conceptualization, M.S.; methodology, M.S., F.L. and L.L.; formal analysis, M.S. and J.L.; investigation, M.S., A.B. and J.L.; resources, M.S.; data curation, M.S., F.L. and J.L.; writing—original draft preparation, M.S.; writing—review and editing, F.L., L.L., P.K., A.B. and J.L.; visualization, M.S.; supervision, P.K.; project administration, M.S., P.K. and A.B.; funding acquisition, P.K. All authors have read and agreed to the published version of the manuscript.

Funding: This project received funding from the Lloyd's Register Foundation, a charitable foundation, helping to protect life and property by supporting engineering-related education, public engagement, and the application of research (www.lrfoundation.org.uk). The PSRV S.A. Agulhas II was instrumented within a project funded by Tekes (Finnish Funding Agency for Technology and Innovation), nowadays Business Finland. The authors would like to acknowledge Tekes for funding.

Acknowledgments: The authors would like to acknowledge Tekes project consortium partners for collaboration, namely, STX Finland (currently Rauma Marine Construction), the University of Oulu, Aker Arctic Technology, Rolls-Royce, Wärtsilä, DNV-GL, and the Department of Environmental Affairs of South Africa.

Conflicts of Interest: The authors declare no conflict of interest.

Appendix A

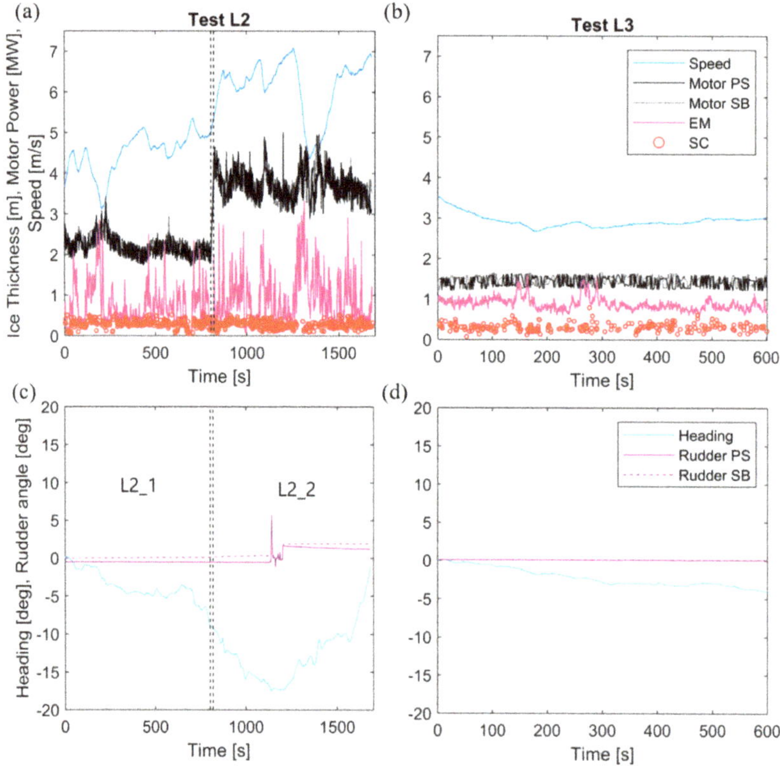

Figure A1. Applied motor power, measured ice thickness with the electromagnetic device (EM) and stereo camera system (SC), and obtained speed during tests (**a**) L2_1 and L2_2, and (**b**) L3. Applied rudder angles with obtained heading during tests (**c**) L2_1 and L2_2, and (**d**) L3.

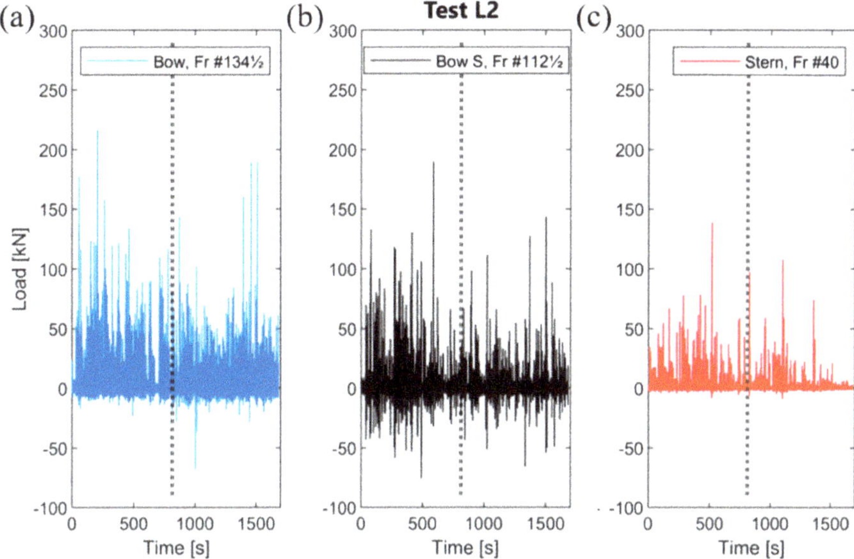

Figure A2. Measured ice-induced loads at (**a**) bow, (**b**) bow shoulder, and (**c**) stern shoulder during tests L2_1 and L2_2.

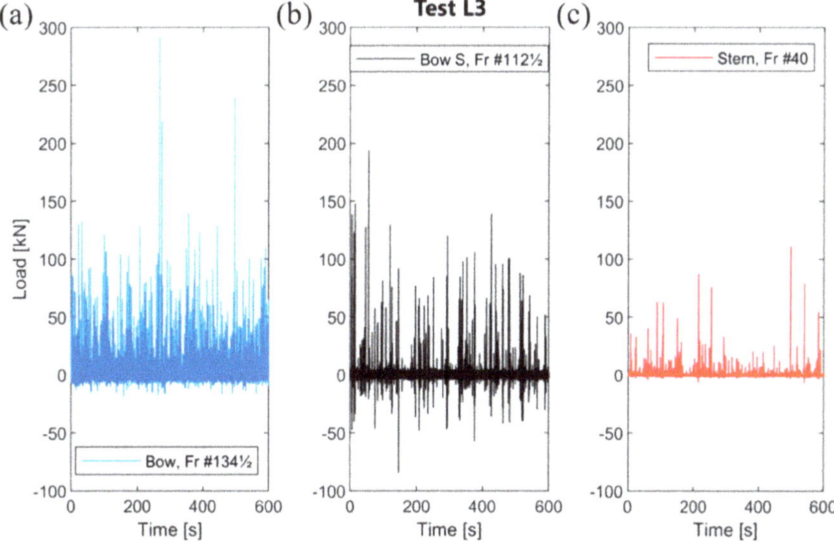

Figure A3. Measured ice-induced loads at (**a**) bow, (**b**) bow shoulder, and (**c**) stern shoulder during test L3.

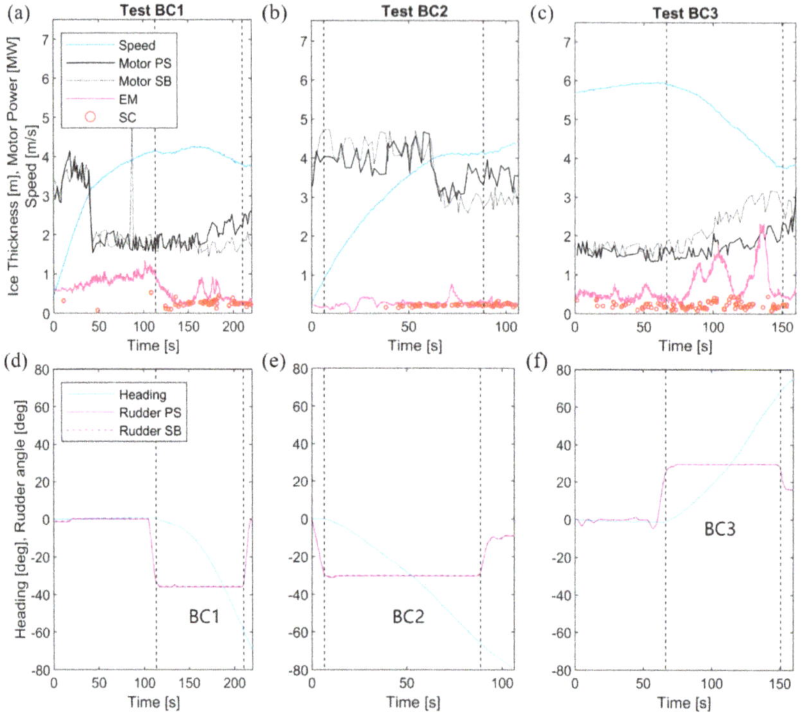

Figure A4. Applied motor power, measured ice thickness with the electromagnetic device (EM) and stereo camera system (SC), and obtained speed during tests (**a**) BC1, (**b**) BC2, and (**c**) BC3. Applied rudder angles with obtained heading during tests (**d**) BC1, (**e**) BC2, and (**f**) BC3.

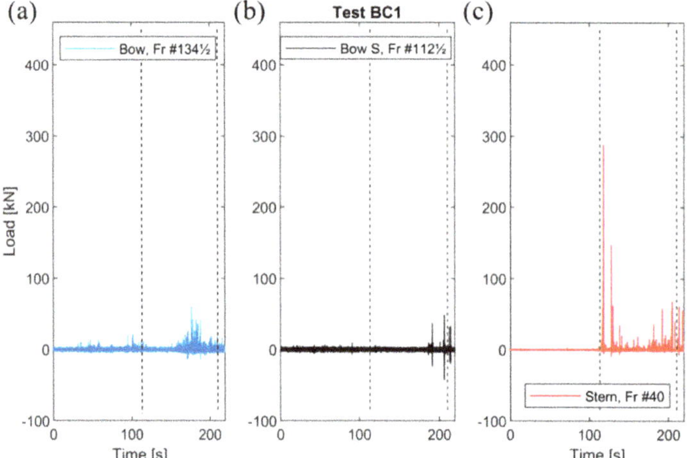

Figure A5. Measured ice-induced loads at (**a**) bow, (**b**) bow shoulder, and (**c**) stern shoulder during test BC1.

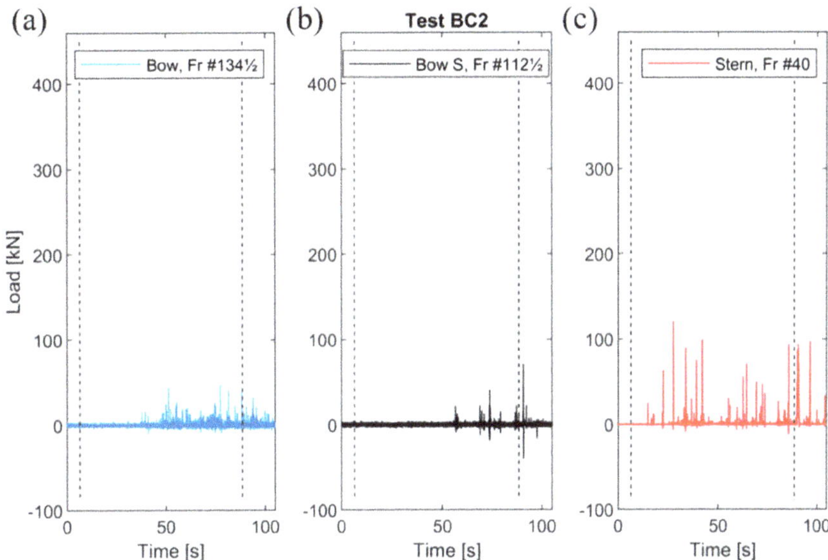

Figure A6. Measured ice-induced loads at (**a**) bow, (**b**) bow shoulder, and (**c**) stern shoulder during test BC2.

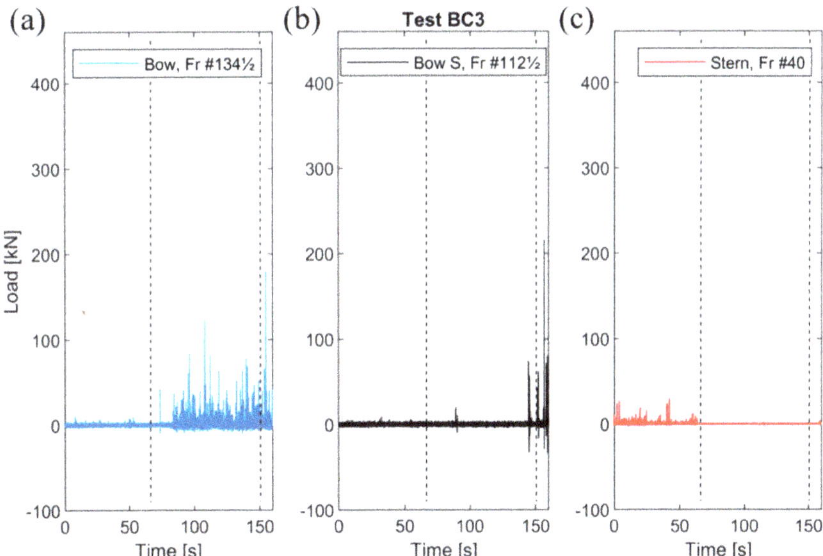

Figure A7. Measured ice-induced loads at (**a**) bow, (**b**) bow shoulder, and (**c**) stern shoulder during test BC3.

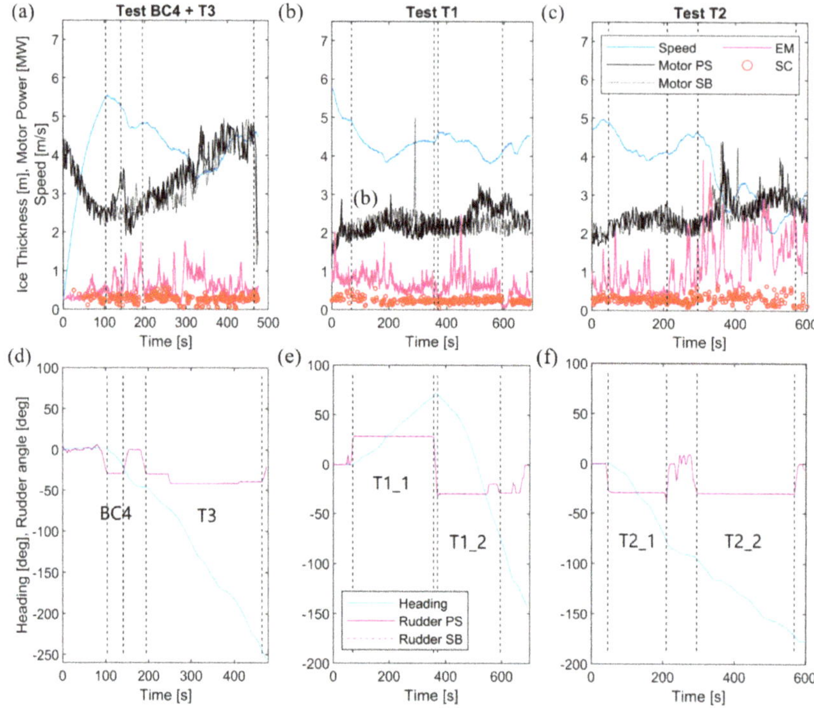

Figure A8. Applied motor power, measured ice thickness with the electromagnetic device (EM) and stereo camera system (SC), and obtained speed during tests (**a**) BC4 and T3, (**b**) T1_1 and T1_2, and (**c**) T2_1 and T2_2. Applied rudder angles with obtained heading during tests (**d**) BC4 and T3, (**e**) T1_1 and T1_2, and (**f**) T2_1 and T2_2.

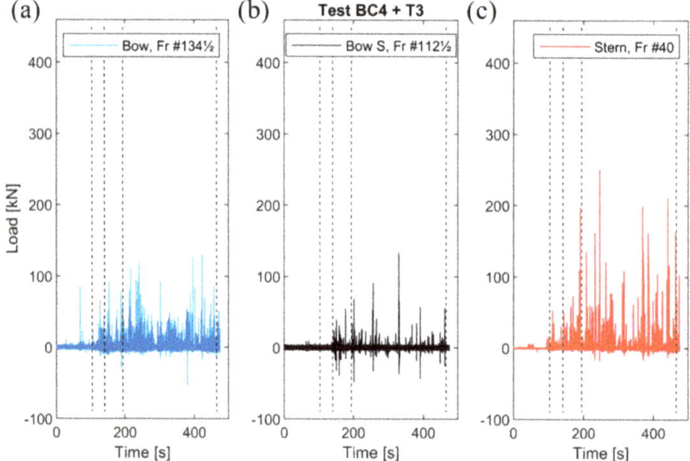

Figure A9. Measured ice-induced loads at (**a**) bow, (**b**) bow shoulder, and (**c**) stern shoulder during tests BC4 and T3.

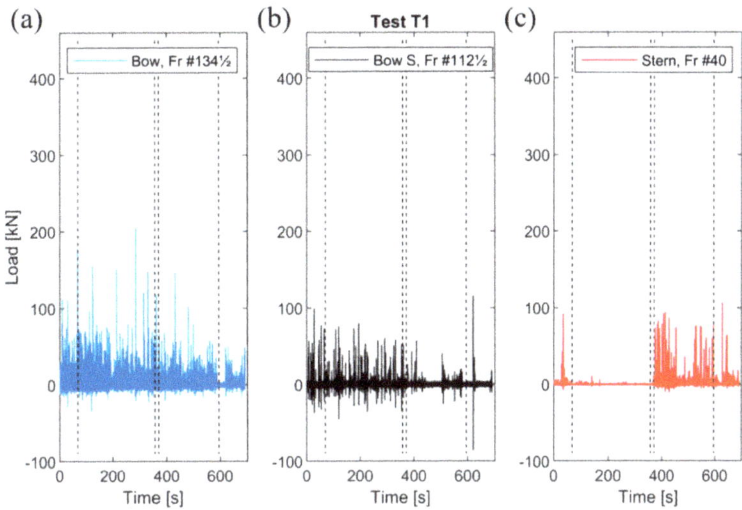

Figure A10. Measured ice-induced loads at (**a**) bow, (**b**) bow shoulder, and (**c**) stern shoulder during tests T1_1 and T1_2.

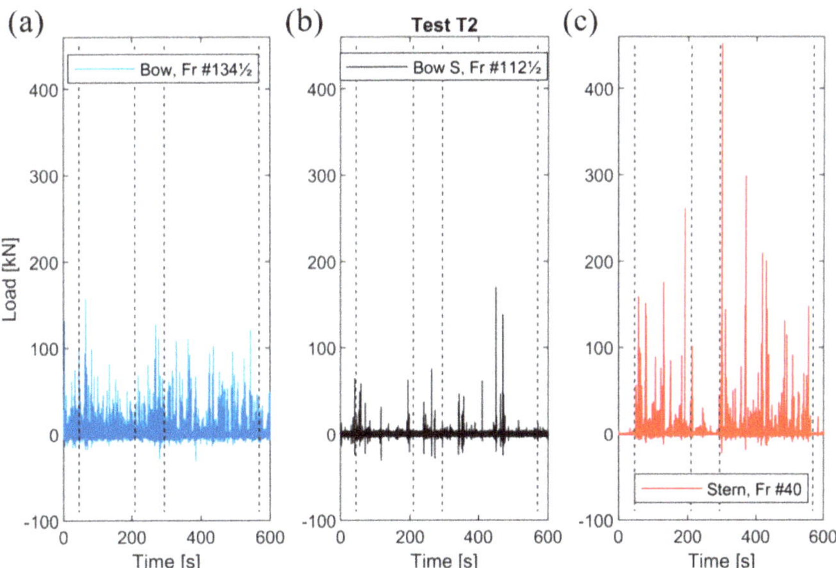

Figure A11. Measured ice-induced loads at (**a**) bow, (**b**) bow shoulder, and (**c**) stern shoulder during tests T2_1 and T2_2.

Appendix B

The measurement data were collected in separate folders in accordance with Table 2. Each of the folders was named based on the test presented in each row. As the measured parameters were sampled with different rates from different sources, the measurement data were saved in -ascii format in different files depending on the source of the measurements. As an example, the folder 'T1' contains files: 'T1_hice_SC.txt', 'T1_hice_EM.txt', 'T1_Navigation.txt', 'T1_Machinery.txt', and 'T1_ Loads.txt'.

The name of a file indicates the content of the file. In each file, the first column gives the synchronized time that is zeroed to the beginning of the tests. As an example, time 0 s in test T1 equals 21.3. 10:46:00. The starting times of the tests are presented in Table 2. The other columns of the measurement files are also given. It should be noted that test T1 is applied as an example here, but the content of the files in different tests is the same.

T1_hice_SC.txt:	Column1—Time [s]; Column2—Ice thickness [m];
T1_hice_EM.txt:	Column1—Time [s]; Column2—Ice thickness [m];
T1_Navigation.txt:	Column1—Time [s]; Column2—Latitude [deg]; Column3—Longitude [deg]; Column4—Speed over ground [m/s]; Column5—Course over ground [deg]; Column6—Heading [deg];
T1_Machinery.txt:	Column1—Time [s]; Column2—PS Motor power [kW]; Column3—SB Motor power [kW]; Column4—PS Rudder position [deg]; Column5—SB Rudder position [deg]; Column6—PS Propeller pitch [%]; Column7—SB Propeller pitch [%]; Column8—PS Shaft RPM [RPM]; Column9—SB Shaft RPM [RPM];
T1_ Loads.txt:	Column1—Time [s]; Column2—Load on frame $134\frac{1}{2}$ [kN]; Column3—Load on frame 134 [kN]; Column4—Load on frame 113 [kN]; Column5—Load on frame $112\frac{1}{2}$ [kN]; Column6—Load on frame 112 [kN]; Column7—Load on frame 41 [kN]; Column8—Load on frame $40\frac{1}{2}$ [kN]; Column9—Load on frame 40 [kN]; Column10—Load on frame $39\frac{1}{2}$ [kN];

In addition to the measurement data, the approximated waterline of S.A. Agulhas II with the frame angles at the draught 7.55 m was provided in excel format. It should be noted that the frame angles were defined from the frame pictures presented in [4] for the locations of instrumented areas. The hull angels for other locations were approximated based on the information on the instrumented area. The shape of the waterline was defined based on the general arrangement picture. Thus, the waterline and the frame angles were coarse approximations from the actual hull form.

References

1. Bekker, A.; Suominen, M.; Peltokorpi, O.; Kulovesi, J.; Kujala, P.; Karhunen, J. Full Scale Measurements on a Polar Supply and Research Vessel During Maneuver Tests in an Ice Field in the Baltic Sea. In Proceedings of the 33rd International Conference on Ocean, Offshore and Arctic Engineering, OMAE33, San Francisco, CA, USA, 8–13 June 2014.
2. Suominen, M.; Kujala, P.; Kotilainen, M. The Encountered Extreme Events and Predicted Maximum Ice Induced Loads on the Ship Hull in the Southern Ocean. In Proceedings of the 34th International Conference on Ocean, Offshore and Arctic Engineering, OMAE34, St. John's, NL, Canada, 31 May–5 June 2015.
3. Valtonen, V. Effect of rudder force on stern shoulder region ice-induced loads. In Proceedings of the Arctic Technology Conference, St. John's, NL, Canada, 24–26 October 2016.
4. Suominen, M.; Kujala, P.; von Bock und Polach, R.; Kiviranta, J. Measured Ice-induced loads and Design Ice-induced loads. Analysis and Design of Marine Structures. In Proceedings of the 4th International Conference on Marine Structures, MARSTRUCT 2013, Espoo, Finland, 25–27 March 2013.
5. Bekker, A.; Suominen, M.; Kujala, P.; De Waal, R.J.O.; Soal, K.I. From data to insight for a polar supply and research vessel. *Ship Technol. Res.* **2019**, *66*, 57–73. [CrossRef]
6. Iyerusalimskiy, A.; Choi, J.; Park, G.; Kim, Y.; Yu, H. The Interim Results of the Long-Term Ice-Induced Loads Monitoring on the Large Arctic Tanker. In Proceedings of the 21st International Conference on Port and Ocean Engineering under Arctic Conditions, POAC11, Montréal, QC, Canada, 10–14 July 2011.
7. Abramian, A.; Vekulenko, S.; van Horssen, W. On a simple oscillator problem describing ice-induced vibrations of an offshore structure. *Nonlinear Dyn.* **2019**, *98*, 151–166. [CrossRef]
8. Korri, P.; Varsta, P. On the Ice Trial of a 14500 DWT Tanker on the Gulf of Bothnia. In Proceedings of the 24th Norwegian Ship Technology Conference, Helsinki, Finland, 4–5 October 1979.
9. Vuorio, J.; Riska, K.; Varsta, P. *Long Term Measurements of Ice Pressure and Ice-Induced Stresses on the Icebreaker Sisu in Winter 1978*; Report No 28; Winter Navigation Research Board: Helsinki, Finland, 1979; p. 50.

10. Glen, I.; Blount, H. Measurement of ice impact pressures and loads onboard CCGS Louis S. St. Laurent. In Proceedings of the 3rd International Conference on Ocean, Offshore & Arctic Engineering, OMAE3, New Orleans, LA, USA, 1984.
11. Kujala, P.; Vuorio, J. On the statistical nature of the ice-induced pressures measured on board I.B. Sisu. In Proceedings of the International Conference on Port and Ocean Engineering under Arctic Conditions, POAC 1985, Narssarssuaq, Greenland, 7–14 September 1985; pp. 823–837.
12. Riska, K.; Rantala, H.; Joensuu, A. *Full Scale Observations on Ship-Ice Contact Results from Tests Series Onboard IB Sampo, Winter 1989*; Report M-97; Helsinki University of Technology, Laboratory of Naval Architecture and Marine Engineering: Espoo, Finland, 1990.
13. St. John, J.; Sheinberg, R.; Ritch, R.; Minnick, P. Ice impact load measurements abroad the Oden during the International Arctic Ocean Expedition 1991. In Proceedings of the SNAME, 5th International Conference on Ships and Marine Structures in Cold Regions, Calgary, AB, Canada, 16–18 March 1994.
14. Izumiyama, K.; Takimoto, T.; Uto, S. Length of Ice-induced load Patch on a Ship Bow in Level Ice. In *10th International Symposium on Practical Design of Ships and other Floating Structures, PRADS 2007*; ABS: Houston, TX, USA, 2007.
15. Gagnon, R. Analysis of data from bergy bit impacts using a novel hull-mounted external Impact Panel. *Cold Reg. Sci. Technol.* **2008**, *52*, 50–66. [CrossRef]
16. Hänninen, S.; Lensu, M.; Riska, K. *Analysis of the Ice Lad Measurements during USCGC Healy Ice Trials, Spring 2000*; Report M-265; Helsinki University of Technology, Ship Laboratory: Espoo, Finland, 2001.
17. Lee, J.-H.; Hwang, M.-R.; Kwon, S.; Kang, J.-G.; Park, H.-M.; Choi, K.; Lee, T.-K. Analysis of Local Ice-Induced Load Signals Measured on an Arctic Voyage in 2013. In Proceedings of the 23rd International Conference on Port and Ocean Engineering under Arctic Conditions, Trondheim, Norway, 14–18 June 2015.
18. Ahn, S.-J.; An, W.-S.; Lee, T.-K.; Choi, K. An Analysis on Strain Gauge Signal Measured from Repetitive Ramming in Heavy Ice Condition. In Proceedings of the 24th International Conference on Port and Ocean Engineering under Arctic Conditions, Busan, Korea, 11–16 June 2017.
19. Ahn, S.-J.; Lee, T.-K. An experimental study on occurrence of intermediate peaks in ice-induced load signals. *Int. J. Nav. Archit. Ocean Eng.* **2020**, *12*, 157–167. [CrossRef]
20. Suominen, M.; Karhunen, J.; Bekker, A.; Kujala, P.; Elo, M.; Von Bock und Polach, R.; Endlund, H.; Saarinen, S. Full scale measurements on board PSRV S.A. Agulhas II in the Baltic Sea. In Proceedings of the 22nd International Conference on Port and Ocean Engineering under Arctic Conditions, POAC13, Espoo, Finland, 9–13 June 2013.
21. Daley, C.G. *Ice Edge Contact—A Brittle Failure Process Model*; Acta Polytechnica Scandinavica, Mechanical Engineering Series No. 100; Finnish Academy of Technology: Helsinki, Finland, 1991.
22. Kotilainen, M.; Suominen, M.; Kujala, P. Rotating ice cusps on ship's bow shoulder: Full-scale study on the cusp sizes and corresponding peak loads in different ice and operational conditions. *Ocean Eng.* **2019**, *189*, 106280. [CrossRef]
23. Kujala, P. *On the Statistics of Ice-Induced Loads on Ship Hull, in the Baltic*; Acta Polytechnica Scandinavica, Mechanical Engineering Series No. 116; Finnish Academy of Technology: Helsinki, Finland, 1994.
24. Kotilainen, M.; Vanhatalo, J.; Suominen, M.; Kujala, P. Predicting ice-induced load amplitudes on ship bow conditional on ice thickness and ship speed in the Baltic Sea. *Cold Reg. Sci. Technol.* **2017**, *135*, 116–126. [CrossRef]
25. Suominen, M.; Kujala, P.; Romanoff, J.; Remes, H. Influence of load length on short-term ice-induced load statistics in full-scale. *Mar. Struct.* **2017**, *52*, 153–172. [CrossRef]
26. Suominen, M.; Kujala, P.; Romanoff, J.; Remes, H. The effect of the extension of the instrumentation on the measured ice-induced load on a ship hull. *Ocean Eng.* **2017**, *144*. [CrossRef]
27. Suominen, M.; Kujala, P. The measured line load as a function of the load length in the Antarctic Waters. In Proceedings of the 23rd International Conference on Port and Ocean Engineering under Arctic Conditions, POAC15, Trondheim, Norway, 14–18 June 2015.
28. Kulovesi, J.; Lehtiranta, J. Level ice thickness measurement using ship-based semi-automatic computer vision. In Proceedings of the 22nd IAHR International Symposium on Ice, Singapore, 11–15 August 2014.
29. Lensu, M.; Kujala, P.; Kulovesi, J.; Lehtiranta, J.; Suominen, M. Measurements of Antarctic Sea Ice Thickness during the Ice Transit of S.A. Agulha II. In Proceedings of the 23rd International Conference on Port and Ocean Engineering under Arctic Conditions, POAC15, Trondheim, Norway, 14–18 June 2015.

30. Suominen, M.; Kulovesi, J.; Lensu, M.; Lehtiranta, J.; Kujala, P. A Comparison of Shipborne Methods for Ice Thickness Determination. In Proceedings of the 22nd IAHR International Symposium on Ice, Singapore, 11–15 August 2014.
31. Lensu, M. The Evolution of Ridged Ice Fields. Ph.D. Thesis, Report M-280. Laboratory of Naval Architecture and Marine Engineering, Helsinki University of Technology, Espoo, Finland, 2003.

Journal of
Marine Science and Engineering

Article

Investigation of the Flow Field of a Ship in Planar Motion Mechanism Tests by the Vortex Identification Method

Zhen Ren, Jianhua Wang and Decheng Wan *

Computational Marine Hydrodynamics Lab (CMHL), School of Naval Architecture, Ocean and Civil Engineering, State Key Laboratory of Ocean Engineering, Shanghai Jiao Tong University, Shanghai 200240, China; renzhen90@163.com (Z.R.); jianhuawang@sjtu.edu.cn (J.W.)

* Correspondence: dcwan@sjtu.edu.cn

Received: 1 July 2020; Accepted: 15 August 2020; Published: 24 August 2020

Abstract: Planar motion mechanism (PMM) tests provide a means of obtaining the hydrodynamic derivatives needed to assess ship maneuverability properties. In this paper, the self-developed computational fluid dynamic (CFD) solver based on the open source code platform OpenFOAM, naoe-FOAM-SJTU, associated with the overset grid method is used to simulate the complex viscous flow field of PMM tests for a benchmark model Yupeng Ship. This paper discusses the effect of several parameters such as the drift angle and period on the hydrodynamic performance of the ship and compares the time histories of the predicted forces and moments with experimental data. To investigate the complex viscous flows with a large separation, four vortex identification methods are used to capture the vortex structures. The results show that the forces and moments are in good agreement in static drift and dynamic tests. By comparing the vortex structures, it is found that the third generation vortex identification methods, OmegaR and Liutex, are able to more accurately capture the vortex structures. The paper concludes that the present numerical scheme is reliable and the third generation vortex identification methods are more suitable for displaying the vortex structures in a complex viscous flow field.

Keywords: planar motion mechanism; vortex identification; OmegaR; Liutex; naoe-FOAM-SJTU solver

---

1. Introduction

The accurate assessment of ship maneuverability in the design stage is essential for the safe navigation and efficient operation of ships. The International Maritime Organization adopted the "Standards for Ship Maneuverability" in 1993, which provided provisions and recommendations for ship maneuverability indicators. The accuracy of ship maneuverability prediction relies on the quality of the hydrodynamic derivatives estimation. The planar motion mechanism (PMM) test is a widely applied method to obtain the hydrodynamic derivatives of a ship, including the static drift, pure sway and pure yaw tests. Nowadays, the ship maneuvering behavior is mainly predicted by direct, experimental and numerical methods. As a powerful and efficient tool, the popularity of computational fluid dynamics (CFD) is increasing rapidly in the prediction of ship maneuverability.

Tahara et al. [1] used self-developed solver to predict the hydrodynamic characteristics of S60 at different drift angles. In their study, the mesh independency was carried out by five grids resolutions. Simonsen and Stern [2] simulated the oblique towing test of Esso Osaka. However, the free surface was ignored in the numerical simulations. Pinto-Heredero et al. [3] adopted the solver, CFD Ship-Iowa, to simulate the flow field of the Wigley ship with the drift angle from 10° to 60°. The free surface was captured by the level-set method. Ismail et al. [4] conducted the numerical simulations of the KVLCC2 model with the drift angle from 0° and 12°. They found that a second-order total variation diminishing

(TVD) scheme was more suitable to predict the forces/moments acting on the hull. Xing et al. [5] also used the solver, CFD Ship-Iowa, to carry out numerical simulations of oblique towing tests of a KVLCC2 tanker at different drift angles (0°, 12°, 30°). In the study, a detached eddy simulation (DES) was applied to capture the flow separation phenomenon and the vortex structures around the hull. In the numerical simulations, the total number of grids was 13.0 million. However, the free surface was not taken into account. Stern et al. [6] summarized the numerical schemes in the simulation of ship maneuverability and found that using the DES method combined with enough grid refinement can obtain a more accurate numerical prediction. Meng and Wan [7] used the unsteady Reynolds averaged Navier–Stokes (RANS) to simulate the flow field of KVLCC2M at different drift angles in deep and shallow water. The connected vessels' effect in shallow water was captured and abundant information in the flow field was presented. Wang et al. [8] simulated the flow field of ship hull at six drift angles by the CFD solver, naoe-FOAM-SJTU. The predicted hydrodynamic derivatives were in good agreement with the experiments.

Broglia et al. [9] conducted numerical predictions for the dynamic pure sway and pure sway tests of KVLCC1 and KVLCC2. They found that the stern of hull was the main region inducing the lateral force. Toxopeus et al. [10] adopted the dynamic mesh to realize the ship motion of KVLCC2 with full appendages. The predicted lateral force and yawing moment were in good agreement with the experiments. Cura-Hochbaum et al. [11] also carried out the numerical simulation of the dynamic test of KVLCC1 tanker based on RANS equations. The predicted hydrodynamic derivatives agreed well with the experiment results. Simonsen and Stern [12] simulated the pure yaw test of KRISO Container Ship (KCS) and analyzed the uncertainty of forces/moments acting on the hull. Sakamoto et al. [13,14] and Yoon et al. [15,16] used the solver, CFD Ship-Iowa, to simulate the static and dynamic PMM tests of DTMB 5415. The predicted hydrodynamic derivatives were compared with the test results, and the highest number of errors of linear hydrodynamic derivatives was less than 10%. Kim et al. [17] simulated the PMM test for a KCS ship by the CFD solver SHIP Motion. The predicted results of the static drift test and the dynamic pure sway test were in good agreement with the tests, but the predicted results of the pure yaw test were not accurate enough. Yang et al. [18] simulated the pure sway tests of the hull in deep and shallow water by solving unsteady RANS equations. However, the free surface was not taken into consideration in the numerical simulations. The effect of speed and water depth on the hydrodynamic performance of the hull in the pure sway test was investigated by Liu et al. [19]. In their studies, the CFD solver, naoe-FOAM-SJTU, was used to solve unsteady RANS equations by coupling it with the shear stress transport (SST) k-ω turbulence model. The lateral force and yawing moment increased rapidly. Wang and Wan [20] also adopted the CFD solver, naoe-FOAM-SJTU, to simulate the dynamic tests of DTMB 5415, and the solver was coupled with the overset grid technology. The predicted hydrodynamic derivatives were in good agreement with the experiments.

So far, the main aim of a ship planar motion mechanism is the hydrodynamic derivatives of the hull. There are few studies to simulate the detailed flow field in the larger separation flow, especially by using the new vortex identification methods. In this paper, reliable numerical schemes are presented, and four vortex identification methods are applied to analyze the separation flow in the PMM tests of a Yupeng ship.

In the present work, the authors perform the numerical simulations of static drift tests and dynamic tests of a Yupeng Ship. The CFD solver, naoe-FOAM-SJTU, is applied in the numerical simulations. Unsteady RANS equations coupled with an SST k-ω turbulence model are adopted to solve the flow field. The main framework of this paper goes as follows. The first part is the basic numerical methods and the vortex identification methods. The second part is the geometry model, grid generation and test conditions. Then comes the results and analysis, which contains the predicted results of the static drift, pure sway and pure yaw tests of the Yupeng Ship. In this section, the forces/moment, free surface, dynamic pressure and vortex structures are presented. Finally, the conclusion of this paper is drawn.

2. Numerical Methods

2.1. Basic Numerical Scheme

In this paper, the self-developed CFD solver, naoe-FOAM-SJTU, is used to perform the numerical simulations of the PMM tests. In the numerical computations, the governing equations are the unsteady Reynolds averaged Navier–Stokes equations, which are as follows:

$$\nabla\cdot\mathbf{U} = 0 \tag{1}$$

$$\frac{\partial\rho\mathbf{U}}{\partial t} + \nabla\cdot\left(\rho(\mathbf{U}-\mathbf{U_g})\mathbf{U}\right) = -\nabla p_d - \mathbf{g}\cdot\mathbf{x}\nabla\rho + \nabla\cdot(\mu_{eff}\nabla\mathbf{U}) + (\nabla\mathbf{U})\cdot\nabla\mu_{eff} + f_\sigma \tag{2}$$

where $\mathbf{U}$ and $\mathbf{U_g}$ are the fluid velocity field and the grid velocity, respectively; p_d represents the dynamic pressure; μ_{eff} is the effective dynamic viscosity. f_σ is the surface tension term; ρ represents the density of the fluid and g is the gravity acceleration.

In the numerical calculations, the shear stress transport turbulence model, SST k-ω model, is selected to enclosure the URANS equations. The SST k-ω model is one of the widely used turbulence models and the model combines the advantages of the standard k-ε model and k-ω model to improve the accuracy when solving the free surface and near-wall region. In addition, it is noted that the rotation/curvature correction [21] of the SST k-ω model is added into the governing equations for capturing the vortex structures better. The free surface is captured by the volume of fluid (VOF) approach. The relative proportion of different fluids in the grid cell is expressed as:

$$\begin{cases} \alpha = 0 & air \\ \alpha = 1 & water \\ 0 < \alpha < 1 & interface \end{cases} \tag{3}$$

where α represents the relative proportion. When it equals 0, it means that the grid cell is completely filled with air. When it equals 1, the cell is filled with water fully.

In the present numerical/experimental setup, the amplitude of sway motion in the dynamic tests can be as large as the beam of the ship model. Therefore, it is necessary to apply the dynamic overset grid method in the present numerical simulations to avoid the numerical error caused by grid deformation and improving the predicted accuracy. The transformation between the different computational domains is achieved by the Suggar++ library [22].

The velocity and pressure equations are decoupled by the merged PIMPLE (PISO-SIMPLE) algorithm. The algorithm allows the iterative procedure and pressure–velocity correction to solve the Navier–Stokes equations. In addition, built-in numerical schemes in OpenFOAM are applied to solve the partial differential equations (PDEs). A second-order TVD limited linear scheme is used to discretize the convection terms and the diffusion terms are approximated by a second-order central difference scheme.

2.2. Vortex Identification Methods

2.2.1. First Generation of Vorticity-Based Vortex Identification Methods

The first generation of vortex identification methods is based on the definition of vorticity proposed by Helmholtz in 1858. The vorticity is defined as velocity curl, $\nabla\times U$. Based on the Cauchy–Stokes decomposition, it is considered to be twice the rotational angular velocity of the rigid body rotation angular velocity of the fluid mass. The Cauchy–Stokes decomposition [23] is expressed as:

$$U(r+dr) = U(r) + Adr + \frac{1}{2}\omega\times dr \tag{4}$$

According to the decomposition, U is the velocity vector and r is radius vector. The movement of fluid mass can be decomposed into three parts: translation, symmetrical tensor A and vorticity ω. In this method, the vorticity ω represents the rigid body rotating part of the fluid mass movement.

2.2.2. Second Generation of Eigenvalue-Based Vortex Identification Methods

The second generation encompasses modifications based on the first generation. Both methods are derived from the Cauchy–Stokes decomposition. The second generation of vortex identification methods includes Q, λ_2, Δ, and λ_{ci}. Here is a brief introduction of the Q criterion. The Q criterion is a widely used vortex identification methods. The expression [23] can be written as:

$$Q = \frac{1}{2}\left(||B||_F^2 - ||A||_F^2\right) \tag{5}$$

where A and B represent the symmetric and antisymmetric parts of the velocity gradient tensor, respectively.

However, in this method, the threshold of Q had to be selected manually, so that the threshold has an important influence on the display of the vortex structure.

2.2.3. Third Generation of Vortex Identification Methods

Liu et al. [23] proposed the third vortex identification methods: OmegaR and Liutex/Rortex.

The OmegaR method was also derived from the Cauchy–Stokes decomposition. In this method, the vorticity ω is further decomposed into a rotating part and a nonrotating part. The Ω [23] can be expressed as:

$$\Omega = \frac{||B||_F^2}{||A||_F^2 + ||B||_F^2 + \varepsilon} \tag{6}$$

Obviously, Ω is between 0 and 1, which can be understood as the concentration of vorticity. $\Omega = 0.51$ or 0.52 are always selected as a fixed threshold to identify the vortex structure. ε is a small positive number for avoiding a division by zero. The OmegaR method avoids the problem of artificially adjusting the threshold.

According to the Cauchy–Stokes decomposition, the antisymmetric tensor B cannot represent the rigid body part of fluid motion. The Liutex/Rortex vector systematically solves the problem to obtain a rigid rotating part from the fluid motion. The Liutex vector [23] is defined as:

$$R = \left(\langle\omega, r\rangle - \sqrt{\langle\omega, r\rangle^2 - 4\lambda_{ci}^2}\right)r \tag{7}$$

where ω is the vorticity vector.

The Liutex vector represents not only the rotation intensity, but also the local rotation axis; compared with the Cauchy–Stokes decomposition, the Liutex vector provides an accurate motion decomposition to capture the rigid rotation part in the flow field.

3. Numerical Simulation Setup

3.1. Geometry Model

In the present study, the benchmark model, Yupeng Ship, is used for all the computations. The geometry model of Yupeng Ship is shown in Figure 1. The principle particulars are listed in Table 1. In the numerical simulations, a model scale is adopted and the scale ratio is 1:49. The length between the perpendiculars of the ship model is 3.857 m. The researchers of Marine Design and Research Institute of China (MARIC) conducted extensive experiments using the model ship. In the experiments, the speed of ship model is 1.323 m/s, corresponding to Fr = 0.215. In the PMM tests, the heave and pitch motion are taken into consideration.

Figure 1. Geometry model of a Yupeng Ship.

Table 1. Principle particulars of a Yupeng Ship.

Main Particulars	Symbols	Unit	Full Scale
Length between perpendiculars	L_{pp}	m	189
Beam	B_{WL}	m	27.8
Draft	T_M	m	10.3
Displacement volume	∇	m^3	39,019
Initial stability	GM	m	1.18
Block coefficient	CB		0.721
Longitudinal inertial radius	Ryy	m	48.206
Speed		kn	18

3.2. Tests and Calculated Conditions

In the experiments, the static drift and dynamic tests were carried out in the towing tank of MARIC. The test conditions are summarized and listed in Table 2. The static drift tests included two conditions, namely drift angle 4° and 20°. The dynamic tests included the pure sway and pure yaw tests. In the dynamic tests, the amplitude of the sway motion is 0.4 m. The periods of pure sway motions are 8 s and 16 s, respectively. The periods of pure yaw motions are the same with pure sway tests. The calculation reference point of forces and moments is the center of gravity.

Table 2. Summary of the tests and calculated conditions.

Conditions	CASE	Amplitude (m)	Period (s)	Drift Angle (°)
Static drift test	DA4	-	-	4
	DA20	-	-	20
Dynamic pure sway test	PS8	0.4	8	-
	PS16	0.4	16	-
Dynamic pure yaw test	PY8	0.4	8	-
	PY16	0.4	16	-

3.3. Grids Generation

In order to simulate the large amplitude motion of the Yupeng ship, the dynamic overset mesh is adopted in the present numerical calculations. The computational domain and boundary conditions are presented in Figure 2. The computational domain is $-1.0\ L_{pp} \le x \le 3.0\ L_{pp}$ in length, $-1.5\ L_{pp} \le y \le 1.5\ L_{pp}$ in width and $-1.0\ L_{pp} \le z \le 0.5\ L_{pp}$ in height. The red line indicates the boundary of the overlap domain. In the present simulations, the grid is generated by the commercial software, Hexpress. The grids of the background and hull domain are generated respectively. To capture the free surface accurately, a block is adopted to refine the grid near the free surface. The total grid number is about 6.28 million. The grid distribution is shown in Figure 3. Figure 3a presents the global grid distribution. The red and blue lines depict the grid distribution of the background and hull domain, respectively. A box is used to refine the mesh around the hull. Figure 3b shows the grid distribution on the free surface. The grid distribution of the hull is depicted in Figure 3c.

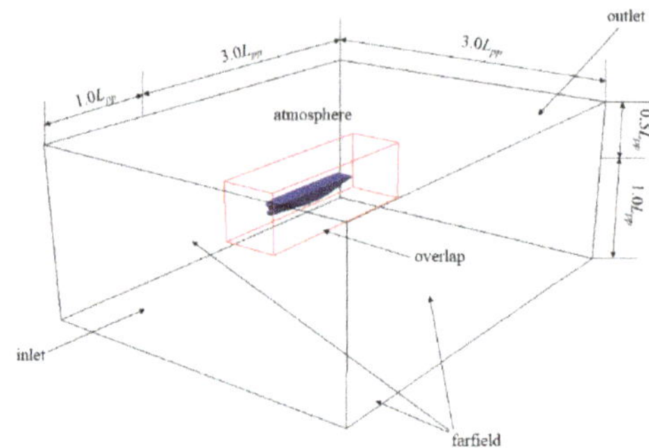

Figure 2. Computational domain and boundary condition.

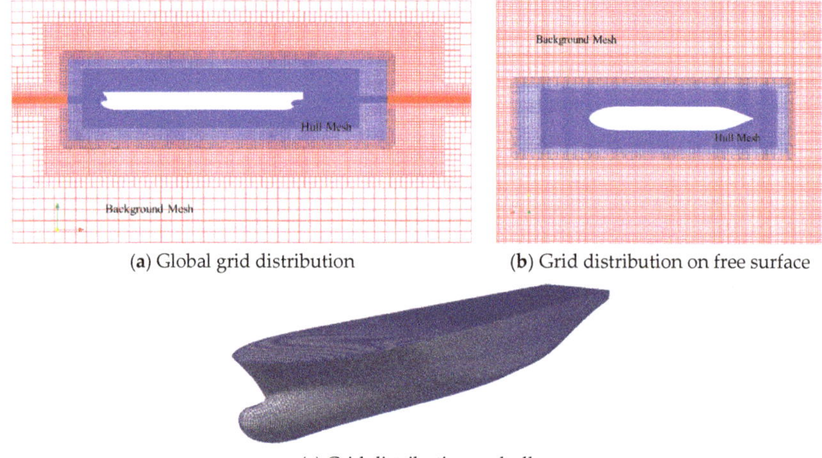

(**a**) Global grid distribution (**b**) Grid distribution on free surface

(**c**) Grid distribution on hull

Figure 3. Grid distribution: (**a**) global grid distribution; (**b**) grid distribution on the free surface; (**c**) grid distribution on the hull.

To analyze the mesh independence, three mesh resolutions, with the refinement ratio of $\sqrt{2}$, were adopted in the numerical simulations. Table 3 summaries the details of the mesh generated with different resolutions. As shown in the table, the mesh of the background and hull domain was refined. The grid number with fine resolution approaches 17.59 million. Figure 4 presents the grid distribution with different mesh resolutions at the longitudinal cutting plane.

Table 3. Details of the mesh generation.

Grid Sets	Background Mesh (Million)	Hull Mesh (Million)	Total (Million)
Coarse	0.77	1.47	2.24
Mid	2.16	4.12	6.28
Fine	6.05	11.54	17.59

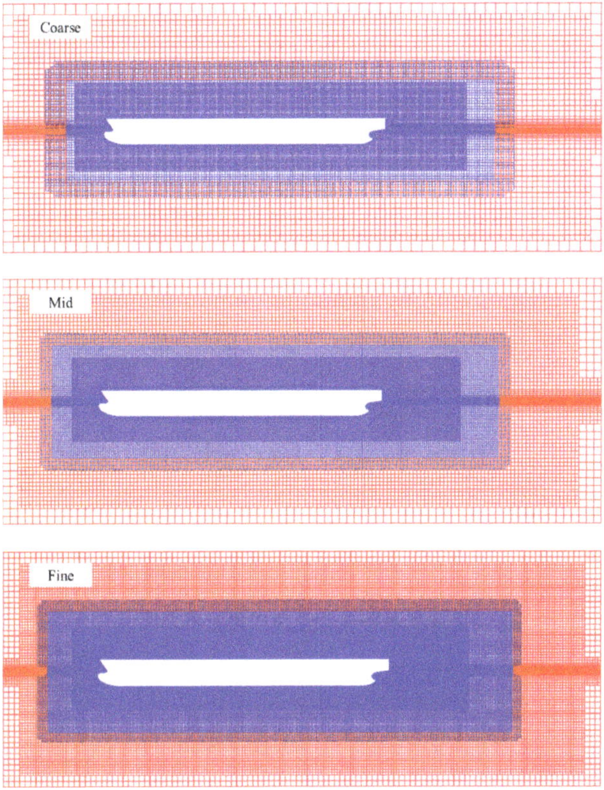

Figure 4. Grids distributions with different mesh resolutions (top: coarse; middle: mid; bottom: fine).

The different grid distributions are adopted to simulate the hydrodynamic performance of the Yupeng Ship in CASE DA4. The predicted results are summarized in Table 4. Since the resistance is affected by many factors and fluctuates greatly in the experiments, the resistance was slightly over-predicted in the numerical simulations. However, the predicted lateral force and yawing moment are in good agreement with the experimental values, and all of their errors achieved by different mesh resolutions were less than 5%. The results of the fine mesh scheme were slightly improved compared to mid mesh scheme.

Table 4. Grid convergence of the resistance in CASE DA4.

	Resistance (N)	Error	Lateral Force (N)	Error	Yawing Moment (NM)	Error
EFD	11.30	-	7.33	-	27.83	-
Coarse	9.89	−12.48%	7.45	1.64%	26.94	−3.20%
Mid	10.12	−10.44%	7.36	0.41%	27.28	−1.98%
Fine	10.23	−9.47%	7.34	0.14%	27.52	−1.11%

Taking the predicted accuracy and time-consumption into consideration, the mid mesh distribution is selected to carry out the next simulations.

4. Results and Discussion

In this section, all of the experimental conditions are simulated by the CFD solver, naoe-FOAM-SJTU. The predicted forces and moments are compared with the experiments. In the static drift tests, the values at the last 18 s of the prediction are plotted. The plotted values in the dynamic tests are the last two periods of the numerical simulations. All the numerical simulations are performed on the high performance computing (HPC) cluster center in Computational Marine Hydrodynamics Lab (CMHL), Shanghai Jiao Tong University. Each node consists of two Central Processing Units (CPUs) with 20 cores per node and 64GB accessible memory (Intel Xeon E5-2680v2 @2.8 GHz). In total, 40 processors are assigned for each case, in which 39 are for the calculation of the flow field and the other processor is for the interpolation calculation of the overset grids. In the numerical simulations, the time step is set to $\Delta t = 0.0005$ s, and the Courant number is less than 1.0 in the current predictions. Each calculation costs approximately 264 h with about 60,000 time steps.

4.1. Static Drift Test

4.1.1. Force and Moments

In the static drift tests, the forces and moments acting on the hull are greatly affected by the drift angle. The comparison between the experiments and the predicted results are presented in Figure 5. The left column is the results of the CASE DA4 tests and the right is the CASE DA20 tests. It can be seen that there are greater fluctuations in the experiment data and the predicted results are more regular. In CASE DA20, the forces and moments are greater than that in CASE DA4. The lateral force and yawing moment at drift angle 20° are much larger than the results at drift angle 4°. The errors between the experimental and predicted results are listed in Table 5. The forces and moment are the mean value of their time histories. There is the largest error in terms of the resistance in CASE DA4. The measured resistance, affected by many factors, leads to a great fluctuation of the time history, so that the predicted resistance is not very accurate. However, all the errors of the lateral force and yawing moment in both cases are less than 6%. The resistance and yawing moment are under-predicted compared to the experiments, while the lateral force is over-predicted. Overall, the predicted results are in good agreement with the experimental data.

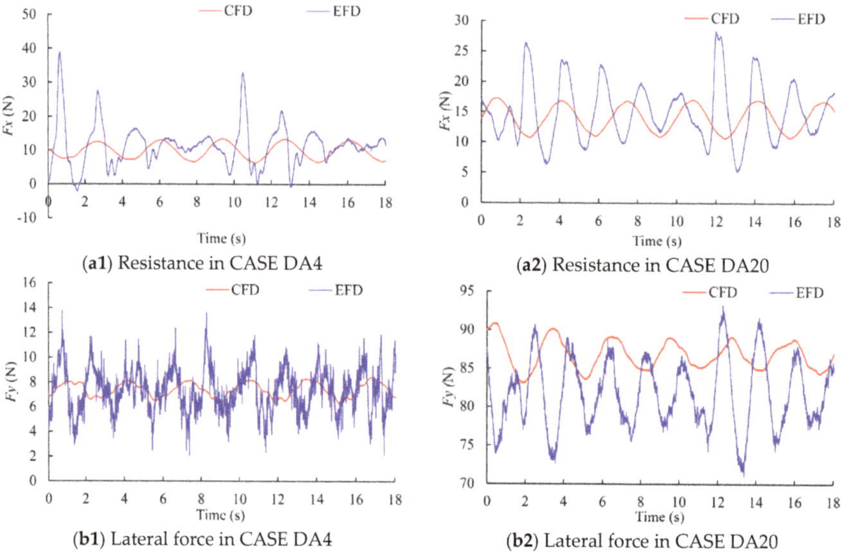

(a1) Resistance in CASE DA4

(a2) Resistance in CASE DA20

(b1) Lateral force in CASE DA4

(b2) Lateral force in CASE DA20

Figure 5. *Cont.*

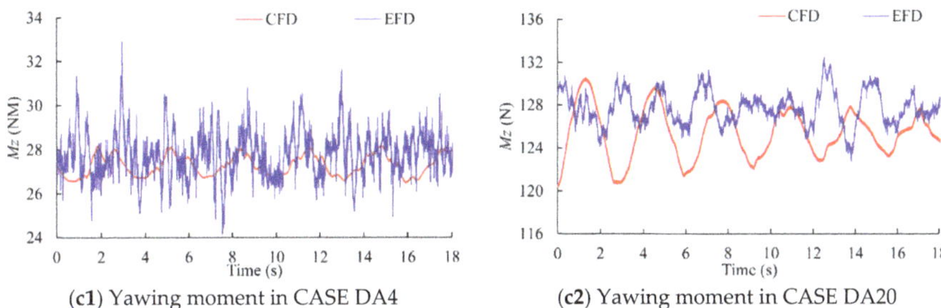

(c1) Yawing moment in CASE DA4 (c2) Yawing moment in CASE DA20

Figure 5. Forces and moments in CASE DA4 (left) and DA20 (right) (**a**: resistance; **b**: lateral force; c: yawing moment).

Table 5. Comparison between the predicted and experimental forces and moment in the static drift test.

CASE	Force/Moment	EFD	CFD	Error
DA4	Resistance (N)	11.30	10.32	−10.44%
	Lateral force (N)	7.33	7.36	0.43%
	Yawing moment (NM)	27.83	27.28	−1.98%
DA20	Resistance (N)	15.08	13.77	−8.69%
	Lateral force (N)	81.89	86.72	5.90%
	Yawing moment (NM)	127.80	125.28	−1.97%

4.1.2. Free Surface

The main advantage of CFDs over the experiment is that, not only can the hydrodynamic characteristics of the ship can be predicted, but a detailed flow field distribution around the hull can also be investigated. Figure 6 shows the comparison of the free surface at different drift angles. The static hydrodynamic characteristics of the hull can be explained to some extent by the wave-making pattern such as the height and length. As shown in Figure 6, the asymmetry of the wave-making wave pattern is more obvious at a larger drift angle. In addition, the range of the bow wave peak (starboard side) is wider and the peak value is larger. The dynamic pressure difference on both sides of the hull is intensified rapidly, caused by the increasing diffraction wave peak on the portside, so that the lateral force and yawing moment increase rapidly at the larger drift angle. In addition, the bow wave breaking is captured in CASE DA20, as shown in Figure 6c. The plunging wave breaking, which leads to the energy loss, is found in the evolution of diffraction waves on the portside.

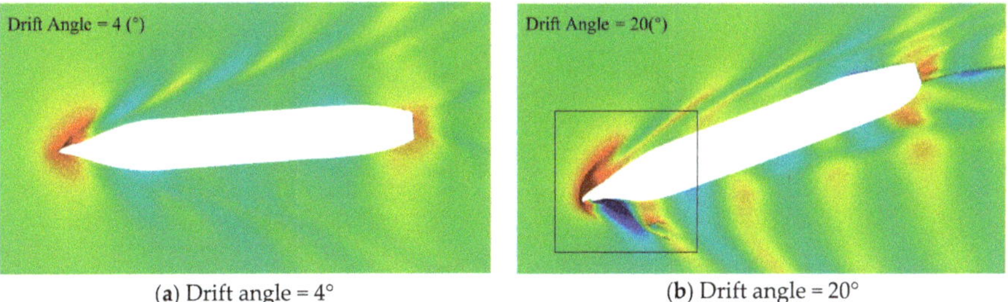

(**a**) Drift angle = 4° (**b**) Drift angle = 20°

Figure 6. *Cont.*

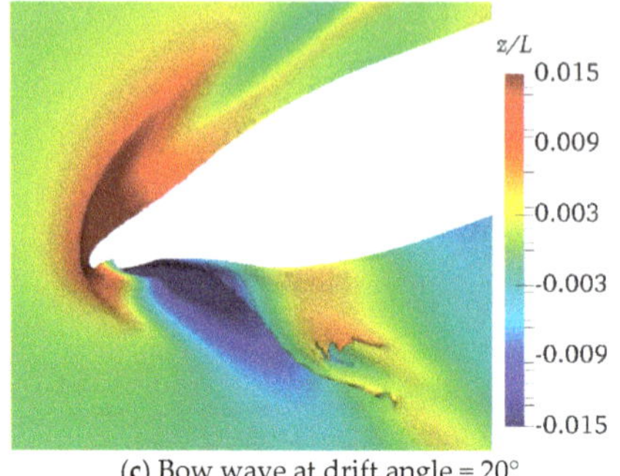

(c) Bow wave at drift angle = 20°

Figure 6. Free surface: (**a**) free surface in CASE DA4; (**b**) free surface in CASE DA20; (**c**) bow wave in CASE DA20.

4.1.3. Dynamic Pressure

Affected by the drift angle, the dynamic pressure distribution on the portside and starboard side is no longer symmetrical. The variation of the pressure distribution leads to the rapid change of the forces and moments. Figure 7 depicts the dynamic pressure distribution on the portside and starboard side of the Yupeng ship. Figure 7a shows the dynamic pressure on the portside. It is clear that the dynamic pressure in CASE DA20 is much lower than it is in other case. For the large drift angle case, the diffraction wave appears near the bow of the portside and the dynamic pressure decreases rapidly. Figure 7b shows the dynamic pressure distribution on the starboard side. The dynamic pressure near the bow of the starboard side at a larger drift angle is much larger than the results in CASE DA4. Compared with the dynamic pressure distribution on the portside, the difference of the dynamic pressure between both sides in CASE DA20 is much greater than the results of CASE DA4. Therefore, it results in a rapid increase in the lateral force and yawing moment.

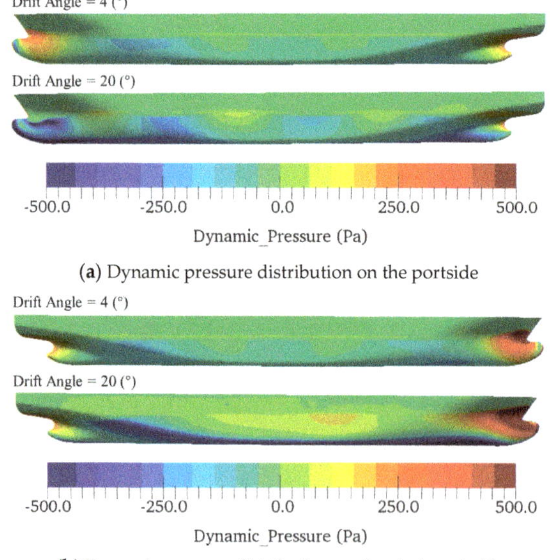

Figure 7. Dynamic pressure distribution on the hull: (**a**) dynamic pressure distribution on the portside; (**b**) dynamic pressure distribution on the starboard side.

4.1.4. Vortex Structure

Vortex structures around the hull, propeller and rudder are in different scales and intensities. In order to compare the behaviors of different vortex identification methods, the vortex structures obtained by four vortex identification methods are presented in Figure 8, where Figure 8a,b show the vortex structure in CASE DA4 and DA20, respectively. Overall, the hull is covered fully by the vortex structures based on vorticity in both cases. Obviously, the results are not acceptable. The Q criterion seems to more suitably capture the vortex structures in the flow field than the first generation method. As a third generation vortex identification method, the OmegaR method is able to capture larger and more vortex structures in the flow field. Compared to the results obtained by the OmegaR method, more broken vortex structures near the hull are captured by the Liutex method. Comparing the vortex structures in CASE DA4 and DA20, the vortex structure is longer and more complex at a larger drift angle. At larger drift angles (DA20), a vortex occurs on the portside of the hull. It is captured by the vortex identification method based on the vorticity, Q and the OmegaR method. It indicates that the large separation of field flow appears near the hull in CASE DA20 in the static drift tests.

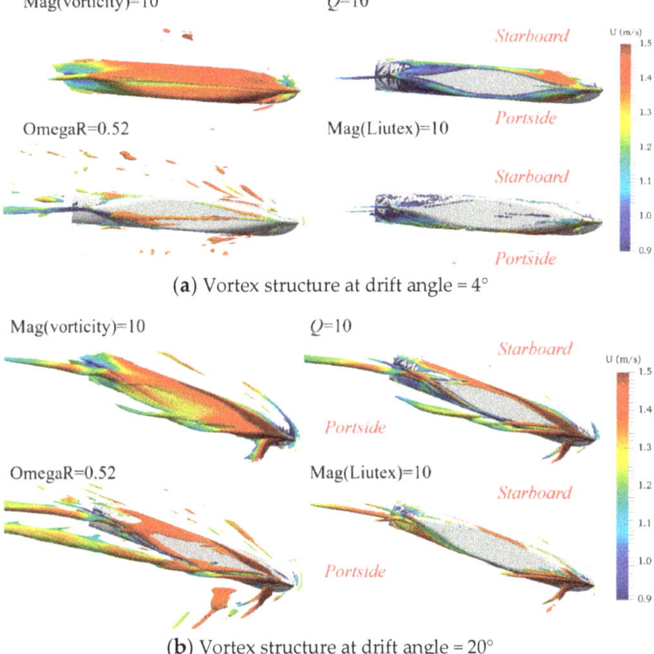

(**a**) Vortex structure at drift angle = 4°

(**b**) Vortex structure at drift angle = 20°

Figure 8. Vortex structure in the simulations of the static drift test: (**a**) vortex structure in CASE DA4; (**b**) vortex structure in CASE DA20.

In order to study the evolution of vortices along the ship longitudinal direction, a series of planes are selected to present the axial vorticity and Liutex. As shown in Figure 9, the vortex always appears in pairs with the alternate positive and negative value. The value of the axial vorticity is much larger than the value of axial Liutex. As shown in Figure 9a, there are ten pairs of vortices, eight of which are near the free surface, and the other two pairs occur near the bilge of the hull. By using the Liutex method, the vortex pairs are also captured clearly. Most of the vortex pairs also appear near the free surface, and there are two pairs of vortex near the bilge of the hull. The axial vorticity and Liutex occur at the same position, indicating that the strong rotating and separating motion appears near the bilge of the Yupeng Ship.

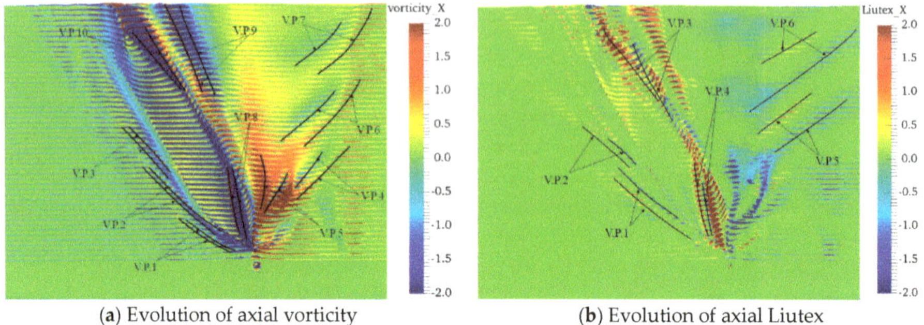

(**a**) Evolution of axial vorticity (**b**) Evolution of axial Liutex

Figure 9. Evolution of the axial vorticity (**a**) and Liutex (**b**) along the ship length direction in CASE DA20.

Figure 10 presents the evolution of vertical vorticity and Liutex at z/Lpp = −0.03572 along the ship length direction in CASE DA4 and DA20. At the small drift angle, the vertical vorticity and Liutex on the portside are very small and the flow separation is not very strong. The vertical vorticity and Liutex on the portside increase rapidly in CASE DA20. The violent flow separation appears on the portside at a large drift angle. Figure 10a presents the violent variation of the vertical vorticity in CASE DA20. The extensive vortices originates from the shoulder of the hull, while a pair of vertical Liutex originates from the mid-hull. Far from the hull, a vertical vorticity pair (V.P.) with an alternate positive and negative value is captured and the negative one is closer to the hull. By using the Liutex method, the vertical Liutex, which is similar to a Karman vortex street (K.V.S.), is displayed, as shown in Figure 10b.

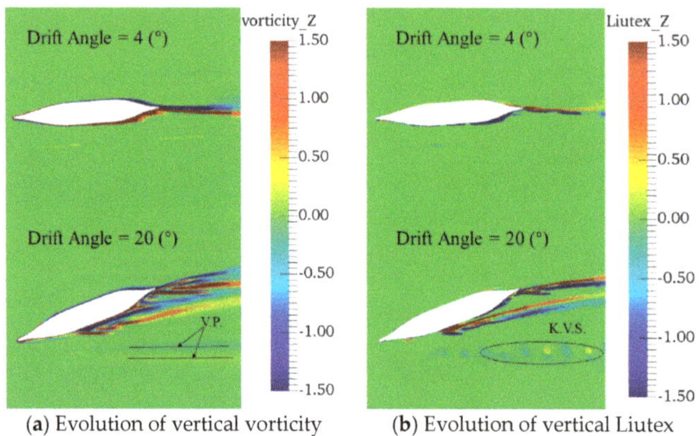

(**a**) Evolution of vertical vorticity (**b**) Evolution of vertical Liutex

Figure 10. Evolution of the vertical vorticity (**a**) and Liutex (**b**) at $z/L_{pp} = -0.03572$ along the ship length direction in static drift tests (top: CASE DA4; bottom: CASE DA20).

4.2. Dynamic Test: Pure Sway

4.2.1. Force and Moments

In the dynamic tests, the forces and moments acting on the hull are smoother—especially the lateral forces and yawing moment. The predicted forces and moments in the pure sway tests are shown in Figure 11. It is obvious that the experimental resistance data fluctuate more sharply than the numerical results. Compared with the resistance in CASE PS8 and PS16, it is found that the pure sway period has little effect on ship resistance. The lateral force and yawing moment are affected significantly by the period of sway motion in the dynamic pure sway tests. The short period leads to a large lateral force and yawing moment. Although the forces and moments at different periods vary greatly, the lateral force and yawing moment in both cases are in good agreement with the experiment.

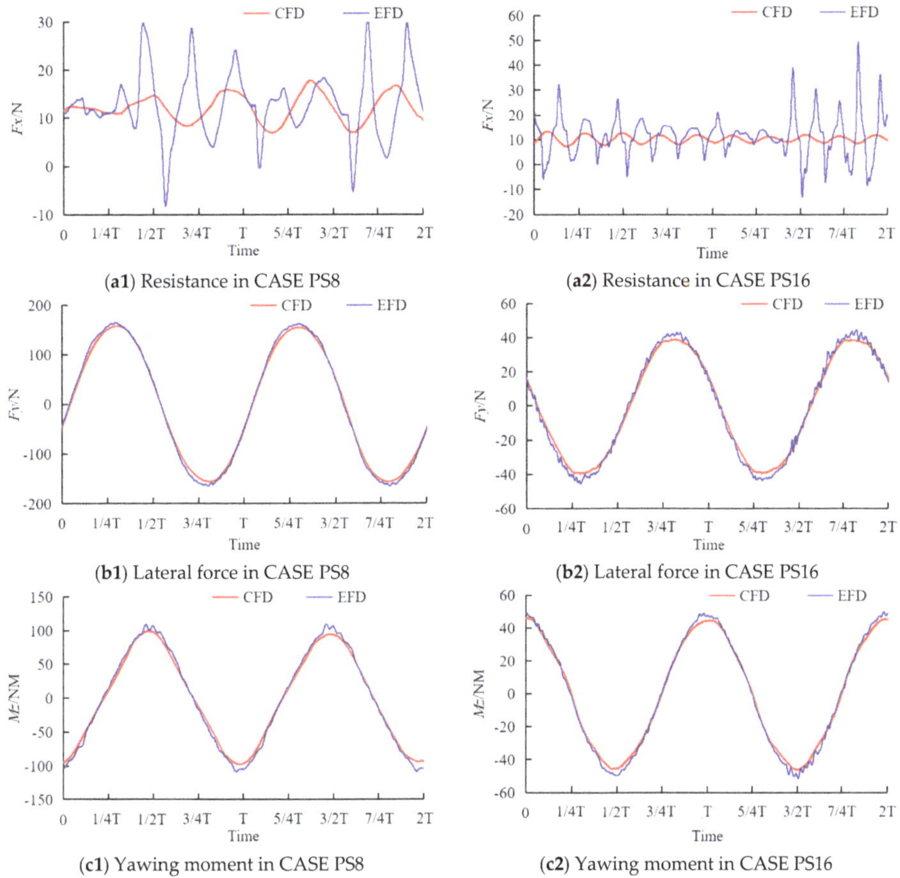

(**a1**) Resistance in CASE PS8 (**a2**) Resistance in CASE PS16

(**b1**) Lateral force in CASE PS8 (**b2**) Lateral force in CASE PS16

(**c1**) Yawing moment in CASE PS8 (**c2**) Yawing moment in CASE PS16

Figure 11. Forces and moment in CASE PS8 (left) and PS16 (right) (**a**: resistance; **b**: lateral force; **c**: yawing moment).

The errors between the experimental and predicted forces and moments are summarized in Table 6. The resistance is the mean value at the last two periods. The lateral force and yawing moment are the amplitudes of the time histories and the amplitude is the mean of the four absolute peak values in the plots. As shown in the table, the resistance in CASE PS8 is slightly larger than that in CASE PS16, while the lateral force and yawing moment in the shorter period are much larger than the results of CASE PS16. Since the amplitude of the pure sway motion is the same, a smaller pure sway motion period leads to a higher sway acceleration, resulting in the larger lateral force and yawing moment. The predicted results of CASE PS18, whose errors are less than 10%, is relatively better than those of CASE PS16. Overall, the errors of forces and moment are less than 15%, which indicates that CFDs is able to accurately predict the forces and moments of the hull in the dynamic PMM tests.

Table 6. Comparison between the calculated and experimental forces and moment in pure sway tests.

CASE	Force/Moment	EFD	CFD	Error
PS8	Resistance (N)	12.65	11.96	−5.46%
	Lateral force (N)	165.61	158.03	−4.57%
	Yawing moment (NM)	109.88	99.66	−9.30%
PS16	Resistance (N)	11.54	10.11	−12.36%
	Lateral force (N)	45.81	40.50	−11.58%
	Yawing moment (NM)	49.74	45.75	−8.01%

4.2.2. Free Surface

Figure 12 shows the evolution of free surface wave-making in a period of pure sway test in CASE PS8. When the hull moves to the portside, the wave-making in the left flow field is squeezed and the amplitudes of wave-making near the bow and stern both increase, which corresponds to the lateral force pointing to the starboard side. Additionally, when the hull moves to starboard side, the trend of change is just the opposite way, which is due to the periodic lateral force and yawing moment, as shown in Figure 11.

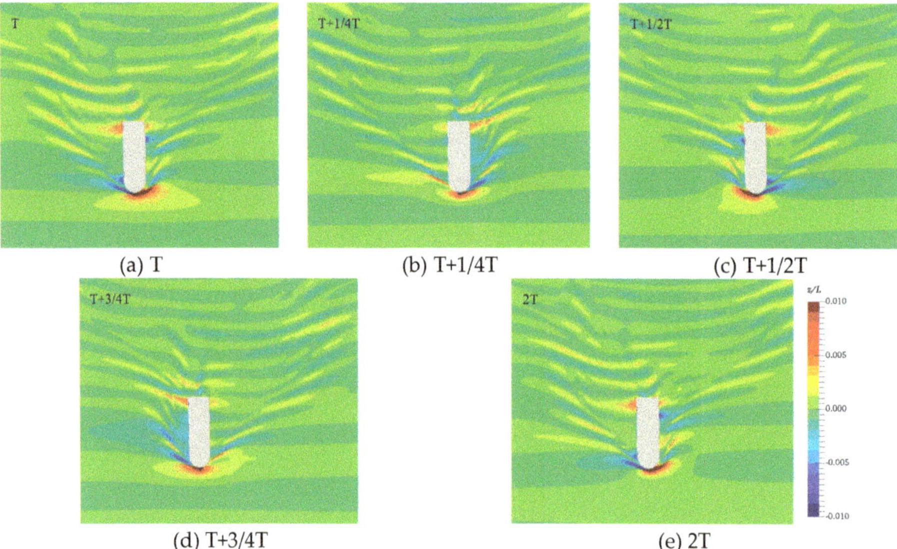

Figure 12. Evolution of free surface in one period in CASE PS8.

4.2.3. Vortex Structure

The vortex structures in the dynamic pure sway test are obviously different from those in the static drift test. Figure 13 shows the vortex structure obtained by four methods in CASE PS8. The vortex structures obtained by four methods are all periodic. The vortex structures appearing on the bottom of the hull at one-quarter and three-quarter periods are also captured by these methods. It indicates that the separated flow occurs near the bottom of the hull. At T+1/4T, the vortex structures gradually approach the starboard side of the hull downstream. The trends of the vortex structures evolution at T+3/4T are opposite to one another. As shown in Figure 13a, the hull is covered fully by the vortex structures obtained based on the vorticity. Obviously, this is impossible in practice. The hull is covered partially by the vortex structures obtained by Q criterion. At T, the vortex structures mainly appear near the bilge of the hull. A vortex originates from the propeller shaft and tilts to the starboard side as it evolves downstream. At T+1/4T, a vortex occurs at the bottom of the hull, which starts from the mid-hull in the initial stage and gradually approaches the starboard side downstream. After separating from the hull at the bilge, it evolves downstream along the ship length direction. At T+1/2T, the pattern of the vortex structures is similar to the results at T, but the vortex that originated from the propeller shaft tilts to the portside downstream. Except for the vortex originated from the mid-hull, the distribution of vortex structures at T+3/4T are analogous to that at T+1/4T. The direction of the evolution of the vortex originated from the mid-hull is opposite to the results at T+1/4T. The vortex structures obtained by OmegaR are more complex and extensive vortices in the flow field are captured, especially near the free surface. However, the evolution of the main vortex is similar to the results obtained by the Q criterion. More broken vortices near the hull are captured by the Liutex method. At T, a vortex is captured at the bottom of the hull and evolves along the mid-hull, as shown in Figure 13d. The vortex is not visible in the results of the other methods.

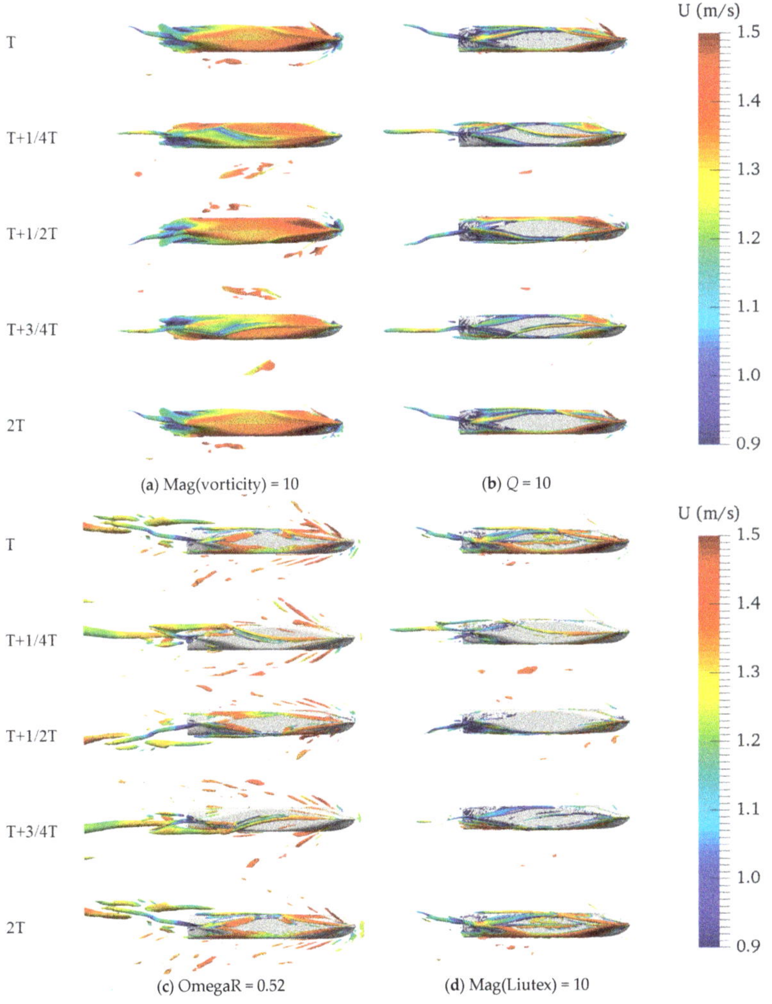

Figure 13. Vortex structure obtained by different vortex identification methods: (**a**) first generation based on vorticity; (**b**) second generation based on eigenvalues; (**c**,**d**) third generation in CASE PS8.

Figures 14 and 15 present the evolution of the vertical vorticity and Liutex at $z/L_{pp} = -0.03572$ along the ship length direction in CASE PS8 and PS16. Since the distribution in one period is basically symmetrical, the distribution in a half-period is only presented here. In CASE PS8, the flow separation is more obvious than that in CASE PS16, which is caused by a shorter period and it will lead to a larger sway velocity. The vertical vortex and Liutex presented in Figure 14 are larger than the results of CASE PS16. As shown in Figure 14, a larger vertical vorticity and Liutex are captured clearly near the stern of the hull, which indicates that a strong flow separation occurs. The violent flow separation causes a big loss of energy, resulting in the rapid increase in the lateral force and yawing moment. Compared with the vertical vorticity, the vertical Liutex is more disordered, and the value of the vertical Liutex is less than that of the vertical vorticity. Since the definition of vorticity and Liutex are different, the positive Liutex appears where a negative vorticity occurs.

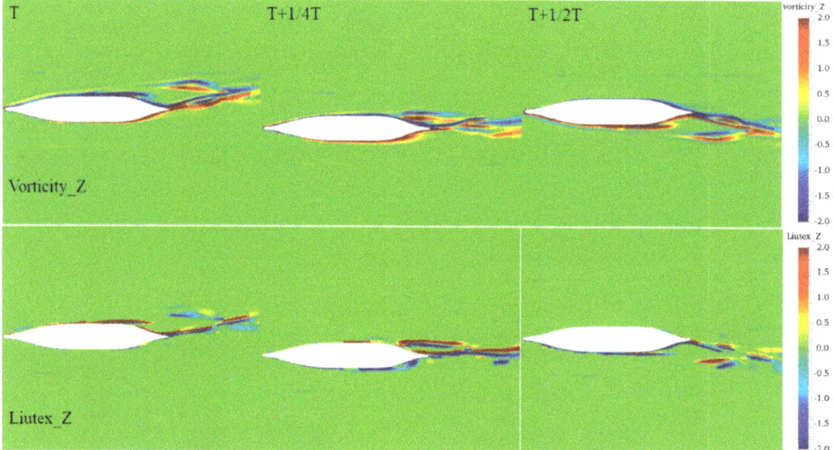

Figure 14. Vertical vorticity (top) and Liutex (bottom) at $z/L_{pp} = -0.03572$ in CASE PS8.

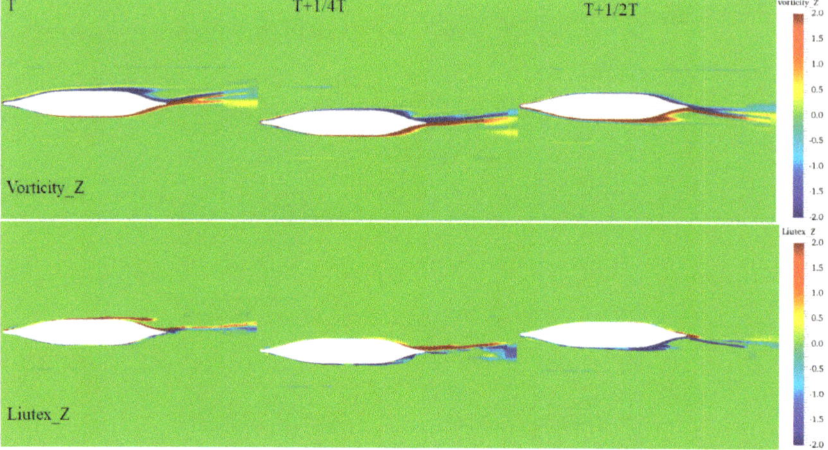

Figure 15. Vertical vorticity (top) and Liutex (bottom) at $z/L_{pp} = -0.03572$ in CASE PS16.

4.3. Dynamic Test: Pure Yaw

4.3.1. Force and Moments

In the dynamic pure yaw tests, the predicted forces and moments are compared with the experiment, as shown in Figure 16. The left column shows the results of CASE PY8 and the right is the results of CASE PY16. Overall, the predicted forces and moment in both cases are in good agreement with the experiment. In CASE PY8, the resistance is also periodic, and there are two peaks in one period. In pure yaw tests, the resistance is also the largest when the drift angle of the hull reaches the maximum. It is more obvious in a shorter period. By comparing with the lateral force and yawing moment in CASE PY8 and PY16, it is found that a shorter period leads to a larger force and moment. The errors between the experimental and predicted forces and moment are listed in Table 7. The resistance is calculated by the mean value of the time history at the last two periods and the lateral force and yawing moment are the average amplitudes of the time history. As shown in the table, the resistance in CASE PY8 is 40% larger than that in CASE PY16, and the lateral force and yawing moment are much larger than the results of CASE PY16. In both cases, the forces and moment predicted in the long period are more accurate than those in CASE PY8. The predicted results indicate that the numerical schemes adopted in the present study are reliable.

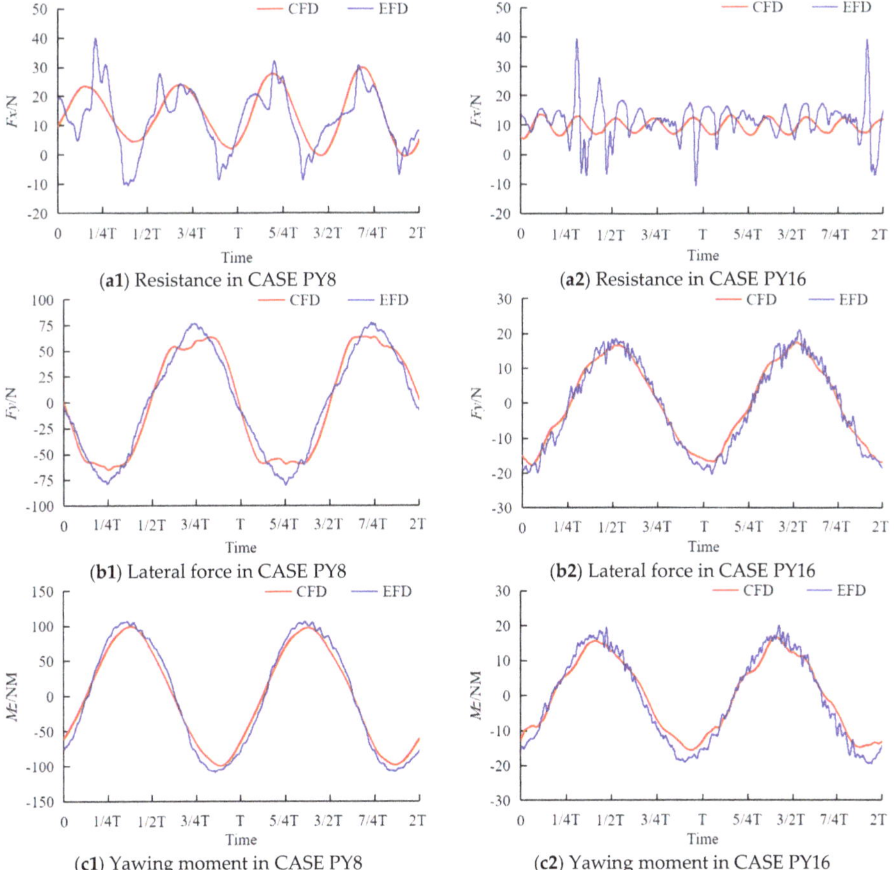

Figure 16. Forces and moment in CASE PY8 (left) and PY16 (right) (**a**: resistance; **b**: lateral force; **c**: yawing moment).

Table 7. Comparison between the predicted and experimental forces and moment in pure yaw tests.

CASE	Force/Moment	EFD	CFD	Error
PY8	Resistance (N)	12.75	14.05	10.21%
	Lateral force (N)	77.78	91.53	17.67%
	Yawing moment (NM)	107.97	99.59	−7.76%
PY16	Resistance (N)	9.84	9.87	0.39%
	Lateral force (N)	16.28	17.48	7.33%
	Yawing moment (NM)	15.73	16.67	5.97%

4.3.2. Free Surface

Figure 17 shows the evolution of the free surface wave-making in one period of CASE PY8. Compared with the results of CASE PS8, the variation of wave-making on the free surface around the hull is more drastic. Meanwhile, the peak of the bow wave is larger than that in the pure sway test due to the large yaw motion. In addition, there is a significant difference between the diffraction wave patterns on both sides of the hull. The bow wave breaking is also captured in CASE PY8. The periodic variation of wave-making leads to the corresponding change in the pressure field around the hull, resulting in the increment in the lateral force and yawing moment.

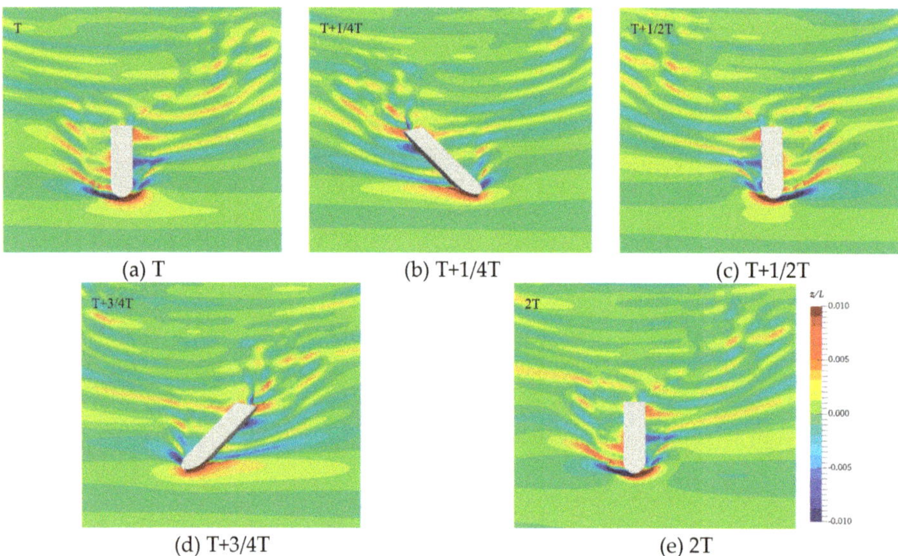

(a) T (b) T+1/4T (c) T+1/2T (d) T+3/4T (e) 2T

Figure 17. Evolution of the free surface in one period in CASE PY8.

4.3.3. Vortex Structure

Since the yaw motion of the hull is taken into consideration, the vortex structures in the dynamic pure yaw test are obviously different from those in the pure sway tests. Figure 18 shows the vortex structures obtained by the different methods in CASE PY8. By observing the vortex structures obtained by the four methods, a vortex, which originated from the propeller shaft, is captured by all of them. The vortex structures near the bow are also visible to different extents. At T, the bow vortex structures near the free surface are captured by the method based on the vorticity, Q criterion and OmegaR, while more broken and fine vortex structures are presented by Liutex. At this time, the hull is moving to the portside, and the separation flow appears on the starboard side, which can be confirmed by the vortex structures near the bilge on the starboard. At T+1/4T, the bottom of the hull is covered by the vortex structures obtained by the Q criterion. Only the OmegaR method presents the bow vortex structures. Since the hull is swinging counterclockwise (see from bottom to top), the vortex structures originated from the mid-hull appear near the stern on the portside of the hull. The vortices are more clearly presented by Q, OmegaR and Liutex. At T+1/2T, the movement of the hull is the opposite to that at T, so the distribution of the vortex structures is basically symmetrical. The vortex structures originated from the propeller shaft approach the starboard side downstream. At T+3/4T, the hull swings clockwise, and the vortex structures occur at the mid-hull and approaches the starboard side downstream. At 2T, the distribution of vortex structures is basically the same as the results at T.

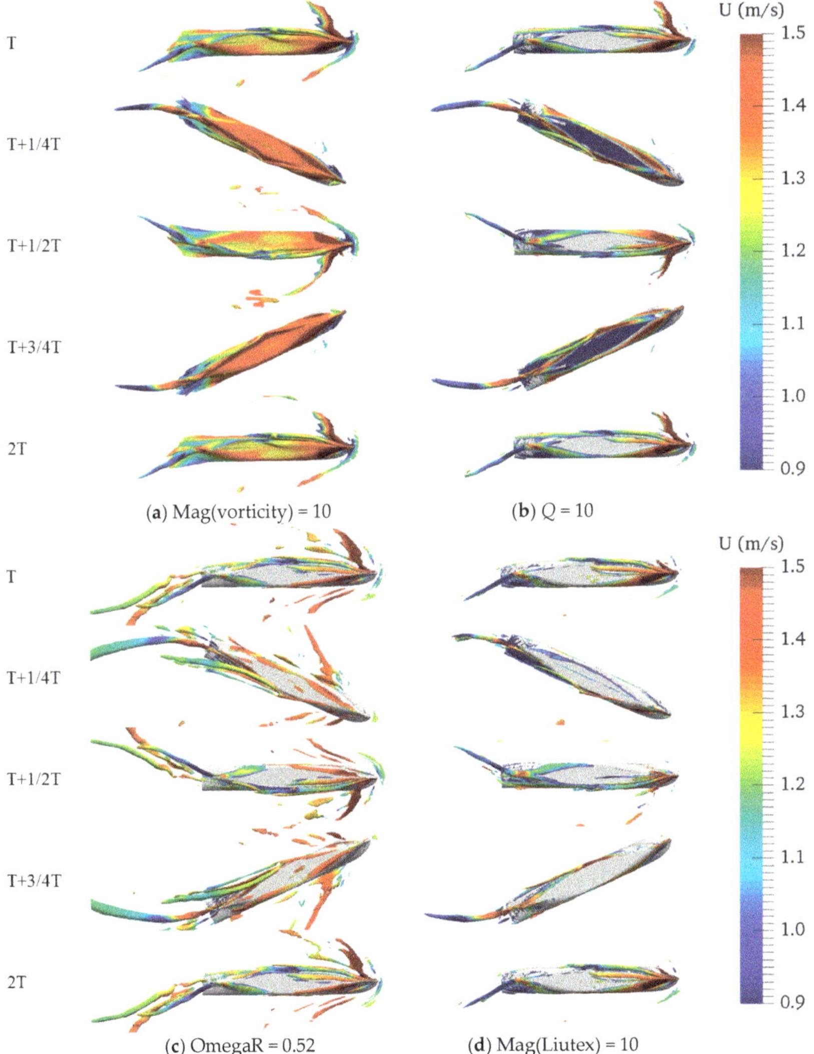

Figure 18. Vortex structures obtained by different vortex identification method: (**a**) first generation based on vorticity; (**b**) second generation based on eigenvalues; (**c**,**d**) third generation in CASE PY8.

Figures 19 and 20 present the evolution of the vertical vorticity and Liutex at z/L_{pp} = −0.03572 along the ship length direction in CASE PY8 and PY16. Similar to the phenomenon presented in the pure sway tests, the flow separation is more obvious in the shorter period. At T+1/4T, the drift angle in CASE PY8 is larger than that presented in Figure 20. Therefore, the flow separation on the portside is more violent in CASE PY8. At T and T+1/2T, the flow separates more obviously. This is because the lateral velocity becomes larger in a short period. It is confirmed by the vertical vorticity and Liutex near the stern of the hull. As shown in Figure 19, the Liutex method is able to capture more broken and finer vortices.

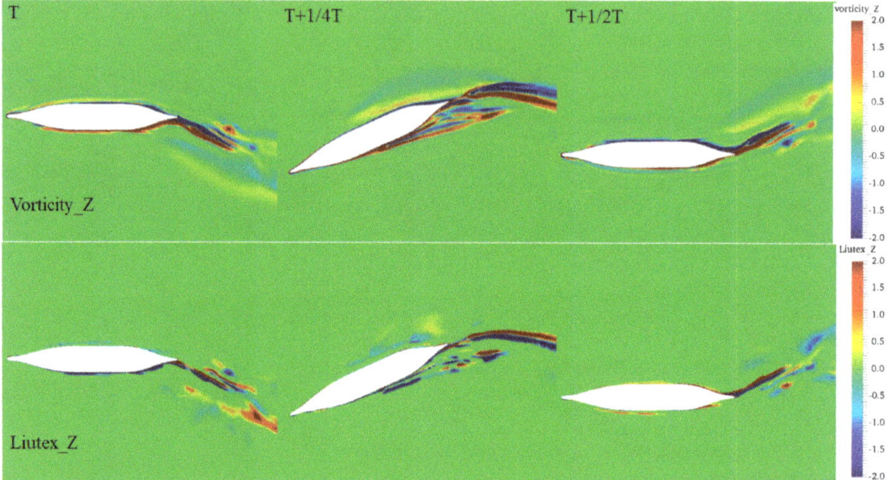

Figure 19. Vertical vorticity (top) and Liutex (bottom) at $z/L_{pp} = -0.03572$ in CASE PY8.

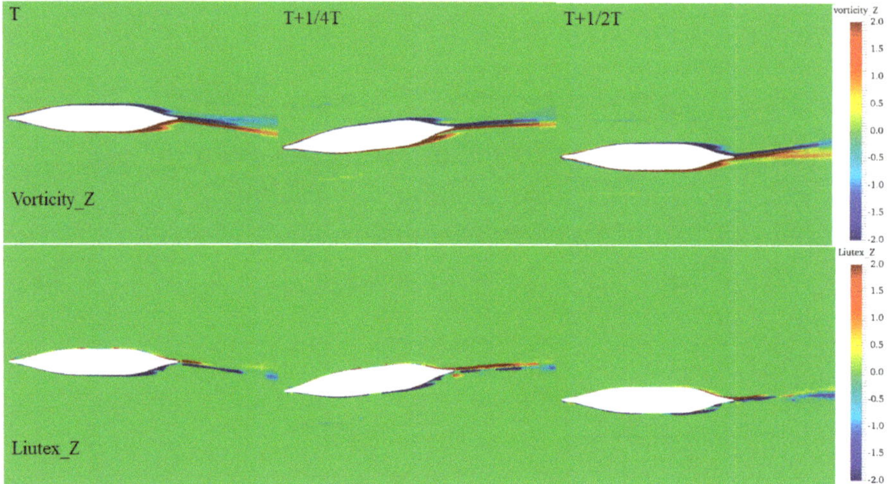

Figure 20. Vertical vorticity (top) and Liutex (bottom) at $z/L_{pp} = -0.03572$ in CASE PY16.

In summary, the characteristics of four vortex identification methods can be obtained by analyzing the capability of capturing the vortex structures in the ship's large separation flow. As shown in Table 8, more information of vortex structures is displayed by the third generation of vortex identification methods, OmegaR and Liutex/Rortex.

Table 8. Comparison of the different vortex identification methods.

Methods	Vortex Boundary	Threshold Sensitivity	Weak Vortex	Vortex Axis
Vorticity	No	sensitive	No	No
Q criterion	Yes	sensitive	No	No
OmegaR	Yes	insensitive	Yes	No
Liutex/Rortex	Yes	sensitive	No	Yes

5. Conclusions

In the present study, an in-house CFD solver, naoe-FOAM-SJTU, is used to perform numerical simulations of planar motion mechanism (PMM) tests of a Yupeng ship. Unsteady RANS equations combined with an SST k-ω turbulence model are adopted to solve the complex viscous flow and a dynamic overset grid method is applied to simulate the large amplitude motion of the hull. The paper not only compares the time histories of predicted and experimental forces and moments acting on the hull, but also uses four vortex identification methods to capture the vortex structures in the large separated flow, which are very helpful for a further analysis on the flow mechanism around the hull.

As for the PMM tests, the predicted forces and moments show a good agreement with the experimental data. It indicates that the present numerical schemes are suitable for the simulations of the large separation flow. In the static drift tests, even though the predicted resistance is not accurate enough since there are big fluctuations in the experimental data, the errors of the lateral force and yawing moment are less than 6%. The evolution of the free surface is also captured well and the bow wave breaking is visible at a larger drift angle (CASE DA 20). In the dynamic tests, the shorter period leads to a larger lateral force and yawing moment. In the pure yaw tests, the lateral force in CASE PY16 is only 16.28N, but it increases to 77.28N rapidly when the period is shortened to 8 s. The periodic variation of the lateral force and yawing moment are confirmed by the evolution of the free surface.

By comparing the vortex structures captured by four vortex identification methods, it is found that the OmegaR method with the insensitive threshold is able to capture more vortex structures in the viscous flows. More breaking and detailed vortex structures are obtained by the Liutex method. The vortex structures obtained by vorticity (first generation) covers the hull fully, which is not reasonable. The Q criterion (second generation) is very sensitive to the artificial threshold. In conclusion, the third generation of vortex identification methods are more suitable for capturing the vortex structures in the large separated flow of the ship.

In the future, DES approaches will be used to solve the viscous flow fields around the ship hull and the results will be compared with the RANS method.

Author Contributions: Funding acquisition, D.W.; writing—original draft preparation, Z.R.; writing—review and editing, J.W.; numerical simulations, Z.R.; post-processing, Z.R. All authors have read and agreed to the published version of the manuscript.

Funding: This work is supported by the National Natural Science Foundation of China (51879159, 51809169). The National Key Research and Development Program of China (2019YFB1704200, 2019YFC0312400), Chang Jiang Scholars Program (T2014099), Shanghai Excellent Academic Leaders Program (17XD1402300), and Innovative Special Project of Numerical Tank of Ministry of Industry and Information Technology of China (2016-23/09), to which the authors are most grateful.

Conflicts of Interest: The authors declare no conflict of interest. The funders had no role in the design of the study; in the collection, analyses, or interpretation of data; in the writing of the manuscript, or in the decision to publish the results.

References

1. Tahara, Y.; Longo, J.; Stern, F. Comparison of CFD and EFD for the Series 60 C B = 0.6 in steady drift motion. *J. Mar. Sci. Technol.* **2002**, *7*, 17–30. [CrossRef]
2. Simonsen, C.D.; Stern, F. Verification and validation of RANS maneuvering simulation of Esso Osaka: Effects of drift and rudder angle on forces and moments. *Comput. Fluids* **2003**, *32*, 1325–1356. [CrossRef]

3. Pinto-Heredero, A.; Xing, T.; Stern, F. URANS and DES analysis for a Wigley hull at extreme drift angles. *J. Mar. Sci. Technol.* **2010**, *15*, 295–315. [CrossRef]
4. Ismail, F.; Carrica, P.M.; Xing, T.; Stern, D. Evaluation of linear and nonlinear convection schemes on multidimensional non-orthogonal grids with applications to KVLCC2 tanker. *Int. J. Numer. Methods Fluids* **2010**, *64*, 850–886. [CrossRef]
5. Xing, T.; Bhushan, S.; Stern, F. Vortical and turbulent structures for KVLCC2 at drift angle 0, 12, and 30 degrees. *Ocean Eng.* **2012**, *55*, 23–43. [CrossRef]
6. Stern, F.; Agdrup, K.; Kim, S.Y.; Hochbaum, A.C.; Rhee, K.P.; Quadvlieg, F.; Perdon, P.; Hino, T.; Broglia, R.; Gorski, J. Experience from SIMMAN 2008—The First Workshop on Verification and Validation of Ship Maneuvering Simulation Methods. *J. Ship Res.* **2011**, *55*, 135–147.
7. Meng, Q.J.; Wan, D.C. Numerical simulations of viscous flow around the obliquely towed KVLCC2M model in deep and shallow water. *J. Hydrodyn.* **2016**, *28*, 506–518. [CrossRef]
8. Wang, J.H.; Liu, X.Y.; Wan, D.C. Numerical Simulation of an Oblique Towed Ship by naoe-FOAM-SJTU Solver. In Proceedings of the 25th International Offshore and Polar Engineering Conference, Big Island, HI, USA, 21–26 June 2015.
9. Broglia, R.; Muscari, R.; Di Mascio, A. Numerical simulations of the pure sway and pure yaw motion of the KVLCC-1 and 2 tanker. In Proceedings of the SIMMAN 2008 Workshop on Verification and Validation of Ship Maneuvering Simulation Methods, Lyngby, Denmark, 14–16 April 2008.
10. Toxopeus, S.L.; van Walree, F.; Hallmann, R. Maneuvering and Seakeeping Tests for 5415M. In Proceedings of the AVT-189 Specialists' Meeting, Portsdown West, UK, 12–14 October 2011.
11. Cura-Hochbaum, A. On the numerical prediction of the ship's manoeuvring behaviour. *Ship Sci. Technol.* **2011**, *5*, 27–39. [CrossRef]
12. Simonsen, C.D.; Stern, F. RANS simulation of the flow around the KCS container ship in pure yaw. In Proceedings of the SIMMAN 2008 Workshop on Verification and Validation of Ship Maneuvering Simulation Methods, Lyngby, Denmark, 14–16 April 2008.
13. Sakamoto, N.; Carrica, P.M.; Stern, F. URANS simulations of static and dynamic maneuvering for surface combatant: Part 1. Verification and validation for forces, moment, and hydrodynamic derivatives. *J. Mar. Sci. Technol.* **2012**, *17*, 422–445. [CrossRef]
14. Sakamoto, N.; Carrica, P.M.; Stern, F. URANS simulations of static and dynamic maneuvering for surface combatant: Part 2. Analysis and validation for local flow characteristics. *J. Mar. Sci. Technol.* **2012**, *17*, 446–468. [CrossRef]
15. Yoon, H.; Simonsen, C.D.; Benedetti, L.; Longo, J.; Toda, Y.; Stern, F. Benchmark CFD validation data for surface combatant 5415 in PMM maneuvers—Part I: Force/moment/motion measurements. *Ocean Eng.* **2015**, *109*, 705–734. [CrossRef]
16. Yoon, H.; Longo, J.; Toda, Y.; Stern, F. Benchmark CFD validation data for surface combatant 5415 in PMM maneuvers—Part II: Phase-averaged stereoscopic PIV flow field measurements. *Ocean Eng.* **2015**, *109*, 735–750. [CrossRef]
17. Kim, H.; Akimoto, H.; Islam, H. Estimation of the hydrodynamic derivatives by RaNS simulation of planar motion mechanism test. *Ocean Eng.* **2015**, *108*, 129–139. [CrossRef]
18. Yang, Y.; Zou, Z.J.; Zhang, C.X. Calculation of hydrodynamic forces on a KVLCC hull in sway motion in deep and shallow water. *Chin. J. Hydrodyn.* **2011**, *26*, 85–93.
19. Liu, X.Y.; Fan, S.; Wang, J.H.; Wang, J.; Wan, D. Hydrodynamic Simulation of Pure Sway Tests with Ship Speed and Water Depth Effects. In Proceedings of the 25th International Offshore and Polar Engineering Conference, Big Island, HI, USA, 21–26 June 2015.
20. Wang, J.H.; Wan, D.C. Numerical simulation of pure yaw motion using dynamic overset grid technology. *Chin. J. Hydrodyn.* **2016**, *31*, 567–574.
21. Omer, M.; Chen, X.; Zhou, C.S.; Lunkun, G. Assessment of the modified rotation/curvature correction SST turbulence model for simulating swirling reacting unsteady flows in a solid-fuel ramjet engine. *Acta Astronaut.* **2016**, *129*, 241–252.

22. Noack, R.W.; Boger, D.A.; Kunz, R.F.; Carrica, P. Suggar++: An improved general overset grid assembly capability. In Proceedings of the 47th AIAA Aerospace Science and Exhibit, San Antonio, TX, USA, 22–25 June 2009; pp. 22–25.
23. Liu, C.Q.; Gao, Y.S.; Dong, X.R.; Wang, Y.-Q.; Liu, J.-M.; Zhang, Y.-N.; Cai, X.-S.; Gui, N. Third generation of vortex identification methods: Omega and Liutex/Rortex based systems. *J. Hydrodyn.* **2019**, *31*, 205–223. [CrossRef]

Journal of
Marine Science and Engineering

Article

Prediction of the Side Drift Force of Full Ships Advancing in Waves at Low Speeds

Shukui Liu [1,*] and Apostolos Papanikolaou [2,3]

1 School of Mechanical and Aerospace Engineering, Nanyang Technological University, 50 Nanyang Ave., Singapore 639798, Singapore
2 Ship Design Laboratory, National Technical University of Athens, 9 Heroon Polytechniou, 15773 Athens, Greece; papa@deslab.ntua.gr
3 Hamburger Schiffbau-Versuchsanstalt, 164 Bramfelder Str., 22305 Hamburg, Germany
* Correspondence: skliu@ntu.edu.sg

Received: 3 May 2020; Accepted: 21 May 2020; Published: 25 May 2020

Abstract: In this study, we analyze the experimental results of the mean sway (side drift) forces of six full type ships at low speeds in regular waves of various directions and compare them with numerical results of the in-house 3D panel code NEWDRIFT. It is noted that the mean sway force is most significant in relatively short waves, with the peak being observed at $\lambda/L_{PP} \approx 0.5$–$0.6$. For $\lambda/L_{PP} > 1.0$, the corresponding value is rather small. We also observe a solid recurring pattern of the mean sway force acting on the analyzed full type ships. On this basis, we proceed to approximate the mean sway force with an empirical formula, in which only the main ship particulars and wave parameters are used. Preliminary validation results show that the developed empirical formula, which is readily applicable in practice, can accurately predict the mean sway force acting on a full ship, at both zero and non-zero speeds.

Keywords: second order steady sway wave force; side drift force; empirical formula; 3D panel method; near-field method; maneuverability in waves

1. Introduction

The prediction of the second order steady wave forces and moment of a ship or a floating structure in waves is a classical ship dynamic problem. The longitudinal component in the direction of a ship's forward speed (added resistance) is of major concern when discussing ship's speed-power performance in realistic seaways, while the transverse and rotational components significantly affect the maneuvering performance of a ship, hence, they are directly related to the safe navigation of the ship in a seaway.

Several theoretical approaches of varying complexity and accuracy have been developed and validated since the early 1940s, when Havelock [1] developed the first theoretical approach to calculate the steady drift force acting on a fixed vertical circular cylinder in waves. Besides a few analytical approaches to the drift forces acting on analytically defined forms, the current well established numerical methods can be classified into two main categories, namely, the far-field and near-field, pressure methods. The far-field methods are based on energy considerations for the diffracted (reflected and transmitted) and radiated wave, the momentum flux at infinity and the work done on the body in the near-field, leading to the acting steady force by the total rate of momentum change. The near-field methods, on the other side, calculate the steady second order forces/moments by direct integration of the hydrodynamic, steady second order pressure acting on the wetted body surface. This pressure can be calculated exactly from first order potential functions and their derivatives, thus, there is no need to solve the exact second order potential theory problem.

In the frame of potential flow theory, the far-field approach was introduced by Maruo [2] leading to the calculation of the steady second order longitudinal forces and then extended by Newman [3] to calculate the steady second order yaw moment as well. Salvesen [4] investigated the added resistance problem by applying the radiated energy approach of Gerritsma and Beukelman [5], which is in line with Maruo's theory, but while using the basic potential flow solution of the Salvesen–Tuck–Faltinsen (STF) seakeeping strip theory [6]; they obtained satisfactory results for the investigated ship hull forms. Following also the far-field approach, Mavrakos [7] presented a solution to the vertical second order forces and pitch moment acting on floating axisymmetric bodies. Finally, Kashiwagi [8] presented results for the second order steady forces based on the far-field method but using a more consistent numerical seakeeping method accounting for 3D effects.

The near-field, direct pressure integration method has been practically developed in parallel with the far-field method. Boese [9] proposed a first pressure integration method, though the associated hydrodynamic pressure distribution was very simplified. More exact 3D formulations and numerical implementations were introduced in the later years, first dealing with the zero speed second order problem [10–14] with good validation results. It was proven that both far- and near-field methods lead to the same results, even though numerical implementations may justify minor differences [15,16].

An important issue in both methods is the treatment of the forward speed effect in the calculation of the ensuing first order potential function and its derivatives. For slender, ship-like forms, slender body theory may be applied for the forward speed effect. It can be shown that this simplification is also well applicable to non-slender ships (full type ships) at low speed of advance [17]. A negative side effect of this approximation is, however, that the steady wave formation at a ship's bow region is not properly accounted for. This is problematic when applying the near-field pressure integration method, because of the dominance of the so-called line integral term. Joncquez [18] further developed the near-field method by including an improved formulation for the steady pressure and the results improved, even though it still shows discrepancy from experimental data in short waves.

In the validation of both methods, the focus has been mainly on the longitudinal component, which is also known as added resistance, as it affects significantly the speed-power performance (and then, the associated fuel and emission performances) of a ship in seaways. Therefore, many studies were devoted to the improvement of various issues in the implementation of such methods, as documented in, but not limited to, the works of Chen [19], Duan & Li [20], Liu et al. [21], Ohkusu [22], Sakamoto and Baba [23], Tsujimoto et al. [24], and Yang et al. [25], etc.

The other force and moment components, such as the steady sway force and yaw moment, mainly influence the design of mooring systems of floating structures (zero speed problem) and the maneuvering performance of ships in waves. Experimental measurements for the validation of numerical methods have to be carried out in ocean basins and the model size is trivially constrained by the size of the basin, hence, it is usually relatively small for the demand of the accurate validation measurements of the concerned second order quantities. For this reason, rather limited results have been reported to the public and the validation of various methods on predicting these components has been much less intensive, as indicated in various studies (for instance, in Skejic & Faltinsen [26]). Nevertheless, over the years, some experimental data of standard designs as well as several standard hull forms have been gradually released to the public. For VLCC tanker designs, Naito et al. [27] reported the experimental steady forces and moment acting on the *Esso Osaka* ship at low speeds using a 4.0 m long model. Iwashita et al. [28] published the experimental data of a VLCC model advancing in oblique waves using a 3.14 m long model. Ueno et al. [29] studied the steady wave forces and moment of a VLCC tanker in short waves at zero and small forward speed. Ueno et al. [30] measured the steady forces and moment of a VLCC tanker at zero speed in waves of various lengths and headings. Within the framework of the EU funded SHOPERA project [31], the steady forces and yaw moment of the benchmark ship KVLCC2 in steep waves at zero speed were measured using a 4 m model [32]; the same ship was the subject of an international benchmark study on contemporary numerical methods [33]. For bulk carrier designs, Kadomatsu et al. [34] published the experimental results of a capesize bulker design using a 2.91 m long model. Yasukawa et al. [35] reported to the public the steady wave forces and yaw moment of the S-Cb84 model at several speeds.

The experimental data accumulated in the public domain over the years make nowadays a qualitative, and to some extent, quantitative validation of numerical methods possible, namely comparing the numerical results generated by a numerical method and software for a specific design and examine the results against available experimental data of the same or similar hull forms. In the present study, the well-established frequency domain 3D boundary element method and associated computer code NEWDRIFT [12,14,17,36] is used to solve the basic seakeeping problem and to calculate the second order steady sway force. Furthermore, we proceeded to approximate the mean sway force with an empirical formula, in which only the main ship particulars and wave parameters are used. Applications to standard ship designs were carried out to validate the applicability and the accuracy of the developed and implemented methods in practice and to make recommendations on the way ahead.

2. Background of the Employed 3D Panel Method

We consider a ship advancing at constant mean forward speed U in regular sinusoidal waves of small amplitude. The ship's heading is defined by angle α measured between the direction of U and the direction of wave propagation (α = 0° represents following waves). The wave frequency ω_0 is related to the ship's frequency of encounter ω by

$$\omega = |\omega_0 - Uk\cos\alpha| \tag{1}$$

where wave number $k = 2\pi/\lambda$, and with λ being the wavelength.

The resulting oscillatory motions of the ship are assumed linear and harmonic. An orthogonal coordinate system Oxyz is assumed at the mean position of the ship, with positive z-axis vertically upwards through the center of gravity of the ship, positive x-axis towards the ship's bow, and O in the plane of undisturbed free surface. The ship is assumed to oscillate as a rigid body in six degrees of freedom with amplitudes η_i, where i = 1, 2, ... , 6 refer to surge, sway, heave, roll, pitch and yaw motions, respectively. The six linear coupled differential equations of motions are:

$$\sum_{l=1}^{6}\left[-\omega^2(M_{il}+A_{il}) - j\omega B_{il} + C_{il}\right]\eta_l = f_i e^{-j\omega t},\ i = 1,2,\ldots 6 \tag{2}$$

where M_{il} are components of the generalized mass matrix of the ship, A_{il} and B_{il} are added mass and damping coefficients, C_{il} are hydrostatic restoring force coefficients, f_i are the complex amplitudes of wave exciting forces and moments, and j is the imaginary unit associated with a harmonic time function. Only the real part is taken in all expressions involving $e^{-j\omega t}$.

Assuming potential flow, the velocity potential $\Phi(\mathrm{x},\mathrm{y},\mathrm{z};\mathrm{t})$ is separated into a time-independent steady part due to ship's speed U and time-dependent part associated with the incident wave and the excited unsteady oscillatory motions, which are assumed small in response to small amplitude incident waves:

$$\Phi(\mathrm{x},\mathrm{y},\mathrm{z};\mathrm{t}) = -Ux + \Phi_s(\mathrm{x},\mathrm{y},\mathrm{z}) + \Phi_T(\mathrm{x},\mathrm{y},\mathrm{z})e^{-j\omega t} \tag{3}$$

Here, $-Ux + \Phi_s$ is the steady part and Φ_T the complex amplitude of the unsteady potential part. We assume that the operator $\frac{U}{\omega}\frac{\partial}{\partial x}$, appearing in the free surface boundary condition of the ensuing BVP for Φ, is small, which holds for

- U small (low speed),
- ω large (high wave frequency/short waves) and/or
- $\frac{\partial}{\partial x}$ small, which holds for slender bodies.

Under these conditions, the problem is linearized by disregarding higher order terms in Φ_s and Φ_T and terms containing their cross products. The steady and unsteady velocity potentials can be determined separately. Φ_T is then decomposed as follows:

$$\Phi_T = \varphi_0 + \varphi_7 + \sum_{i=1}^{6} \eta_i \varphi_i \tag{4}$$

where φ_0 is the incident wave potential, φ_7 the diffracted wave potential, and φ_i the velocity potential for the i-th mode of motion. All φ_i, i = 0, ..., 7 satisfy Laplace's equation in the fluid domain and appropriate boundary conditions at the wetted surface, the free surface and at infinity.

We introduce the zero speed Green's function $G(p, q)$ representing the potential at a field point $p(z, y, z)$ of a pulsating source of unit strength at point $q(\xi, \eta, \zeta)$. Applying Green's 3rd theorem to the harmonic functions $G(p, q)$ and φ_i, a set of Fredholm integral-equations is derived, to be solved for the complex source strengths, based on which the potential values φ_i can be obtained.

Having obtained φ_i, the hydrodynamic pressure p due to the unsteady velocity potential can be calculated by Bernoulli's equation:

$$p = -\rho\left(\frac{\partial \varphi}{\partial t} + \frac{1}{2}|\nabla\varphi|^2 + gz\right) \tag{5}$$

Integrating the pressure over the wetted hull surface yields the hydrodynamic forces and moments, which finally lead to the values of the hydrodynamic coefficients in the equations of motion (2), namely A_{il}, B_{il} and f_i. These coefficients are corrected for the effect of forward speed on the basis of expressions derived from the slender body theory [17]. Thus, the six DOF equations of motions can be set up and solved for the motions of the ship.

Both the far-field method and near-field (pressure integration) method have been implemented in the code NEWDRIFT to calculate the quasi second order forces/moments [12,37]. The classic near-field (pressure integration) method for the calculation of the transverse steady second order force at low speed of advance has the following form:

$$\begin{aligned}
\overline{F}_Y = & -\tfrac{1}{2}\rho g \int_C \zeta_r^2 \sec\alpha_{WL} n_2 dl - \omega^2 M\eta_3\eta_4 + \omega^2 M(\eta_1 - z_G\eta_5)\eta_6 \\
& + \rho \iint_{S_B} \Big\{ (\eta_2 + x\eta_6 - z\eta_4)\tfrac{\partial}{\partial y}\Big(\tfrac{\partial\phi}{\partial t} + U\tfrac{\partial\phi}{\partial x}\Big) \\
& + (\eta_3 + x\eta_5 - y\eta_4)\tfrac{\partial}{\partial z}\Big(\tfrac{\partial\phi}{\partial t} + U\tfrac{\partial\phi}{\partial x}\Big)\Big\} n_2 ds \\
& - \tfrac{1}{2}\rho \iint_{S_B} \Big[\Big(\tfrac{\partial\phi}{\partial x}\Big)^2 + \Big(\tfrac{\partial\phi}{\partial y}\Big)^2 + \Big(\tfrac{\partial\phi}{\partial z}\Big)^2\Big] n_2 ds
\end{aligned} \tag{6}$$

where ζ_r is the first order relative wave height along the ship waterline and η_i is the ship motion amplitude in i-th direction, ϕ is the total velocity potential, M is the mass of the ship, and z_G the z-coordinate of the center of gravitation of the ship, and α_{WL} is the sectional flare angle (0 means vertical wall) at the ship's SWL waterline.

3. Subject Models

Table 1 presents the main particulars, associated hull coefficients and ratios, and testing conditions of six full ships that are used in the present analysis. The associated experimental results are available for Froude numbers Fr between 0.0 and 0.10, which correspond to low and up to moderate speeds and are typical conditions in ship maneuvering. Four sets of the experimental data are obtained in free motion conditions, while the data for VLCCa and *Esso Osaka* are in motion restrained conditions. All of these experimental data are in the public domain and the references for each set of data are given.

Table 1. Main particulars, associated coefficients and ratios, and testing conditions of several full ships.

Item	Unit	VLCCa	VLCCb	KVLCC2	S-Cb84	Bulk Carrier	Esso Osaka
Model Length	m	3.14	2.97	4.00	3.10	2.91	4.00
L/B	-	5.08	5.52	5.52	5.52	5.70	6.13
B/T	-	3.05	3.01	2.79	2.79	2.70	2.44
L/T	-	15.49	16.62	15.40	15.40	15.39	14.96
C_B	-	0.81	0.81	0.81	0.84	0.83	0.83
Speed, Fr	-	0.0–0.2	0–0.069	0.0	0.0–0.05–0.10	0.0–0.10	0.0–0.055
Motion	-	restrained	free	free	free	free	restrained
Reference	-	[28]	[29,30]	[32,33]	[35]	[34]	[27]

Regarding the available information of the hull forms, the lines of the KVLCC2, S-Cb84, the bulk carrier and the *Esso Osaka* are available in the public domain, while that of the VLCCa and VLCCb are not. For VLCCa, the block coefficient is the same as KVLCC2, while the L/T is close to that of KVLCC2, and the L/B smaller and B/T larger. These characteristics give us good reason to apply a simple affine distortion method to the KVLCC2 offsets to generate an approximate lines plan [38]. The scaling factors for the three axes in this simple affine distortion are as follows:

$\alpha_1 = 3.100/320.0 = 0.00969$; $\beta_1 = 0.610/58.0 = 0.01052$; $\gamma_1 = 0.200/20.8 = 0.00962$

For VLCCb, a similar procedure is applied to generate her hull form from KVLCC2's lines:

$\alpha_2 = 2.970/320.0 = 0.00928$; $\beta_2 = 0.538/58.0 = 0.00928$; $\gamma_2 = 0.179/20.8 = 0.00861$

The predicted mean second order force for the two approximated hull forms will be denoted as VLCCa′ and VLCCb′, respectively.

4. Numerical Results and Discussion

Figure 1 shows the used panelization of the wetted part of the hull form of S-Cb84 model, with 739 panels for half of the ship's surface. Similar panelizations have been applied to all the investigated ship models.

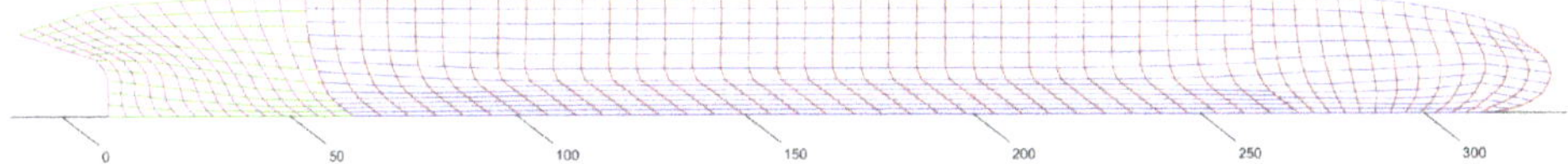

Figure 1. Panelization of the underwater hull of S-Cb84 model for 3D panel code NEWDRIFT.

Figure 2 shows the calculated mean sway forces of the six ships in beam waves at zero speed as obtained both from experiments and from numerical predictions using NEWDRIFT. Overall, it is noted that the four sets of experimental results are close to each other, except for the results of KVLCC2, which show some oscillatory behavior in short waves. This might be due to the fact that the generated waves in this region were quite steep for KVLCC2. Note, also, that the model length of the KVLCC2 was the largest among the six test ships. The obtained numerical results show a similar trend for the mean sway force of the six ships. The maximum non-dimensional mean sway force is observed at $\lambda/L_{PP} \approx 0.5$–$0.7$. In short waves, the non-dimensional values all tend to approach a certain value between 0.4–0.5, which is expected, when considering that the limiting value of the mean sway force on a vertical wall is 0.5. The limiting value is a bit less than 0.5 because the hull is not vertical near the waterline, particularly at both ends. Nevertheless, some increased uncertainty appears in very short waves. This can be due to experimental and numerical issues. On the numerical side, in short waves, very small panels are needed and in addition, the phenomenon of irregular frequencies may trigger oscillations in the obtained results [19,39].

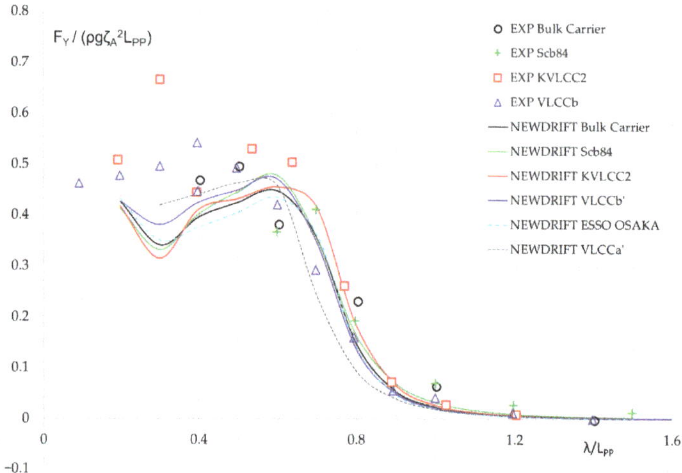

Figure 2. Mean sway force of six ships in regular beam waves at zero speed.

Figure 3 shows the induced mean sway force of five of the test ships in bow waves from two incident directions at zero speed. For $\alpha = 135\ deg$, the experimental data of Esso Osaka and a bulk carrier are available. Reasonable agreement is observed in short waves (λ/L_{PP} = 0.4–0.8). In longer waves, where ship motion plays a role, there are significant deviations as the experiments of Esso Osaka were executed in the motion restrained condition. For $\alpha = 150\ deg$, the experimental data of S-Cb84, KVLCC2 and VLCCb are available. While the measured values are smaller than in the $\alpha = 135\ deg$ case, a close agreement is observed among the three sets of data. For both cases, the numerical results for the studied ships show a high degree of similarity except in very short waves, where the local hull form features appear to play an important role.

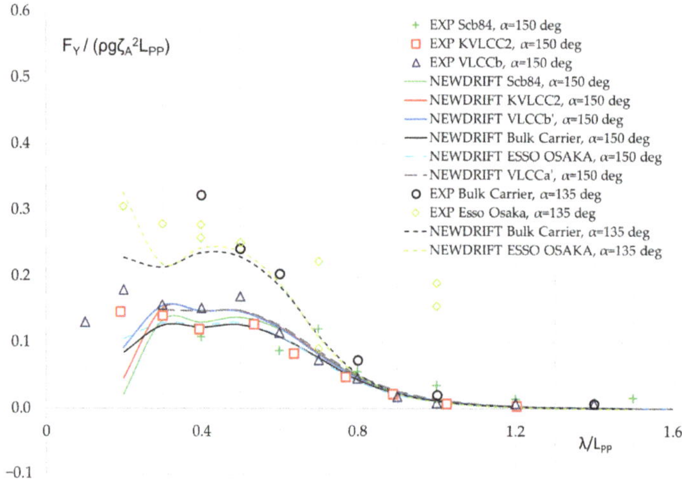

Figure 3. Mean sway force of several ships in regular bow waves at zero speed.

Figure 4 shows the mean sway force of four of the tested ships in stern oblique waves in two incident directions at zero speed. For α = 45 *deg*, the experimental data of a bulk carrier are available, while for α = 30 *deg* the experimental data for S-Cb84, KVLCC2 and VLCCb are available. The measured values for α = 30 *deg* are smaller than in the α = 45 *deg* case and generally, all three sets of data are close together, despite some scattering that might be due to the different stern forms. In both cases, the numerical results for all studied ships are close, except in very short waves, where deviations are notable. Particularly, the experimental results for the bulk carrier at λ/L_{PP} = 0.4 have a peak and are much larger than the numerically predicted value. This is also observed in the associated work of Kadomatsu et al. [34].

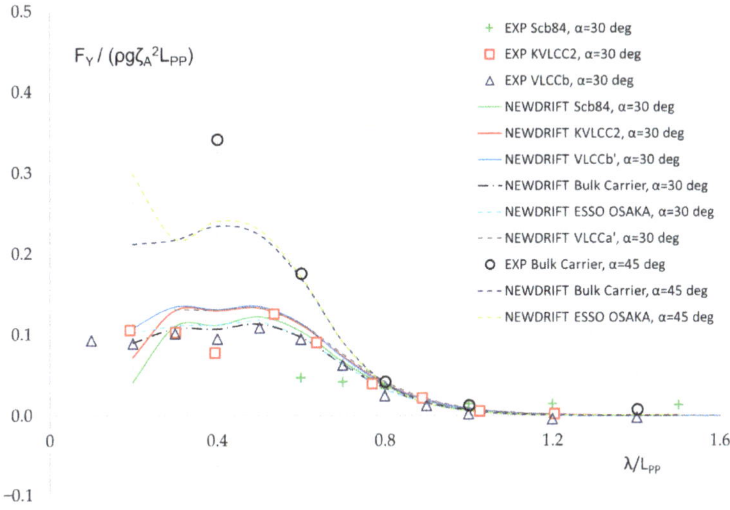

Figure 4. Mean sway force of several ships in regular stern waves at zero speed.

Though the degree of uncertainty of the analyzed set of experimental data is not available to the public, it is useful to briefly comment on the quality of the measurements on the basis of made observations. In this respect, the experimental results depicted in Figures 2–4 seem to be in general reliable, except for a few dubious points significantly deviating from trend lines. In terms of the depicted numerical results, it appears that prediction results by NEWDRIFT agree fairly well with experimental results for all wave directions at zero speed.

For the cases with forward speed, much less experimental data are available, due to the very limited availability of proper facilities to conduct such tests. Figure 5 shows the experimental results for the *Esso Osaka* at Fr = 0.055 in waves of α = 135 *deg* and for Scb-84 at Fr = 0.05 in waves of α = 150 *deg*; numerical results for both ships in both testing conditions are also added. Good agreement between experimental results and numerical results has been achieved, though it seems that for α = 135 *deg*, the numerical prediction is slightly lower than the experimental results, while for α = 150 *deg*, the numerical prediction is slightly higher than the experimental data.

Figure 6 shows the experimental results for S-Cb84 at Fr = 0.05 in beam waves and stern oblique waves, together with the numerical results for the other tested ships for the same testing condition. For α = 90 *deg*, the numerical prediction is slightly higher than the experimental results. The maximum non-dimensional mean sway force is observed at $\lambda/L_{PP} \approx 0.6$. For α = 30 *deg*, the numerical prediction is lower than the experimental results, which may be related to insufficiency of the ensuing slender body theory when dealing with a ship sailing in stern oblique waves and at low frequency of encounter.

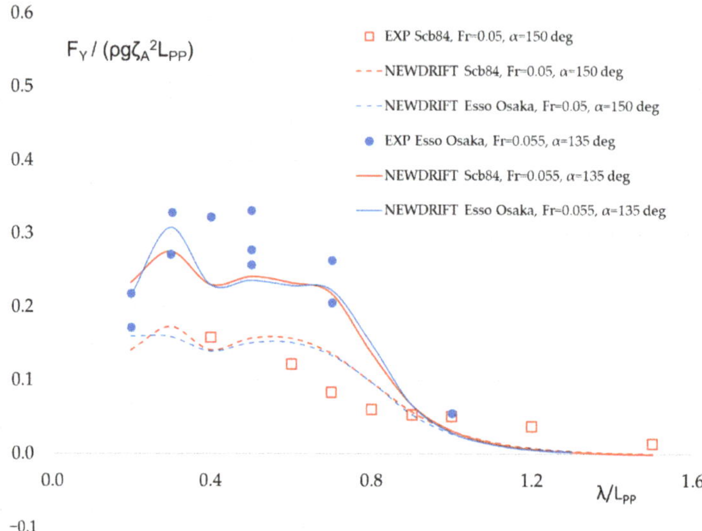

Figure 5. Mean sway force of two ships in regular bow waves at low speeds.

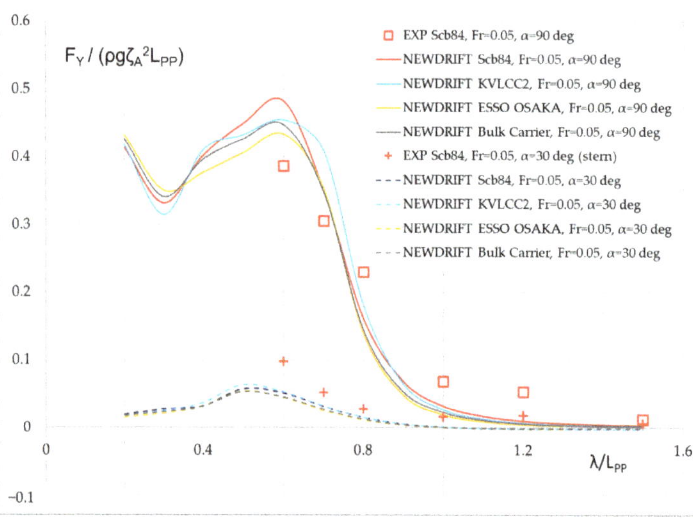

Figure 6. Mean sway force of various ships in regular beam and stern oblique waves at low speeds.

Figures 7–9 present a comparison between the experimental and numerical results for the mean sway force of S-Cb84 and a bulk carrier in beam waves, bow waves and stern oblique waves, at $Fr = 0.10$. It is observed that in bow and beam waves, the numerical results agree very well with experimental results. However, at this moderate speed, the numerical method significantly underpredicts the mean sway force for both ships in stern quartering waves, which indicates the limitations for the applied method (slender body theory) to account for the forward speed effect at lower frequency of encounter.

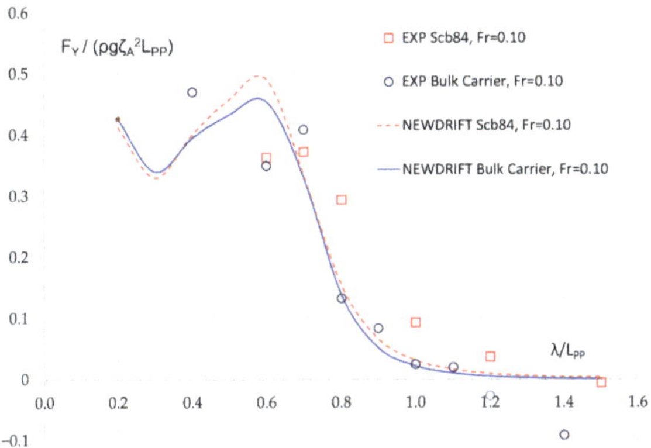

Figure 7. Mean sway force in regular beam waves, $Fr = 0.10$.

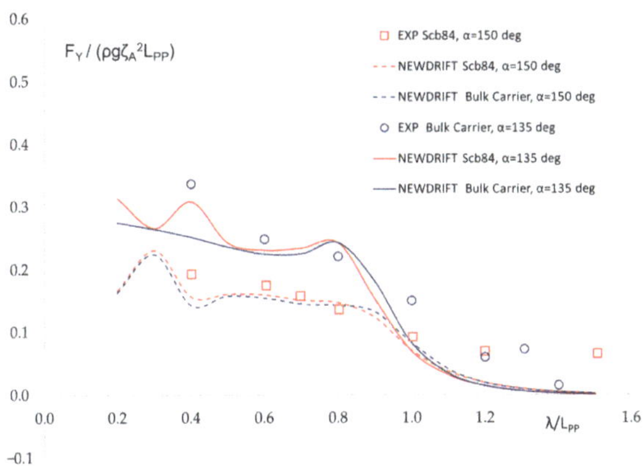

Figure 8. Mean sway force in regular bow quartering waves, $Fr = 0.10$.

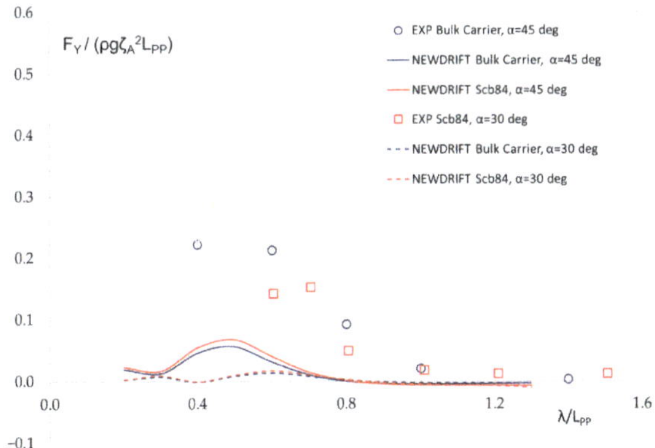

Figure 9. Mean sway force in regular stern oblique waves, $Fr = 0.10$.

5. Development of a Semi-Empirical Formula for the Side Drift Force of Full Type Ships at Low Speeds

Examining both the experimental and numerical results in Figures 2–9, we may detect a strong recurring functional pattern for the concerned mean sway force of the tested full ships at low speeds. This encourages us to attempt the development of an empirical formula to approximate the mean sway force in waves, as we did for the mean surge force or added resistance [40–42].

We start from the zero speed case, namely, to consider the mean sway force on the VLCCb′ in beam waves. As shown in Figure 10, the black curve shows the numerical results predicted by NEWDRIFT and it agrees reasonably well with the experimental results. Assuming that the total mean sway force consists of a reflection induced part and a motion induced part, and the diffraction part can be approximated by:

$$\overline{F}_{Y,\,R}=\frac{1}{2}\rho g{\zeta_A}^2 B B_F(\alpha)R(\alpha)^2(1-e^{-2kT}) \tag{7}$$

where ρ is density of water, g the gravitational acceleration, ζ_A the incident wave amplitude, B the beam of a ship, $B_F(\alpha)$ is the bluntness coefficient of a ship in oblique waves and it is defined at zero speed as follows:

$$B_F(\alpha) = \frac{1}{B}\int_C \sin^2(\theta-\alpha)\mathrm{d}l \tag{8}$$

and $R(\alpha)$ is the generalized *reflection coefficient* given by Evans and Morris [43]. Liu [44] elaborated Equation (7) in detail.

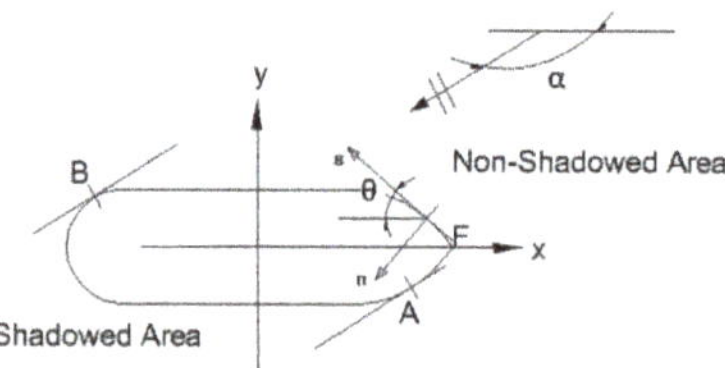

Figure 10. Coordinate system in calculating drift force due to reflection.

The red curve in Figure 11 shows the mean sway force exerting on the VLCCb′ approximated by Equation (7). Then, the shown dashed blue curve represents the difference between the total mean sway force calculated by NEWDRIFT and the approximate diffraction part of the side drift force, thus, the missing motion induced part. If the dashed blue curve can be properly represented by a formula, then an empirical formula can be formed.

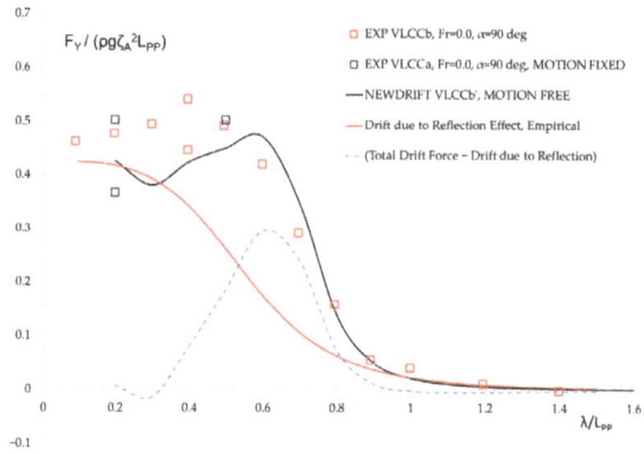

Figure 11. Various components of the mean sway force of a ship in regular beam waves, $Fr = 0.0$.

In a previous study [42], it was demonstrated that the component of the mean surge force due to motion effect $X_{AW,M}$ can be approximated as follows:

$$X_{AW,M} = 4\rho g \zeta_A{}^2 \frac{B^2}{L_{pp}} a_1 a_2 a_3 \overline{\omega}^{b_1} e^{\frac{b_1}{d_1}(1-\overline{\omega}^{d_1})} \tag{9}$$

At zero speed, the mean surge force and the mean sway force are equivalent physical quantities. Thus, the mean sway force can be approximated similarly, though the various form coefficients will need some tuning. Indeed, the following formula is deduced after some trial-and-error procedures:

$$\overline{F}_{Y,M} = \rho g \zeta_A{}^2 L_{pp} a_1 \overline{\omega}^{b_1} e^{\frac{b_1}{d_1}(1-\overline{\omega}^{d_1})} \tag{10}$$

with

$$\overline{\omega} = 2.142 \sqrt[3]{k_{yy}} \sqrt{\frac{L_{pp}}{\lambda}} \left[\left(-1.377Fr^2 + 1.157Fr\right)|\cos\alpha| + \frac{0.618(13+\cos 2\alpha)}{14} \right]$$

$$a_1 = 0.3|\sin\alpha|$$

$$b_1 = \begin{cases} 11.0 \text{ for } \omega < 1 \\ -8.5 \text{ elsewhere} \end{cases}$$

$$d_1 = \begin{cases} 566\left(\frac{L_{pp}C_B}{B}\right)^{-2.66} \text{ for } \omega < 1 \\ -566\left(\frac{L_{pp}}{B}\right)^{-2.66} \text{ elsewhere} \end{cases}$$

where λ is the length of the incident wave, C_B the block coefficient of the ship, and k_{yy} the non-dimensional radius of gyration of pitch.

Figure 12 shows the preliminary results of the approximated mean sway force of the VLCCb′ based on Equations (7) and (10) (sum of reflection and motion contributions):

$$\overline{F}_Y = \overline{F}_{Y,R} + \overline{F}_{Y,M} \tag{11}$$

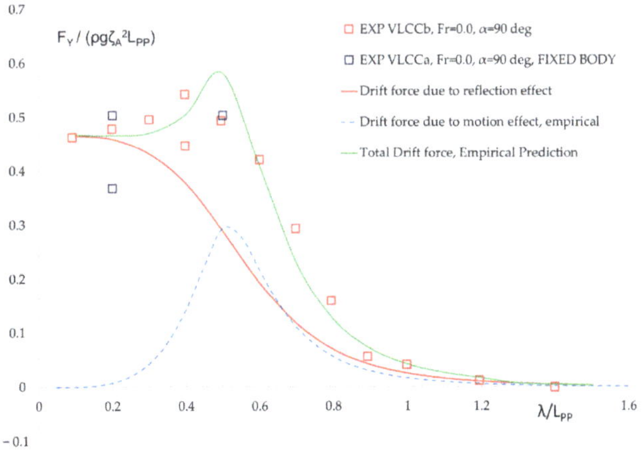

Figure 12. Empirical prediction of the mean sway force of a ship in regular head waves, $Fr = 0.0$.

The results based on Equation (11) are herein denoted as "Total drift force, empirical prediction". Obviously, the obtained results agree very well with the experimental results. It should be noted that the a_1 coefficient is presented in a rather simplified form at the moment. It can be further tuned by introducing the effect of the main ship particulars, such as, C_B, k_{yy}, $\frac{B}{T}$, Fr, etc., so that the formula can be more versatile. This will require, however, more experimental and/or numerical data, as well as details of the test hull forms.

Figure 13 shows the prediction of the mean sway force of the same ship in waves at zero speeds in waves of arbitrary headings by the proposed empirical method in comparison with experimental results.

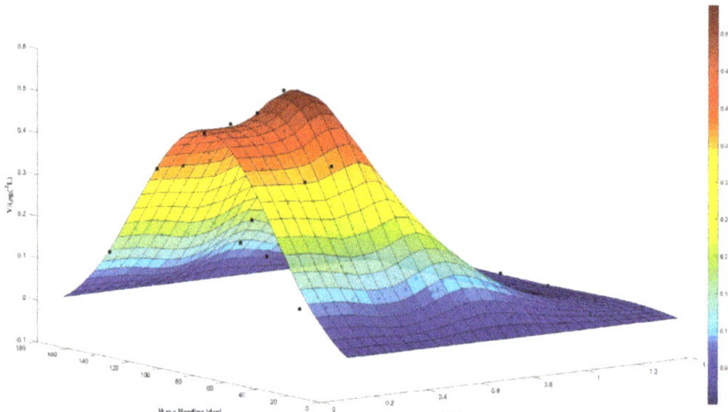

Figure 13. Empirical prediction of the mean sway force of a ship in waves, $Fr = 0.0$.

Figure 14 shows the prediction of the mean sway force for two other full ships in waves at low speeds by the proposed empirical method and model experimental values. Examining the results of VLCCb at two speeds, it is noted that the empirical results agree fairly well with the experimental data and that the forward speed effect is also properly predicted. The empirical prediction of the mean sway force for *Esso Osaka* at $Fr = 0.055$ in regular waves of $\lambda/L_{PP} = 0.3$ also shows an excellent agreement with experimental data.

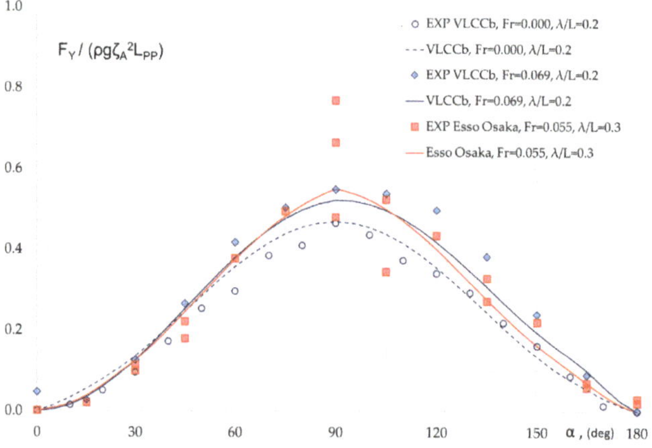

Figure 14. Prediction of the mean sway force by proposed empirical formula for two full type ships in regular waves at low speeds.

Figure 15 shows the experimental and empirical prediction of the mean sway force of a bulk carrier at a higher speed ($Fr = 0.1$) in waves of various directions. A reasonable agreement has been observed for all three wave headings. For head waves, the empirical prediction appears to be a bit higher than the experimental results in short waves. As stated, more validation and calibration of the formula can be carried out when more data is available.

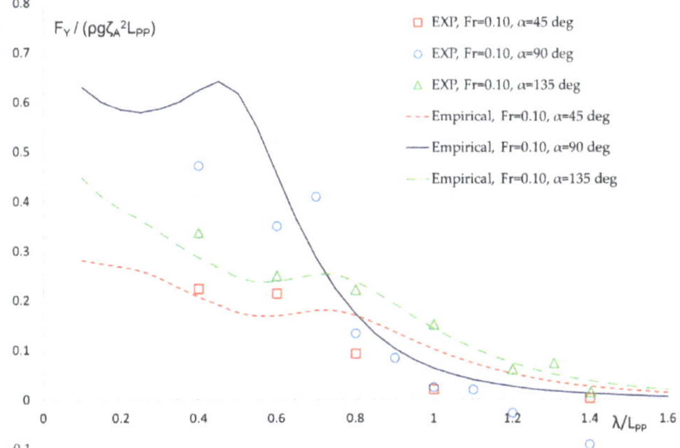

Figure 15. Empirical prediction of the mean sway force of a bulk carrier in waves, $Fr = 0.10$.

6. Summary and Conclusions

In this study, we first analyze a series of published experimental data of the mean sway wave induced forces (side drift) for six full ships at low speeds and various wave headings. A strong recurring pattern was observed in the presented results. In relatively short waves, namely $\lambda/L_{PP} < 0.3$, a spreading of the experimental data is observed. This is because the measured values are rather small, while it is also very challenging to generate stable waves of very small lengths. For this reason, in tank tests, the generated waves are rather steep for small wavelengths, which in turn, leads to complicated nonlinear wave–structure interaction phenomena and further exaggerates the scatter. The influence of steepness on the concerned quantity needs further systematic investigation.

In a second step of the presented study, a 3D panel method based on the in-house software NEWDRIFT is further validated by comparing the numerical results for the second order mean sway force in waves at low speeds with the corresponding experimental data for the tested set of six full type ships. It is noted that the mean sway force is most significant in relatively short waves, with the peak being observed at $\lambda/L_{PP} \approx 0.5$–$0.6$. For $\lambda/L_{PP} > 1.0$, the quantities are rather small. This feature is different from that of mean surge force, or the added resistance, which exhibits its non-dimensional maxima, when wavelength is comparable to ship length (close to the heave/pitch motion resonance). Overall, NEWDRIFT delivers satisfactory predictions for the mean sway force, as for the other drift force and moment components.

Finally, we proceed to the development of an empirical formula to approximate the mean sway force using only the main ship particulars and wave parameters, which will enable a wide application in various tasks. Preliminary validation results show that the empirical formula can capture satisfactorily the mean sway force acting on full type ships, at both zero and non-zero, low speeds. In future work, we plan to further develop the proposed empirical formula so that it can deal with various ship types, forward speeds and wave conditions.

Author Contributions: Conceptualization, S.L. and A.P.; methodology, S.L. and A.P.; software, S.L. and A.P.; validation, S.L.; formal analysis, S.L.; investigation, S.L.; data curation, S.L.; writing—original draft preparation, S.L.; writing—review and editing, A.P.; visualization, S.L.; supervision, A.P. All authors have read and agreed to the published version of the manuscript.

Funding: This study was conducted in the framework of the internally funded bilateral research project "Assessment of the Manoeuvering Performance of a Ship in a Seaway" of NTU and NTUA.

Conflicts of Interest: The authors declare no conflict of interest.

References

1. Havelock, T.H. The Pressure of Water Waves upon a Fixed Obstacle. *Proc. R. Soc. Lond. Ser. A Math. Phys. Sci.* **1940**, *175*, 409–421.
2. Maruo, H. The drift of a body floating on waves. *J. Ship Res.* **1960**, *4*, 1–10.
3. Newman, J.N. The drift force and moment on ships in waves. *J. Ship Res.* **1967**, *11*, 51–60.
4. Salvesen, N. Second-Order Steady State Forces and Moments on Surface Ships in Oblique Regular Waves. In *International Symposium on Dynamics of Marine Vehicles and Structures in Waves*; Mechanical Engineering Publications Limited, Univ. College: London, UK, 1974; pp. 212–226.
5. Gerritsma, J.; Beukelman, W. Analysis of the resistance increase in waves of a fast cargo ship 12. *Int. Shipbuild. Prog.* **1972**, *19*, 285–293. [CrossRef]
6. Salvesen, N.; Tuck, E.O.; Faltinsen, O. Ship motions and sea loads. *Trans. Soc. Nav. Archit. Mar. Eng.* **1970**, *78*, 250–287.
7. Mavrakos, S. The vertical drift force and pitch moment on axisymmetric bodies in regular waves. *Appl. Ocean Res.* **1988**, *10*, 207–218. [CrossRef]
8. Kashiwagi, M. Added Resistance, Wave-Induced Steady Sway Force and Yaw Moment on an Advancing Ship. *Ship Technol. Res. (Schiffstech.)* **1992**, *39*, 3–16.
9. Boese, P. Eine einfache Methode zur Berechnung der Widerstandserhöhung eines Schiffes im Seegang. *Ship Technol. Res.* **1970**, *258*, 17. [CrossRef]
10. Faltinsen, O.M.; Minsaas, K.J.; Liapis, N.; Skjordal, S.O. Prediction of Resistance and Propulsion of a Ship in a Seaway. In Proceedings of the 13th Symposium on Naval Hydrodynamics, Tokyo, Japan, 6–10 October 1980; pp. 505–529.
11. Papanikolaou, A.; Nowacki, Z. Second-Order Theory of Oscillating Cylinders in a Regular Steep Wave. In Proceedings of the 13th ONR Symp., Tokyo, Japan, 6–10 October 1980; pp. 303–333.
12. Papanikolaou, A.; Zaraphonitis, G. On an Improved Method for the Evaluation of Second-Order Motions and Loads on 3D Floating Bodies in Waves. *J. Schiffstech.-Ship Technol. Res.* **1987**, *34*, 170–211.
13. Pinkster, J. Mean and low frequency wave drifting forces on floating structures. *Ocean Eng.* **1979**, *6*, 593–615. [CrossRef]
14. Zaraphonitis, G.; Papanikolaou, A. Second-order theory and calculations of motions and loads of arbitrarily shaped 3D bodies in waves. *Mar. Struct.* **1993**, *6*, 165–185. [CrossRef]
15. Higo, Y.; Ha, M.K. A Study on Added Resistance of a Restrained Body with Forward Speed in Waves. *J. Soc. Nav. Arch. Jpn.* **1991**, *1991*, 75–83. [CrossRef]
16. Dai, Y.S.; Huang, D.B. The near-field and far-field formulae for the added resistance of ships. *J. Harbin Shipbuild. Eng. Inst.* **1994**, *15*, 1–5.
17. Papanikolaou, A.; Schellin, T.E. A Three Dimensional Panel Method for Motions and Loads of Ships with Forward Speed. *Ship Technol. Res. (Schiffstech.)* **1992**, *39*, 147–156.
18. Joncques, S.A.G. Second Order Forces and Moments Acting on Ships in Waves. Ph.D. Thesis, Technical University of Denmark, Mechanical Engineering, Lyngby, Denmark, 2009.
19. Chen, X.B. On the Irregular Frequencies Appearing in Wave Diffraction-Radiation Solutions. *Int. J. Offshore Polar Eng.* **1998**, *8*, 110–114.
20. Duan, W.; Li, C. Estimation of added resistance for large blunt ship in waves. *J. Mar. Sci. Appl.* **2013**, *12*, 1–12. [CrossRef]
21. Liu, S.; Papanikolaou, A.; Zaraphonitis, G. Practical Approach to the Added Resistance of a Ship in Short Waves. In Proceedings of the 25th International Offshore and Polar Engineering Conference, Kona, HI, USA, 21–26 June 2015; Volume 3, pp. 11–18.
22. Ohkusu, M. Added Resistance in Waves of Hull Forms with Blunt Bow. In Proceedings of the 15th Symposium on Naval Hydrodynamics, Hamburg, Germany, 2–7 September 1984; pp. 135–147.
23. Sakamoto, T.; Baba, E. Minimization of Resistance of Slowing Moving Full Hull Forms in Short Waves. In Proceedings of the 16th Symposium on Naval Hydrodynamics, Berkeley, CA, USA, 13–18 July 1986; pp. 598–612.

24. Tsujimoto, M.; Kuroda, M.; Shiraishi, K.; Ichinose, Y.; Sogihara, N. Verification on the resistance test in waves using the actual sea model basin. *J. Jpn. Soc. Nav. Archit. Ocean Eng.* **2012**, *16*, 33–39. [CrossRef]
25. Yang, K.K.; Kim, Y.; Jung, Y.W. Enhancement of Asymptotic Formula for Added Resistance of Ships in Short Waves. *Ocean Eng.* **2018**, *148*, 211–222. [CrossRef]
26. Skejic, R.; Faltinsen, O.M. A unified seakeeping and maneuvering analysis of ships in regular waves. *J. Mar. Sci. Technol.* **2008**, *13*, 371–394. [CrossRef]
27. Naito, S.; Mizoguchi, S.; Kagawa, K. Steady forces acting on ships with advance velocity in oblique waves. *J. Kansai Soc. Nav. Archit.* **1990**, *213*, 45–50.
28. Iwashita, H.; Ito, A.; Okada, T.; Ohkusu, M.; Takaki, M.; Mizoguchi, S. Wave Forces Acting on a Blunt Ship with Forward Speed in Oblique Sea. *J. Soc. Nav. Arch. Jpn.* **1992**, *1992*, 109–123. [CrossRef]
29. Ueno, M.; Nimura, T.; Miyazaki, H.; Nonaka, K. Steady Wave Forces and Moment Acting on Ships in Manoeuvring Motion in Short Waves. *J. Soc. Nav. Arch. Jpn.* **2000**, *2000*, 163–172. [CrossRef]
30. Ueno, M.; Nimura, T.; Miyazaki, H.; Nonaka, K.; Haraguchi, T. Model Experiment on Steady Wave Forces and Moment Acting on a Ship a Rest. *J. Kansai Soc. Nav. Archit.* **2001**, *235*, 69–77.
31. SHOPERA Project (2013–2016). Energy Efficient Safe Ship Operation, EU Funded FP7 Project. Grant Agreement Number 605221. Available online: http://www.shopera.org (accessed on 15 March 2020).
32. Sprenger, F.; Maron, A.; Delefortrie, G.; Van Zwijnsvoorde, T.; Cura-Hochbaum, A.; Lengwinat, A.; Papanikolaou, A. Experimental Studies on Seakeeping and Manoeuvrability in Adverse Conditions. *J. Ship Res.* **2017**, *61*, 131–152. [CrossRef]
33. Shigunov, V.; El Moctar, O.; Papanikolaou, A.; Potthoff, R.; Liu, S. International benchmark study on numerical simulation methods for prediction of manoeuvrability of ships in waves. *Ocean Eng.* **2018**, *165*, 365–385. [CrossRef]
34. Kadomatsu, K.; Inoue, Y.; Takarada, N. On the Required Minimum Output of Main Propulsion Engine for Large Fat Ship with Considering Manoeuverability in Rough Seas. *J. Soc. Nav. Arch. Jpn.* **1990**, *1990*, 171–182. [CrossRef]
35. Yasukawa, H.; Hirata, N.; Matsumoto, A.; Kuroiwa, R.; Mizokami, S. Evaluations of wave-induced steady forces and turning motion of a full hull ship in waves. *J. Mar. Sci. Technol.* **2018**, *24*, 1–15. [CrossRef]
36. Papanikolaou, A. On integral-equation-methods for the evaluation of motions and loads of arbitrary bodies in waves. *Arch. Appl. Mech.* **1985**, *55*, 17–29. [CrossRef]
37. Liu, S.; Papanikolaou, A.; Zaraphonitis, G. Prediction of added resistance of ships in waves. *Ocean Eng.* **2011**, *38*, 641–650. [CrossRef]
38. Papanikolaou, A. *Ship Design-Methodologies of Preliminary Design*; Springer: Berlin/Heidelberg, Germany, 2014; p. 628. ISBN1 978-94-017-8751-2.
39. Dafermos, G.K.; Zaraphonitis, G.N.; Papanikolaou, A.D. On an Extended Boundary Method for the Removal of Irregular Frequencies in 3D Pulsating Source Panel Methods. In Proceedings of the 18th International Congress of the Maritime Association of the Mediterranean (IMAM 2019), Varna, Bulgaria, 9–11 September 2019; pp. 53–59.
40. Liu, S.; Papanikolaou, A. Fast approach to the estimation of the added resistance of ships in head waves. *Ocean Eng.* **2016**, *112*, 211–225. [CrossRef]
41. Liu, S.; Papanikolaou, A. Approximation of the added resistance of ships with small draft or in ballast condition by empirical formula. *Proc. Inst. Mech. Eng. Part M J. Eng. Marit. Environ.* **2017**, *233*, 27–40. [CrossRef]
42. Liu, S.; Papanikolaou, A. Regression analysis of experimental data for added resistance in waves of arbitrary heading and development of a semi-empirical formula. *Ocean Eng.* **2020**, *206*, 107357. [CrossRef]
43. Evans, D.V.; Morris, C.A.N. The Effect of a Fixed Vertical Barrier on Obliquely Incident Surface Waves in Deep Water. *IMA J. Appl. Math.* **1971**, *9*, 198–204. [CrossRef]
44. Liu, S. Revisiting the influence of a ship's draft on the drift force due to diffraction effect. *Ship Technol. Res. (Schiffstech.)* **2020**. in Press.

Journal of
Marine Science and Engineering

Article

Numerical Investigation on Nonlinear Dynamic Responses of a Towed Vessel in Calm Water

Bo Woo Nam

Department of Naval Architecture and Ocean Engineering, Seoul National University, Seoul 08826, Korea; bwnam@snu.ac.kr; Tel.: +82-2-880-7324; Fax: +82-2-880-9298

Received: 28 February 2020; Accepted: 18 March 2020; Published: 20 March 2020

Abstract: In this study, we numerically investigated the nonlinear dynamic responses of an autonomous towing system where a vessel is passively towed by a tug via a towline. A three-degrees-of-freedom maneuvering mathematical model is utilized to describe the nonlinear dynamics of the towed vessel in calm sea. The hydrodynamic force acting on the towed vessel is modelled as a modular-type hull force model, which includes linear and nonlinear (third order) damping forces in sway and yawing directions. The towline force is simply modeled as a linear spring. First, the motion responses of a towing system, showing large sway-yaw coupled motions due to unstable towing characteristics, are studied by applying phase plane analysis. For the validation of the present numerical method, the simulation results are directly compared with the model test data. Then, simulation parameters, such as towing speed, initial positions and hull force coefficients, are changed and their resulting limit cycles are investigated. Finally, the effects of towline and tug motion are discussed based on the simulation results. It is found that the dynamic characteristics of the towed vessel come closer to being chaotic due to the nonlinear stiffness effect of the towline and tug motion effect.

Keywords: towing stability; towed vessel; nonlinear dynamics; towline; limit cycle

---

1. Introduction

In towing operations, vessels with no self-propulsion, such as transportation barges, can be towed by single or multiple tug systems. In this case, if the towing stability conditions are not satisfied, the towed vessel may undergo a transition from stable towing to an unstable slewing phenomenon, whereby the towed vessel experiences large sway and yaw coupled motion, such as fishtailing. Due to the fact that the unstable slewing motion of the towed vessel can result in collision with sea-going vessels or moor installations, it is very important to investigate towing stability prior to towing operations. The stability of the towed vessel is mainly influenced by parameters such as its hull geometry, towing conditions, towline and tug system [1–4].

Research on towing stability started with an analysis based on characteristics equations derived from mathematical models. Strandhagen et al. [1], Abkowitz [2], Bernitsas and Kekerdis [3] derived a characteristic equation for determining the stability of a towed vessel in calm sea. They explained that the towing stability is ensured only if two conditions are met. The first one is the restoring moment condition, which means that the towed vessel is hydrodynamically stable if the restoring moment, due to the lateral hydrodynamic force, is larger than the hydrodynamic yawing moment. The second condition is the minimum resistance condition, meaning that the towline tension corresponding to ship resistance should be larger than a critical value. In general, when the first condition is not satisfied, single or multiple rudders can be utilized as a passive control device for the towing system, in order to change the hydrodynamic characteristics of the hull. The introduction of rudders, which are normally installed at the rear of the towed vessel, may increase the restoring moments, due to the lateral hydrodynamic force when the drift angle occurs.

Practically, model tests can be performed to check the stability characteristics of the towed vessel. By using a scaled model under the dynamic similitude law, these tests can provide information about the overall behavior of the towed vessel under various conditions. Latorre [5] showed a series of model test results for determining the stability of a towed ship, discussing the scale effect. He pointed out that special care is needed for the application of model test results to real ship situations, because the model scale result gives more optimistic predictions than the real scale values. Yasukawa et al. [6] presented the results of towing experiments with single and dual barges, showing the slewing motion phenomenon of the barge. Hong et al. [7] carried out captive and towing model tests for a transportation barge, with focus on the scale and wave effects on towing stability.

Recently, the direct numerical simulation method has also been widely used to evaluate the towing performance. In this case, the numerical model is normally based on the maneuvering equation of the towed vessel. The towing simulation can give good predictions for various operation conditions, provided that the input hydrodynamic coefficients are accurately estimated. Yasukawa et al. [6] presented simulation results based on the MMG (maneuvering modeling group) model, comparing them with model test data. Fitriadhy and Yasukawa [8] and Fitriadhy et al. [9] presented towing stability results, considering the wind effects and numerical simulation results based on the MMG model. Fitriadhy and Yasukawa [10] proposed a numerical model for the course stability of a towed ship, that is coupled to a tow ship through a towline. Nam et al. [11] compared a cross-flow model [12] and the MMG model for the towing simulation of barges in calm water. They found a reasonable agreement between these models for the towing simulation. Nam et al. [13] studied the towing characteristics of a barge under various wind conditions. They applied the wind coefficients from CFD calculations. Discussions were made on the effects of the wind direction and skeg on the slewing motion. Fitriadhy et al. [14] presented a numerical simulation of the turning operation, considering the interaction effect between the tug and towed ship. They also suggested the slack towline criteria and validated the theoretical approaches qualitatively, by comparing the model test results.

In this study, nonlinear dynamic responses of a towing system are numerically investigated, based on a three degree-of-freedom maneuvering model. The hydrodynamic force acting on the towed vessel is modelled as an MMG-type hull force model, while the towline force is simply modeled as a linear spring. The dynamic motion responses of the towed vessel are analyzed by using phase plane analysis. Phase trajectories are presented to show the dynamic characteristics of the autonomous towing system. In addition, a series of numerical simulations are carried out to investigate the effect of various simulation parameters on their resulting limit cycles. Additionally, discussions are made on the effects of towline and tug motion, based on the simulation results.

2. Mathematical Model

When a vessel is passively towed by a tug via a towline, it can be exposed to the possibility of unstable slewing motion, which is a periodic fishtailing motion of the towed vessel. The towed vessel in calm sea may normally experience large horizontal motion responses of surge, sway and yaw. In this case, the slewing motion of the towed vessel has long natural periods, due to small restoring forces, so the memory effects from free-surface are considerably small. Therefore, in this study, a three-degrees-of-freedom maneuvering mathematical model is utilized to describe the nonlinear dynamics of the towed vessel in calm sea. Two different coordinate systems are introduced, shown in Figure 1; one is the space-fixed coordinate system (O-XYZ), and the other is the vessel-fixed coordinate system (o-xyz). The positive x-, y- and z-axes point in the forward, starboard and downward directions, respectively.

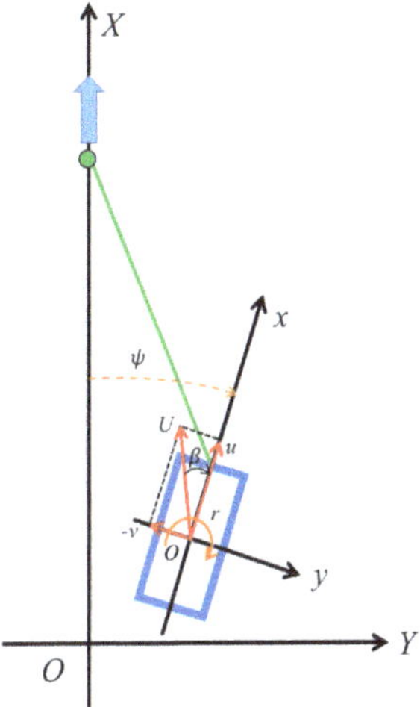

Figure 1. Coordinate system.

The state vector of this system is defined as $\mathbf{x} = \{u, v, r, x, y, \psi\}^T$. Here, u, v, r are surge velocity, sway velocity and yaw rate of the towed vessel in the vessel-fixed coordinate system, respectively. Moreover, x, y, ψ are surge displacement, sway displacement and yaw angle of the towed vessel in the space-fixed coordinate system, respectively. The dynamical system of the towed vessel can be represented by the following state equations:

$$\dot{\mathbf{x}} = \mathbf{F}(\mathbf{x}) = \{f_1(\mathbf{x}), f_2(\mathbf{x}), f_3(\mathbf{x}), \ldots\ldots, f_6(\mathbf{x})\}^T \quad (1)$$

where

$$f_1(\mathbf{x}) = \left\{-\left(m + m_y\right)vr + \frac{1}{2}\rho LdU^2\left(X_{\beta\beta}\beta^2 + X_{\beta r}\beta r + X_{rr}r^2\right) - \frac{1}{2}\rho Ldu^2 X_{uu} + F_x^T\right\}/(m + m_x) \quad (2)$$

$$f_2(\mathbf{x}) = \left\{-(m + m_x)ur + \tfrac{1}{2}\rho LdU^2\left(Y_\beta\beta + Y_r r + Y_{\beta\beta\beta}\beta^3 + Y_{\beta\beta r}\beta^2 r + Y_{\beta rr}\beta r^2 + Y_{rrr}r^3\right) + F_y^T\right\}/\left(m + m_y\right) \quad (3)$$

$$f_3(\mathbf{x}) = \left\{\frac{1}{2}\rho L^2 dU^2\left(N_\beta\beta + N_r r + N_{\beta\beta\beta}\beta^3 + N_{\beta\beta r}\beta^2 r + N_{\beta rr}\beta r^2 + N_{rrr}r^3\right) + M_z^T\right\}/(I_{zz} + J_{zz}) \quad (4)$$

$$f_4(\mathbf{x}) = u\cos\psi - v\sin\psi \quad (5)$$

$$f_5(\mathbf{x}) = u\sin\psi + v\cos\psi \quad (6)$$

$$f_6(\mathbf{x}) = r \quad (7)$$

Here, m and I_{zz} denote the mass and yaw moment of inertia of the towed vessel, respectively. The parameters m_x and m_y are added masses in surge and sway, respectively. J_{zz} is the added yaw moment of inertia; ρ is water density; L and d are the length and draft of the towed vessel, respectively; U is the total speed of the towed vessel, defined as $\sqrt{u^2 + v^2}$; β is a drift angle, which can be defined as $tan^{-1}(-v/U)$; $X_{\beta\beta}$, $X_{\beta r}$ and X_{rr}, are the hydrodynamic force coefficients regarding longitudinal force in surge direction; Y_β, Y_r, $Y_{\beta\beta\beta}$, $Y_{\beta\beta r}$, $Y_{\beta rr}$ and Y_{rrr} are the hydrodynamic force coefficients of lateral force based on a 3rd-order polynomial model [6]; similarly, N_β, N_r, $N_{\beta\beta\beta}$, $N_{\beta\beta r}$, $N_{\beta rr}$ and N_{rrr} are the hydrodynamic force coefficients of lateral force, based on the 3rd-order polynomial model. The dynamic responses of the towed vessel are obviously affected by these hydrodynamic coefficients of the hull. Finally, F_x^T, F_y^T and M_z^T are the towline forces and moment. During a normal towing operation, the axial stiffness of the towline plays a dominant role, and it can be assumed that towline behavior is in a linear regime. Thus, in this study, a simple spring model can be introduced without considering any nonlinear and damping effects in the towline. It can be concluded that the main damping and nonlinearity for the total towing system come from hydrodynamic forces acting on the hull. Due to the fact that the external towline forces are functions of the vessel displacement, but not of time, the motion equations under discussion here are nonlinear and autonomous. In this study, based on the above-mentioned mathematical model, numerical simulations were carried out for a towed barge with length (L) of 1.219 m, breadth (B) of 0.213 m and draft (d) of 0.0548 m in a 1/50 model scale, which was introduced by Yasukawa et al. [6] The principal particulars and the hydrodynamic coefficients can be found in [6].

3. Nonlinear Dynamics of a Towed Vessel

3.1. Dynamic Motion Responses of a Towing System

The first case examined here is the motion responses of a towing system, in which a vessel is tugged though a towline, showing large sway-yaw coupled motions due to unstable towing characteristics. The time histories of the towed vessel motion and towline tension are presented in Figure 2. For the validation of the present numerical method, the simulation results are directly compared with the model test data of Yasukawa et al. [6]. In this case, it is assumed that the towing operation is performed at a constant towing speed (V), following a straight path. Here, the towing speed corresponds to 0.145 as the Froude number which is defined as $\mathrm{Fn} = V/\sqrt{gL}$. No tug control effects are taken into account. Thus, the tug can be assumed as an imaginary towing point, which is constantly moving. It is to be noted that the experimental data are only available up to approximately 60 s in model scale, due to the limitation of basin size. Therefore, only 4–5 periods of slewing motion can be observed. As shown in the figures, large oscillatory yawing motion is clearly observed in terms of yaw angle and yaw rate. In this case, the slewing motion period is approximately 22 s in model scale, and the maximum yawing angle is approximately 40°. In the case of towline tension, high tension peaks occur every 11 s, half of the slewing motion period, because the towline is periodically stretched when the towed vessel changes its heading at both lateral end positions of the slewing trajectory. Due to these fishtailing motions of the towed vessel, the yaw rate and tension time histories clearly show highly nonlinear characteristics with higher harmonic components.

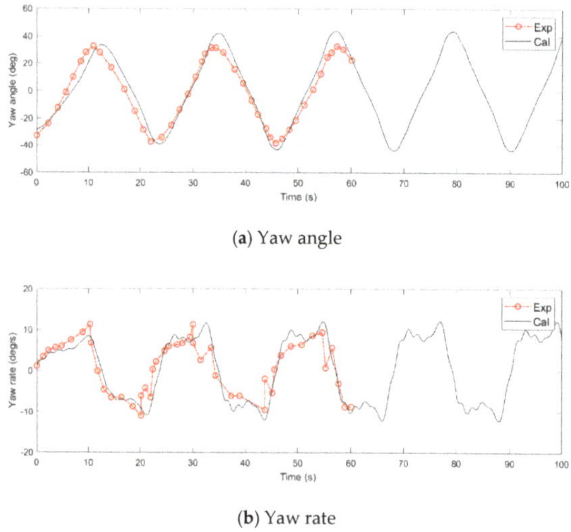

(**a**) Yaw angle

(**b**) Yaw rate

Figure 2. *Cont.*

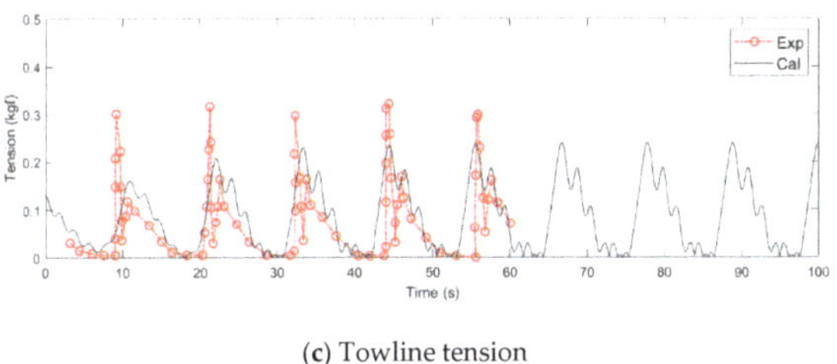

(c) Towline tension

Figure 2. Time histories of the towed vessel motion and towline tension.

To reveal the dynamic characteristics of this towing system, four phase trajectories, which are based on the velocity and motion variables of the towed vessel, are shown in Figure 3. In each figure, the blue dashed line denotes the phase trajectory from 20 to 70 s, which corresponds to the initial transient motion response of the towed vessel. The black solid line represents the phase trajectory from 70 to 500 s, where the steady periodic motions follow the closed oscillation. Therefore, this periodic phase trajectory is 'a single stable limit cycle' of the dynamic system under discussion here. It can be stated that after a short transient response depending on the initial conditions, the dynamic states of the towed vessel are quickly attracted to the limit cycle. When considering the (u, v) and (u, r) projections of the system's state space, the phase trajectories appear to have the shape of a butterfly. This is because the lateral velocity (v) and yaw rate (r) have almost twice the oscillation period of surge velocity (u). As shown in Figure 1, these surge oscillations are also directly related to the towline tension responses. It can be seen that the phase trajectories of sway-yaw velocities (v, r) and motions $(y,\ \psi)$ constitute closed circular curves, which look like skewed ellipses. This indicates that the sway and yaw motion have the same periods of slewing motion. Figure 4 shows the three-dimensional phase trajectories of the towed vessel with the velocity variables (u-v-r). In this figure, the whole dynamic structure is more clearly shown. In the three-dimensional phase space, it can be seen that no intersections are found in the limit cycle, which means that only one path is determined at a certain state on the limit cycle.

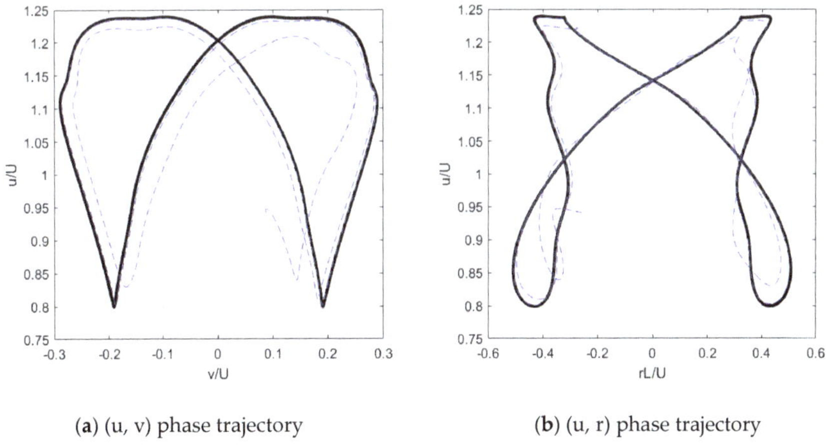

(a) (u, v) phase trajectory

(b) (u, r) phase trajectory

Figure 3. *Cont.*

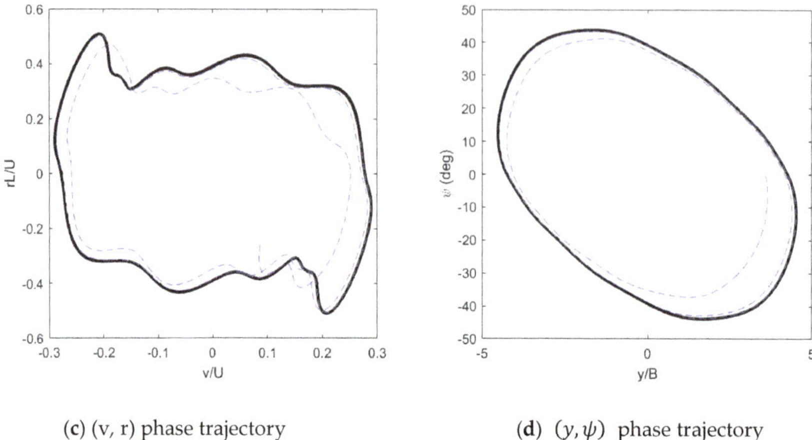

(**c**) (v, r) phase trajectory (**d**) (y, ψ) phase trajectory

Figure 3. Phase trajectories of the towed vessel (blue dashed: 20–70 s, black solid: 70–500 s).

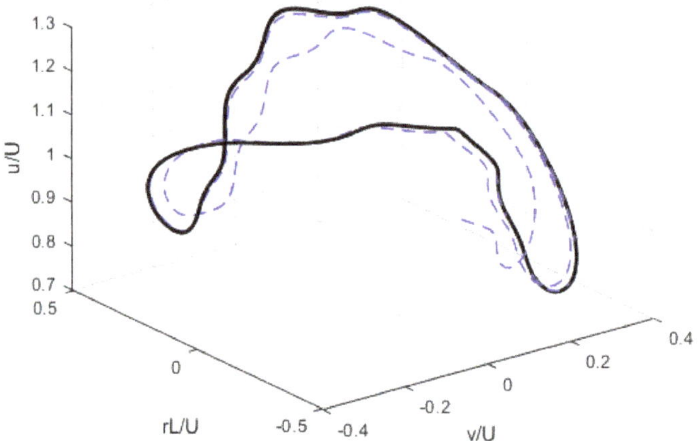

Figure 4. Phase trajectories of the towed vessel in the three-dimensional phase space (blue dashed: 20–70 s, black solid: 70–500 s).

3.2. Simulation Parameters and Limit Cycle

The dynamic characteristics of the towed vessel can be affected by various simulation parameters. Some critical parameters can change the resulting limit cycle of the towed vessel, while others mainly affect the transient response, without changing the limit cycle. For example, the transient response would be affected by the initial conditions of the towed vessel. However, the stable limit cycle of the dynamic system should be the same, regardless of the initial conditions. To check the characteristics of the limit cycle for the towing system under discussion here, various phase trajectories with different initial conditions are suggested in Figure 5. Even though the initial conditions are quite different in terms of the lateral position (y_0) and yaw angle (psi_0) of the towed vessel, it can be clearly observed that the dynamic states of the towed vessel are quickly attracted to the same limit cycle within a short time period. These results demonstrate that the dynamic system of the towed vessel has a stable limit cycle.

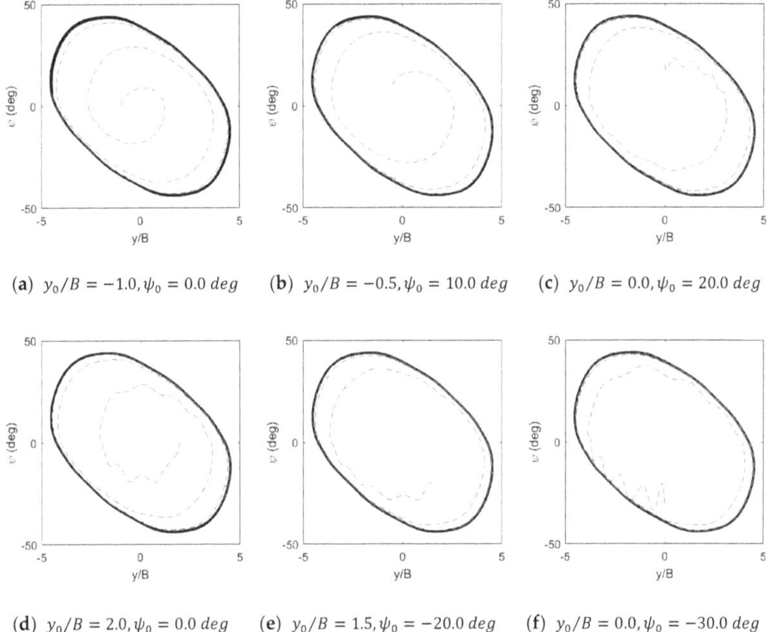

(**a**) $y_0/B = -1.0, \psi_0 = 0.0\ deg$ (**b**) $y_0/B = -0.5, \psi_0 = 10.0\ deg$ (**c**) $y_0/B = 0.0, \psi_0 = 20.0\ deg$

(**d**) $y_0/B = 2.0, \psi_0 = 0.0\ deg$ (**e**) $y_0/B = 1.5, \psi_0 = -20.0\ deg$ (**f**) $y_0/B = 0.0, \psi_0 = -30.0\ deg$

Figure 5. Phase trajectories of the towed vessel with different initial conditions (Fn $= 0.145$, $L_T/L = 1.0$).

Towing speed is another important control parameter of the towing operation, which affects the limiting cycle. Figure 6 compares three phase trajectories under different towing speed conditions. In this figure, only the limit cycles from periodic motion trajectories are plotted together. As the towing speed increases, the resistance of the towed vessel also increases, thereby increasing the towline tension. It is known that the stable towing requires sufficient towline tension [1–3]. This means that a low towing speed makes the towing dynamics more unstable. Conversely, an increase of towing speed can improve the towing characteristics. As shown in the figures, the shapes and sizes of the phase trajectories are very similar, regardless of the towing speed. Since the velocity variables are normalized by the towing speed, similar trajectory sizes mean that the velocity responses of the towed vessel are linearly increased with the towing speed. Under the non-dimensional phase space, it can be confirmed that the limit cycles are not greatly changed according to the towing speed. Therefore, the general nonlinear dynamic characteristics of the towed vessel can be discussed, with normalized phase planes.

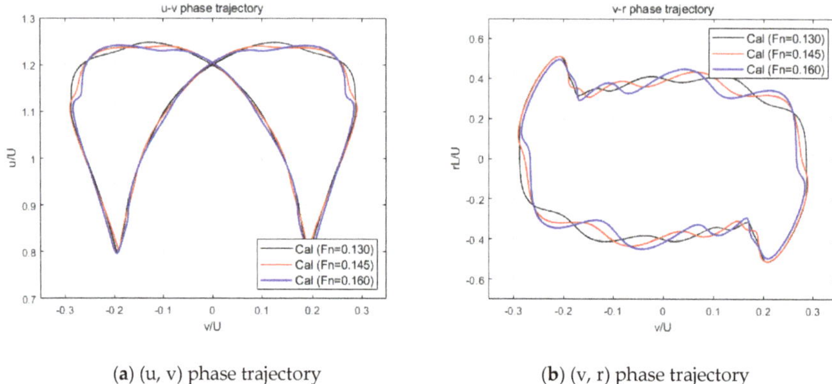

(**a**) (u, v) phase trajectory (**b**) (v, r) phase trajectory

Figure 6. Phase trajectories of the towed vessel under various towing speed conditions ($L_T/L = 1.0$).

Figure 7 shows the effects of hull hydrodynamic force coefficients on towing phase trajectories. In the first row, the simulation results with variations of linear yawing moment coefficient (N_β) are presented. It is known that towing stability is ensured when the restoring moment due to lateral force is bigger than yawing moment [1–3]. Therefore, as the yawing moment decreases, better towing stability may be expected. As shown in the figures, the phase trajectories become smaller with the decrease of the yawing moment. In the second row of Figure 7, the simulation results with resistance (X_{uu}) variation are presented. The resistances are increased by 20% and 40%. As the resistance increases, the towline tension becomes stronger, improving towing stability. Thus, it can be observed that the increase of resistance makes the phase trajectories smaller. Figure 8 shows the effect of numerical models on the dynamics of the towed vessel. Here, two numerical models are taken into account. The original nonlinear MMG model includes higher-order damping force terms while the linearized MMG model only includes linear damping terms for hull force modeling. As shown in the figures, as the linear model is used, the phase trajectories become slightly larger than those of the nonlinear model. This confirms that the linear model generates lower damping forces, so that the motion responses of the towed vessel become larger.

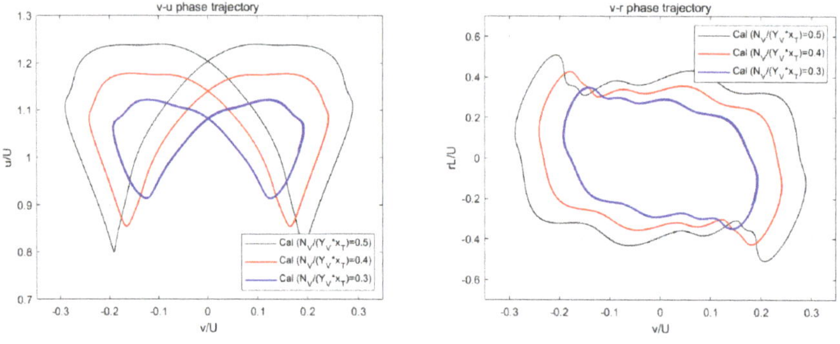

(**a**) Yawing moment (N_β) variation

Figure 7. *Cont.*

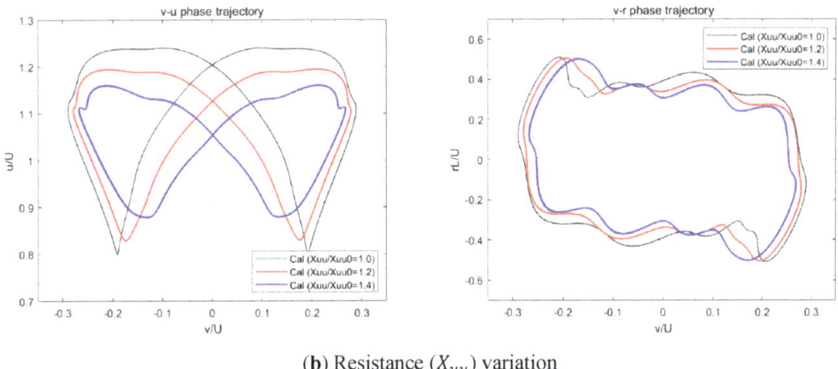

(**b**) Resistance (X_{uu}) variation

Figure 7. Phase trajectories of the towed vessel with hull force variations ($L_T/L = 1.0$).

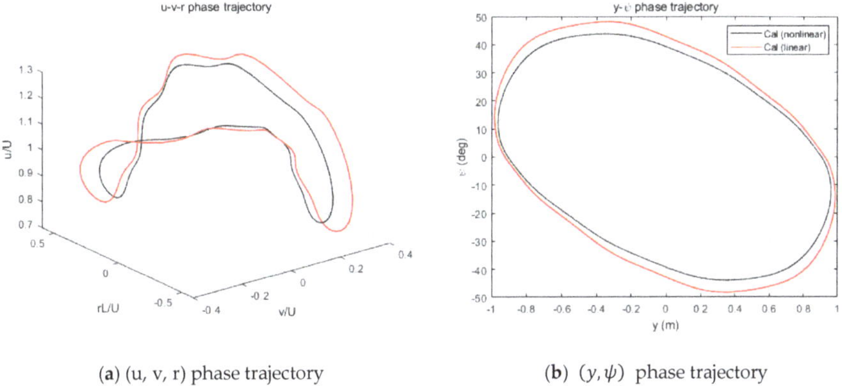

(**a**) (u, v, r) phase trajectory

(**b**) (y, ψ) phase trajectory

Figure 8. Effect of nonlinear hydrodynamic force model on phase trajectories of the towed vessel.

3.3. Effect of Towline and Tug Motion

For the autonomous towing system discussed here, the towing forces can be delivered via towline to the towed vessel. Therefore, characteristics such as towline length (L_T), stiffness (k), and configurations are also important parameters which directly affect the towed vessel dynamics. In particular, the towline length is related to the restoring forces in the sway direction, which has an analogy with a typical pendulum system. If the towline length increases, the restoring force in sway direction is reduced, thereby increasing the slewing motion period. Figure 9 shows the phase trajectories of towed vessels with three different towline lengths. It can be seen that as the towline length increases, the overall velocity phase trajectories become smaller with similar shapes. However, as shown in the phase trajectories of (y, ψ), the lateral distance of the sway motion significantly increases with the increase of towline length. For example, when the towline length is the same as the vessel length, the maximum lateral position is about 4.5 times that the vessel breadth. However, when a longer towline, such as one measuring three times the vessel length, is applied, the maximum lateral position increases up to about 6.5 times the vessel breadth. Interestingly, the heading angle is limited to approximately 45° in this case, and the yaw rate tends to decrease due to the large slewing motion in lateral direction, as shown in the (v, r) phase trajectory.

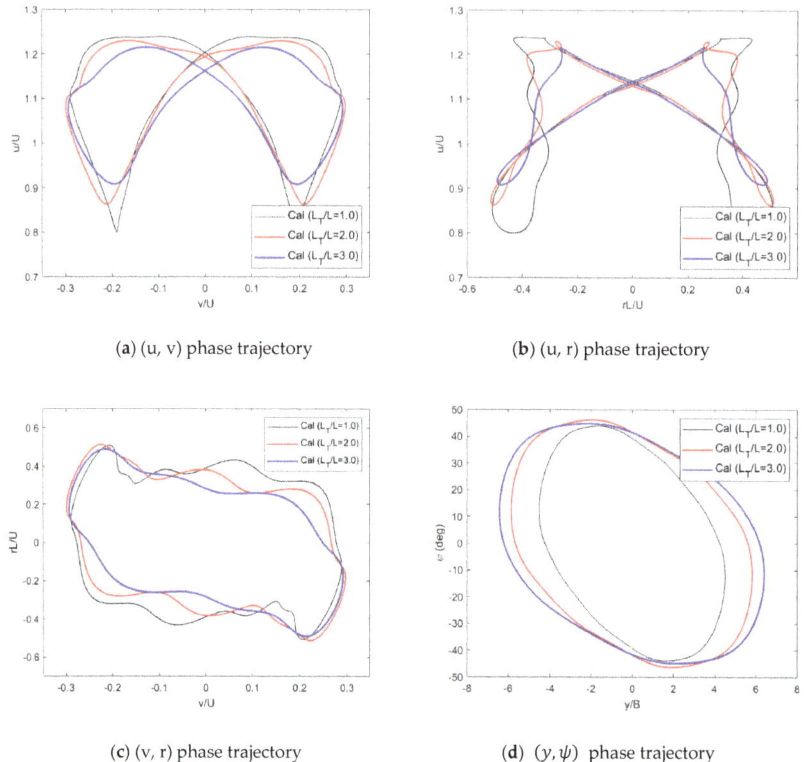

(a) (u, v) phase trajectory (b) (u, r) phase trajectory

(c) (v, r) phase trajectory (d) (y, ψ) phase trajectory

Figure 9. Phase trajectories of the towed vessel under various towline conditions (Fn = 0.145).

In real towing operation, towline has a nonlinear stiffness with horizontal catenary configuration. For simplicity, in this study, to check the nonlinear effect of the towline, bilinear towline models were used. Figure 10 compares the dynamic responses of the towed vessel when different bilinear towline models shown in Equation (8) are applied. Here, two spring constants, k_1 and k_2 denote the spring constants, when the towline (L_T) is stretched and slacked, compared to the initial unstretched length (L_{T0}), respectively.

$$k = \begin{cases} k_1 \ if \ L_T \geq L_{T0} \\ k_2 \ if \ L_T < L_{T0} \end{cases} \tag{8}$$

It is expected that the towline spring will decrease significantly as the towline is slacked. As shown in present numerical simulations, until the stiffness ratio of the towline is bigger than 10%, the limiting cycle is retained, showing almost a single trajectory. However, as the stiffness ratio of the towline becomes about 5%, the trajectories become thicker, meaning that the limiting cycle has a long period. When the stiffness ratio is 0.01, very complex trajectories can be found. This means that the dynamic characteristics of the towed vessel come closer to being chaotic, due to the nonlinear stiffness effect of the towline.

Up to now, we have neglected the effect of tug motion on the towed vessel dynamics. However, in real situations, the towed vessel and tug motions are fully coupled in the total towing system. In other words, the tug motion can also affect the dynamics of the towing system. In this section, in order to check the tug motion effect on the dynamic characteristics of the towing system, additional simulations, including prescribed tug motion, are discussed. Figure 11 shows the time series of the towed vessel motion considering tug longitudinal disturbances. In this case, the tug motion disturbances are assumed to be harmonic motion responses, by applying Equation (9), and the tug motion amplitude (A_{tug}) is assumed as 5% of the towed vessel length.

$$x_{tug} = V_{tow}t + A_{tug}\cos\left(\omega_{tug}t\right) \tag{9}$$

As shown in the figure, the towed vessel exhibits more irregular slewing motion response, due to the tug motion disturbance. In particular, at high frequency condition ($\omega_{tug}(L/g)^{1/2} = 0.25$), it can be observed that not only slewing motion amplitudes, but also oscillation periods, become more irregular. Figure 12 compares the phase trajectories under the various tug motion frequency conditions. Owing to the tug motion effect, two distinct features are found when comparing with the phase trajectories of no tug motion. First, asymmetric patterns are found, especially when the nondimensionalized oscillation periods ($\omega_{tug}(L/g)^{1/2}$) are 0.10, 0.15 and 0.30. The other feature is the limit cycle of multiple lines, which means that the limiting cycle has much longer periods, especially when the nondimensionalized oscillation periods ($\omega_{tug}(L/g)^{1/2}$) are 0.15 and 0.25.

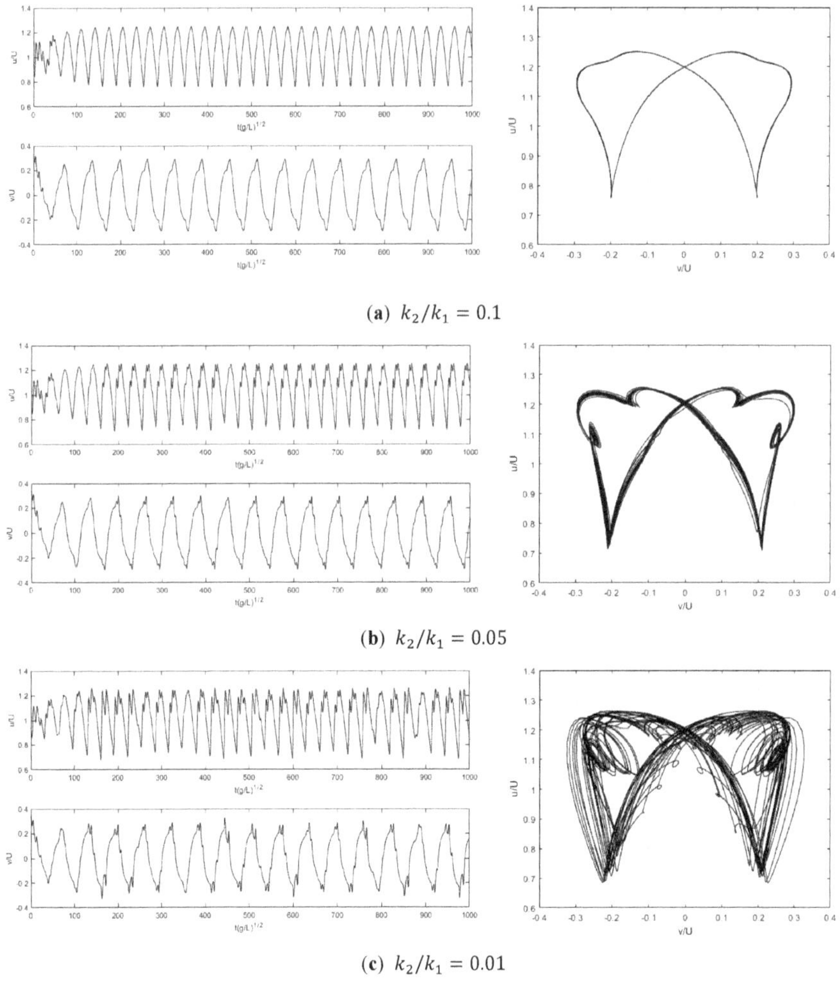

Figure 10. Time series and phase trajectories of the towed vessel with different bilinear towline models.

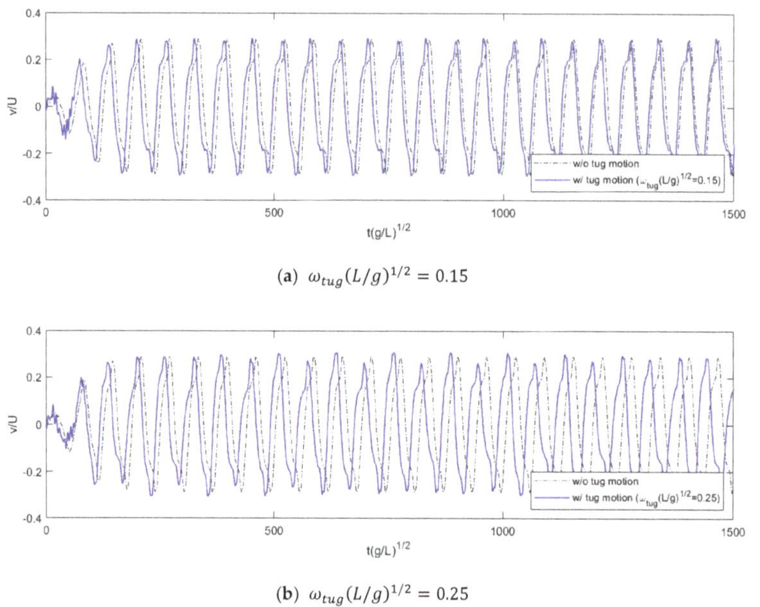

(a) $\omega_{tug}(L/g)^{1/2} = 0.15$

(b) $\omega_{tug}(L/g)^{1/2} = 0.25$

Figure 11. Time series of the towed vessel motion with tug longitudinal motions ($A_{tug}/L = 0.05$).

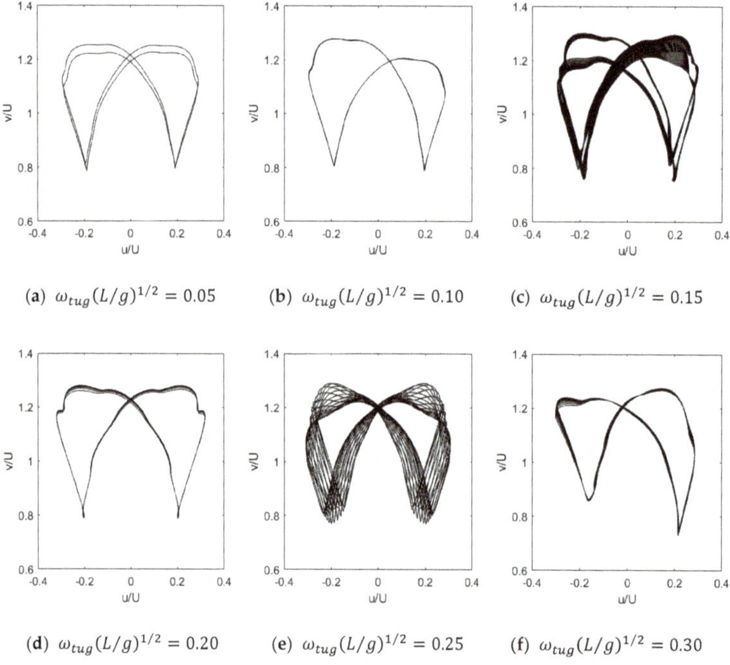

(a) $\omega_{tug}(L/g)^{1/2} = 0.05$ (b) $\omega_{tug}(L/g)^{1/2} = 0.10$ (c) $\omega_{tug}(L/g)^{1/2} = 0.15$

(d) $\omega_{tug}(L/g)^{1/2} = 0.20$ (e) $\omega_{tug}(L/g)^{1/2} = 0.25$ (f) $\omega_{tug}(L/g)^{1/2} = 0.30$

Figure 12. Phase trajectories of the towed vessel with considering tug longitudinal motions ($A_{tug}/L = 0.05$).

4. Conclusions

In this study, various numerical studies were carried out to investigate the nonlinear dynamic characteristics of an autonomous towing system. The following is a summary of the study, the findings, and the above discussions:

(1) The present simulation results with phase plane analysis demonstrate that the dynamic system of the towed vessel has a stable limit cycle. It can be clearly observed that the dynamic states of the towed vessel are quickly attracted to the same limit cycle within a short time of transient response. The limit cycle shows that the lateral velocity (v) and yaw rate (r) have almost twice the oscillation period of surge velocity (u). In the case of towline tension, high tension peaks occur when the towed vessel changes its heading, at both lateral end positions of the slewing trajectory.

(2) The towing phase trajectories are significantly affected by hull hydrodynamic force coefficients, i.e. hull force models, especially yawing moment and fluid resistance. It can be observed that the phase trajectories become smaller with the decrease of the yawing moment, while the increase of resistance makes the phase trajectories smaller. Regarding the towing speed, it can be confirmed that the limit cycles are not greatly changed under the non-dimensional phase space.

(3) The effects of towline and tug motion are discussed based on simulation results. It is found that the dynamic characteristics of the towed vessel come closer to being chaotic due to the nonlinear stiffness effect of the towline, as well as the tug motion effect.

Funding: This work was supported by the New Faculty Startup Fund from Seoul National University.

Conflicts of Interest: The authors declare no conflict of interest.

References

1. Strandhagen, A.; Schoenherr, K.; Kobayashi, F. The stability on course of towed ship. *Trans. SNAME* **1950**, *58*, 32–46.
2. Abkowitz, M.A. *Stability and Motion Control of Ocean Vehicles*; The MIT Press: Cambridge, MA, USA, 1972.
3. Bernitsas, M.; Kekridis, N. Simulation and stability of ship towing. *Int. Shipbuild. Prog.* **1985**, *32*, 112–123. [CrossRef]
4. Hancox, M. *Towing*; Clarkson Research Services Ltd.: London, UK, 2011; Available online: https://www.cargotec.com/4a27af/globalassets/files/investors/presentations/cmd-2011-ships-presentation.pdf (accessed on 20 March 2020).
5. Latorre, R. Scale effect in towed barge course stability tests. *Ocean Eng.* **1988**, *15*, 305–317. [CrossRef]
6. Yasukawa, H.; Hirata, N.; Nakamura, N.; Matsumoto, Y. Simulations of Slewing Motion of a Towed Ship. *J. Jpn. Soc. Nav. Arch. Ocean Eng.* **2006**, *4*, 137–146.
7. Hong, S.Y.; Nam, B.W.; Kim, J.H.; Park, J.Y. A Study on Towing Characteristics of a Transportation Barge in Waves. In Proceedings of the 23th International Offshore and Polar Engineering Conference, Anchorage, AK, USA, 30 June–4 July 2013; Volume 1, pp. 777–783.
8. Fitriadhy, A.; Yasukawa, H. Slewing Motion Characteristics of a Towed Ship in Steady Wind. In Proceedings of the 29th International Conference on Ocean, Offshore and Arctic Engineering, Shanghai, China, 6–11 June 2010; pp. 411–418.
9. Fitriadhy, A.; Yasukawa, H.; Koh, K.K. Course stability of a ship towing system in wind. *Ocean Eng.* **2013**, *64*, 135–145. [CrossRef]
10. Fitriadhy, A.; Yasukawa, H. Course Stability of a Ship Towing System. *Ship Technol. Res.* **2011**, *58*, 4–23. [CrossRef]
11. Nam, B.W.; Park, J.Y.; Hong, S.Y.; Sung, H.G.; Kim, J.-W. Numerical Simulation of Towing Stability of Barges in Calm Water. *J. Ocean Eng. Technol.* **2013**, *27*, 67–73. [CrossRef]
12. Wichers, J.E.W. *A Simulation Model for a Single Point Moored Tanker*; Maritime Research Institute Netherlands Publication, No. 797: Wageningen, The Netherlands, 1988. Available online: https://pdfs.semanticscholar.org/6b2e/81cf7a443caa12680375765181b0710db014.pdf (accessed on 20 March 2020).

13. Nam, B.W.; Choi, Y.-M.; Hong, S.Y. A Study on Towing Characteristics of Barge Considering Wind Force. *J. Ocean Eng. Technol.* **2015**, *29*, 283–290. [CrossRef]
14. Fitriadhy, A.; Yasukawa, H.; Maimun, A. Theoretical and experimental analysis of a slack towline motion on tug-towed ship during turning. *Ocean Eng.* **2015**, *99*, 95–106. [CrossRef]

Article

A Comparison of Numerical Simulations and Model Experiments on Parametric Roll in Irregular Seas

Geert Kapsenberg [1,*], Clève Wandji [2], Bulent Duz [1] and Sungeun (Peter) Kim [3]

1. Maritime Research Institute Netherlands (MARIN), 6708PM Wageningen, The Netherlands; b.duz@marin.nl
2. Bureau Veritas (BV), 92937 Paris La Défense CEDEX, France; cleve.wandji@bureauveritas.com
3. American Bureau of Shipping (ABS), Spring, TX 77389, USA; sukim@eagle.org

* Correspondence: g.k.kapsenberg@marin.nl; Tel.: +31-317-493-911

Received: 29 May 2020; Accepted: 22 June 2020; Published: 27 June 2020

Abstract: The recently finalised Second Generation Intact Stability Criteria (SGISC), produced by the International Maritime Organisation (IMO), contain a level 3 assessment, the so-called Direct Stability Assessment (DSA). This assessment can be carried out using either model experiments or simulations. The fact that such a choice is given implies that the methods are equivalent in accuracy. This assumption has been verified, for one case, by the Cooperative Research Ships (CRS) community. The verification was based on new model experiments and calculated results, using four different programs owned by different CRS members. Results of the verification of the parametric roll failure mode in regular waves were published before, but this study concerns results in irregular seas. The experimental and numerical results are compared in both probabilistic and deterministic manners. The probabilistic comparison showed that the simulation programs considered are sometimes conservative and sometimes non-conservative in the prediction of the probability of an extreme value. The deterministic comparison in head seas showed that parametric roll events were predicted in the simulations in a wave train that showed no sign of important roll events in the measurement. The deterministic comparison in the following seas, on the other hand, showed an accurate fit of experimental and numerical results. It is suggested that predictions could possibly be improved by adding non-linear diffraction forces to the numerical model.

Keywords: parametric roll; numerical simulations; direct stability assessment; statistical comparison; deterministic validation

---

Nomenclature

Symbol	Unit	Description
$A_{\varphi\varphi}$	ton.m^2	Roll added moment of inertia
B	m	Beam
B_1	kNms	Linear component of roll damping
B_2	kNms^2	Quadratic component of roll damping
B_3	kNms^3	Cubic component or roll damping
B_{CR}	kNms	Critical roll damping
$C_{\varphi\varphi}$	kNm	Roll restoring moment
Fn	-	Froude number, $Fn = v/\sqrt{gL}$
F4	kNm	4th component of force (roll moment)
F4a	kNm	Amplitude of roll moment

Symbol	Unit	Description
g	m/s^2	Acceleration due to gravity
GM	m	Transverse metacentric height
Hs	m	Significant wave height
$I_{\varphi\varphi}$	$ton.m^2$	Roll moment of inertia
KG	m	Height Centre of Gravity (CoG) above keel
Lpp	m	Length between perpendiculars
t	s	time
$t_{P\text{-}ENV}$	s	Time instant of the peak of a parametric roll event in a roll angle signal
T	m	Draft
Tp	s	Peak period of the wave spectrum
T_{φ}	s	Roll natural period
k_{XX}	m	Roll gyradius
k_{XX} *	m	Roll gyradius including added mass
k_{YY}	m	Pitch gyradius
p	-	Linear roll damping coefficient
q	-	Quadratic roll damping coefficient
r	-	Cubic roll damping coefficient
Vs	kn	Ship speed
Vs-av	kn	Average ship speed in a sea state
γ	-	Peakedness parameter of Jonswap wave spectrum
Δ	ton	Displacement
ε	rad	Phase angle
ζ_a	m	Wave amplitude
φ	rad	Roll angle
φ_a	rad	Roll angle amplitude
ω_0	rad/s	Earth fixed wave frequency
ω_e	rad/s	Wave-encounter frequency

1. Introduction

As the Second Generation Intact Stability Criteria (SGISC) are now in the final phase, it is now the appropriate time to verify if existing simulation tools are indeed ready for a Level 3 assessment, also called Direct Stability Assessment. A number of basis criteria have been defined by the International Maritem Organization (IMO) Intact Stability Correspondence Group [1], and finalised during a session of the IMO sub-committee on Ship Design and Construction (SDC-7) [2], but these criteria do not guarantee the certain accuracy of the simulations versus the results of experiments. Work has been done in the Cooperative Research Ships (CRS) [3] consortium, which has focused on three out of five stability failure modes: parametric roll, loss of stability and dead ship. This paper focusses on the results for parametric roll in irregular seas. Results in regular waves have been presented before [4]. As in the earlier publication, the results of different simulation programs have been compared to the results of experiments.

Amongst the identified dynamic stability failure mode, parametric roll is probably the most emblematic and the most well-known. Parametric rolling (a parametric resonance phenomenon) is an amplification of roll motions caused by the periodic variation of transverse stability in waves [4]. It is known that parametric roll is mostly expected to occur in head, following, bow and stern quartering seas, when the wave-encounter frequency is approximately twice the natural roll frequency, and the stability variations are large. The consequence of this last requirement is that the wave-length is in the order of the ship length. Additional to this, the roll damping of the ship should be low, so that it is insufficient to dissipate the additional energy accumulated because of the parametric resonance. Since roll damping increases significantly with speed, the phenomenon is more likely to occur at low speeds.

This work adds to existing benchmark cases, like those published by France et al. [5], Spanos and Papanikolaou [6], and Reed [7]. The added value of this work is the effort put into accurately determining roll damping, for larger amplitudes as well, and in the availability of results concerning both regular waves and irregular seas. In particular, the experimental work done to determine the roll damping is valuable, in our opinion.

The prediction of roll damping is notably problematic. Estimation formulas are usually based on the work by Ikeda and colleagues [8–13]. This method has known shortcomings, but it is the best method available. More recently, CFD has been used to estimate the roll damping, and this method has been shown to give good results [14]. It is noted here that an inaccurate estimation of the roll damping dominates the results from further simulations on most of the stability failure modes defined by the IMO. This problem has been circumvented in this publication by using the experimentally determined roll damping in the numerical simulations. Roll decay simulations have been carried out to check on double counts.

The vessel used in this study is the KCS hull form. This vessel has already been used in several parametric roll studies [15–17]. Yu et al. [15,16] worked on the sensitivity of the KCS to parametric roll in regular waves for a large number of speeds. Experiments were carried out and analysed to test an early detection algorithm for parametric roll. Interestingly, they showed that parametric roll occurred mainly for speeds $0.05 < Fn < 0.13$, but this depends largely on the chosen natural roll period (which was 21.6 s for the full-scale vessel). Ruiz et al. [17] studied parametric roll in shallow water by means of experiments. They checked the ABS prediction formula [18] and found it to be conservative; it incorrectly predicted parametric roll in 3 out of 12 cases, and 9 predictions were correct. The 3 degree of freedom (DoF) numerical model developed by Neves et al. [19] performed slightly better, with 2 "false positives" and 10 correct predictions.

The focus of this work is on a comparison of the simulation programs "as is"; no effort has been put into improving the numerical models implemented in the various simulation programs. However, suggestions for improvements are made in this publication.

2. Experiments

The hull form of the KCS is fully specified on the SIMMAN2008 website [20]. A wooden model has been built at a scale ratio of 37.89 according to these lines. This model was equipped with bilge keels, height—0.40 m, length—68.82 m in full scale (St 6–14), and a rudder (span—9.90 m, mean chord—5.54 m in full scale). The main dimensions of the ship and the loading conditions are given in Table 1. Two loading conditions (LC) have been used: LC-1 for the experiments and simulations in head seas, LC-12 for the following seas cases.

Table 1. Main dimensions and loading conditions of the KCS for the parametric roll experiments. LC-1 has been used for experiments in head seas, LC-2 for experiments in following seas.

Parameter	Symbol	LC-1	LC-12	Units
Length perp.	Lpp	230.00		m
Beam	B	32.20		m
Draft	T	10.80		m
Depth	D	19.00		m
Displacement	Δ	53,389		ton
Vertical CoG	KG	13.67	14.34	m
Metacentric height	GM	1.22	0.60	m
Roll nat. period	T_{φ}	23.6	34.1	s
Roll gyradius	k_{XX}	11.90		m
Pitch gyradius	k_{YY}	57.50		m
Yaw gyradius	k_{ZZ}	57.50		m

The experiments were carried out in the Seakeeping and Manoeuvring Basin of MARIN (170 × 40 × 5 m), with a free running model; only thin electric cables connected the model to the carriage. The model maintained heading using an autopilot that controlled the rudder. The duration of the experiments in each of the presented conditions amounted to 3 h. To realise this, a long time trace for the wave generator was made. The actual experiments with the model were done in parts, due to the limited length of the basin. Subsequent runs used subsequent parts of the 3-h wave signal. The length of the individual runs was typically 5300 m (full scale value).

The wave spectra generated for the experiments were based on the JONSWAP formulation [21] for the distribution of the energy over the wave frequency. These spectra are characterised by the period of the peak of the spectrum Tp, the significant wave height Hs and the peak enhancement factor γ. The first two parameters were varied for the different tests, but the peak enhancement factor was fixed at $\gamma = 3.3$. An overview of the conditions in which the tests were done is presented in paragraph 6.

A deterministic comparison requires reliable information on the incoming wave, that is, without disturbances from the model. For experiments in head seas, the wave probe in front of the model can be used for this purpose, but the wave probes before and aft of the model measure the presence of the model in the following seas. Therefore, a few runs were selected for a re-measurement of the wave without the model being present. These are the runs that have been used in the deterministic comparison.

3. Numerical Methods

Four different simulation programs have been used in this paper; two owned by different class societies, and two by different research organisations. The programs have identical basics: the hydrodynamics are calculated by a linear potential flow theory, and the linear restoring and excitation due to the incoming wave are replaced by non-linear Froude–Krylov and restoring forces. Other characteristics of the programs have been summarised in Table 2.

Table 2. Characteristics of the simulation programs (Sim 1–4) used in this study.

Characteristic	Sim-1	Sim-2	Sim-3	Sim-4
Wave model	L	L	L	L
DoF	6	6	6	6
Speed control	S	C	C	C
Hydrodynamics	R	ZG	S	ZG
Rel. motion	I	I	I	I
Pressure for z > 0	H	H	HW	HW
Pressure integration	M	M	M	M
Course control	SD	F	R	R

All programs have used linear waves (L) and assumed 6 Degrees of Freedom (DoF). For all simulations, the speed was kept constant (C), or first order surge motions were allowed by means of a soft spring system (S). The hydrodynamic models used Rankine source panels (R), zero speed Green functions with an encounter frequency correction (ZG) or strip theory (S). To determine the wetted surface, the relative motion was based on ship motions and the incoming wave only (I). The pressure above the calm water surface ($z > 0$) is usually determined by the hydrostatic pressure (H). In two cases (Sim 3, 4) Wheeler stretching [22] was added to the dynamic pressure component (HW). All programs used a mesh (M) for the pressure integration. Course control was realised by springs and dampers (SD), by freezing the yaw degree of freedom (F), or by a rudder controlled via an autopilot (R).

A critical aspect is normally the determination of the roll damping. An estimate is usually based on Ikeda's method; nowadays, CFD is also being used. One program can use a translation of the Ikeda method to the time domain to better capture non-linear effects. For the work presented in this article, the results of experiments were used in a second or third order damping model by all simulation programs.

Simulations in the specified sea states were carried out for all four programs; statistical results are compared in further paragraphs. The additional work to perform the deterministic comparison was only done using programs Sim-2 and Sim-3.

4. Results of Roll Decay and Forced Roll Tests

Quite some effort was spent on measuring the roll damping, since this is a critical parameter in parametric roll predictions, and, in fact, in most of the SGISC failure modes. Roll decay experiments were carried out at different speeds and different initial angles, repeat experiments were done for critical cases, and forced roll experiments were done. This latter experiment was carried out by fitting an electrical motor with a flywheel inside the model. This motor was mounted on a 6 DoF force balance. The motor had a rotation axis in the longitudinal direction of the model, and was forced in a harmonically changing rotation rate. The rotational acceleration of the flywheel provided the roll moment. Experiments were done with various amplitudes, all at the natural roll frequency for the two loading conditions.

Roll decay tests were performed for different initial angles (6°, 12° and 15°), and several repeat tests were done. They were analysed using a fitting procedure for a third order, 1 DoF roll damping model [Equation (1)], an extension of the second order model proposed by Lewandowski [23]. Note that the restoring moment in Equation (1) is defined by just the linear (hydrostatic) coefficient.

The forced roll tests were performed with different values for the roll moment, all at the roll resonance frequency. The experiments were analysed using the roll moment measured by the 6 DoF force balance, and using the phase angle between the roll motion and the moment produced by the motor.

$$\left(I_{\varphi\varphi} + A_{\varphi\varphi}\right)\ddot{\varphi} + B_1\dot{\varphi} + B_2\dot{\varphi}\left|\dot{\varphi}\right| + B_3\dot{\varphi}^3 + C_{\varphi\varphi}\varphi = 0 \quad (1)$$

The damping parameters B_1, B_2 and B_3 are expressed in non-dimensional coefficients p, q and r. These coefficients are defined in Equation (2). These definitions make use of the critical roll damping B_{CR} that is defined in Equation (3). Values for the non-dimensional damping coefficients as derived from the experiments are listed in Table 3. Note that, for some conditions, the coefficients for a second order as well as for a third order model are given.

$$B_1 = \frac{p}{2\pi}B_{CR},\ B_2 = \frac{3qT_\varphi}{32\pi}B_{CR},\ B_3 = \frac{rT_\varphi{}^2}{6\pi^2}B_{CR} \quad (2)$$

$$B_{CR} = 2\Delta k_{XX}\sqrt{gGM} \quad (3)$$

Table 3. Choice of p, q and r coefficients for the simulations.

Vs [kn]	Loading Condition	T_φ [s]	B_{CR} [kNms]	p [-]	q [1/°]	r [1/°²]
0	LC-1	23.6	4.77×10^6	0	2.35×10^{-2}	0
				5.00e−02	0	8.5×10^{-4}
8		23.0	4.69×10^6	1.07e−01	2.30×10^{-2}	0
				2.40e−01	0	7.0×10^{-4}
8	LC-12	32.7	3.33×10^6	9.40e−02	2.50×10^{-2}	0

The results of the roll damping experiments at Vs = 8 kn are shown in Figure 1. This figure shows the results of the roll decay tests up to a maximum roll angle of 15°, and the results of forced oscillation tests in the range 12° < φ < 29°. The two methods give consistent results. It was concluded that the forced roll experiment is a very suitable method for determining the damping at large amplitudes.

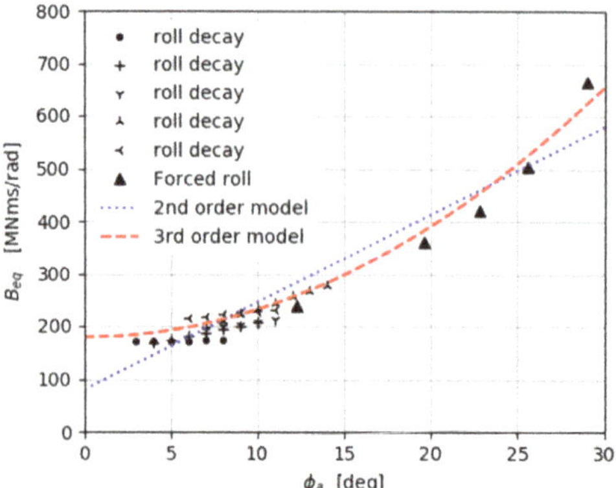

Figure 1. Roll damping for LC-1 at ω_{φ} = 0.273 rad/s and vs. = 8 kn. Results of forced roll tests (full triangles) and of roll decay tests (other symbols). The best-fit quadratic model is displayed as a dashed red line, a cubic model is displayed as a blue dashed line.

Figure 1 also demonstrates a fundamental problem; it is not possible to accurately model the roll damping over the full amplitude range with just a quadratic model. The plot shows the equivalent linear damping as a function of the roll amplitude, so a quadratic roll damping model, as defined in Equation (1), is displayed as a straight line. A third order model gives a much better fit to the experimental data, as is also shown in the figure. Note that the use of either the second or the third order model has a large effect on the limit of the roll damping for $\varphi \to 0$. It was expected that the onset of parametric roll would be affected by this limit.

5. Probabilistic Analysis of Results

5.1. Analysis of Parametric Roll Events

A first analysis aimed at considering the number of parametric roll events and characterising each event by its maximum roll angle. In order to identify these events, a Hilbert transform was made of the roll signal, and this function was filtered by a low-pass filter with frequencies 0.05 and 0.10 rad/s. Signals with a frequency below the first value fully pass the filter; signals with a frequency in-between the two values are partly filtered, and signals with a frequency higher than the highest values do not pass the filter. The filtered Hilbert transform can be considered as the envelope of the roll signal, thus identifying parametric roll events. These events are further identified by the time instant of the peak of the envelope $t_{P\text{-}ENV}$. The actual peak is the peak of the roll signal in the interval defined in Equation (4). The result of this procedure is illustrated in Figure 2.

$$t_{P-ENV} - T_{\varphi} < t < t_{P-ENV} + T_{\varphi} \tag{4}$$

Using this analysis, the ensemble statistics have been made from parametric roll events classed in bins of 5°, starting at 10° roll amplitude. As an example, the results for the simulations done by program Sim-2 are given in Figure 3. The results show that convergence, in terms of the number of events/h, is slow. The figure suggests that only the last five bars show little changes, meaning that 15 simulations of 3 h each are necessary for convergence. It is concluded that results from one 3-h simulation (and hence also from the 3-h experimental results) have a large uncertainty margin.

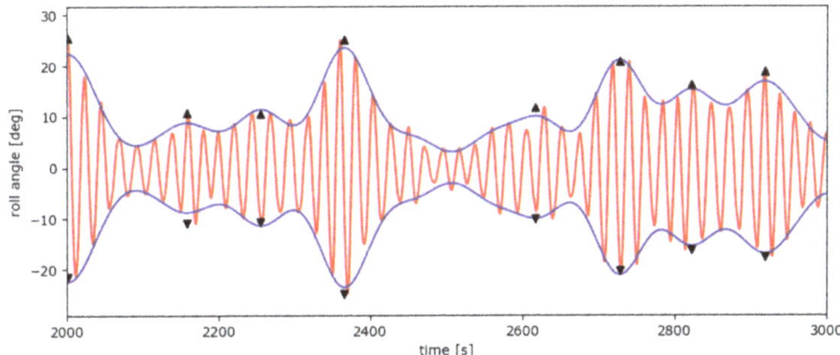

Figure 2. Visualisation of the analysis procedure of parametric roll events: plot of the roll motion (red line) and the envelope (blue lines), with the peak of each event in the interval defined by Equation (4), (black symbols).

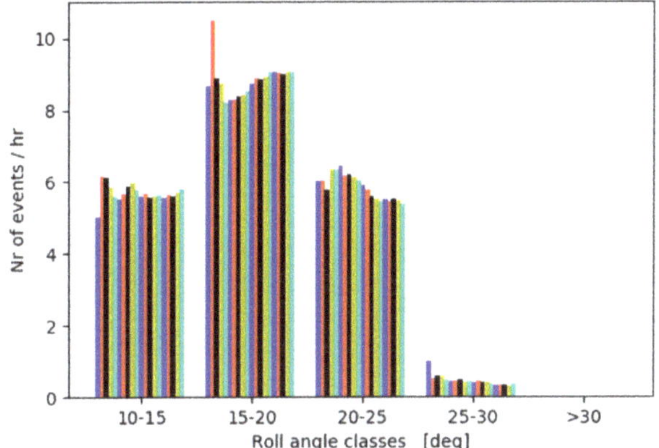

Figure 3. Ensemble statistics of the number of parametric roll events classified in different groups on the basis of the maximum roll angle in each event. The plot shows successive results 3 h, 6 h, 9 h, etc of simulations, with a maximum of 60 h. Sea state: Hs = 7 m, Tp = 13.5 s, speed Vs = 7.6 kn.

5.2. Analysis of Extreme Roll Angles

A probability of exceedance plot can be made from the local extreme values of the roll motion. An extreme value has been defined as the maximum positive value between two subsequent zero up-crossings. Such a plot contains both related events (belonging to the same parametric roll event) and unrelated events. From a statistical point of view, this is not a nice property, but it has been argued by Wandji [24] that the peak values are unrelated.

Figures 4 and 5 show the extreme value distributions of the roll motions for two sea states, from the programs Sim-1 and Sim-4. The figures show the extreme value distributions for 10 realisations of each sea state. The mean (solid black circles with dotted lines) and the 95% confidence intervals (solid red lines) were obtained from the 10 realisations per case. Extreme roll values smaller than 3° were ignored while obtaining the probabilities. These two examples show cases with a small and a larger confidence interval.

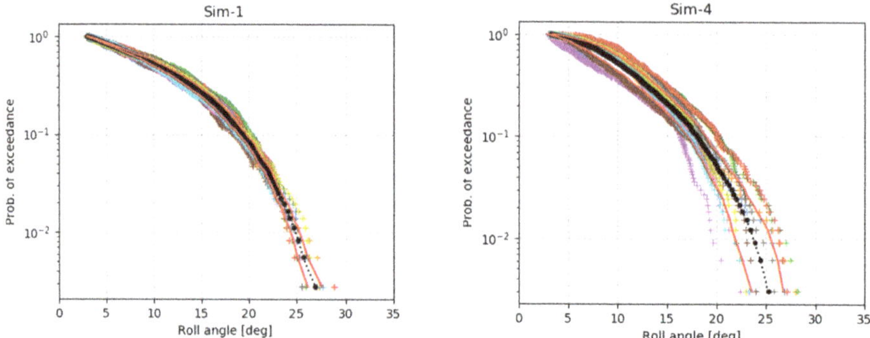

Figure 4. Extreme value distributions of the roll motions from programs Sim-1 (left) and Sim-4 (right) for 10 realisations of the wave. Simulations were run in head seas with Hs = 7 m, Tp = 13.5 s and average speed Vs = 7.9 kn.

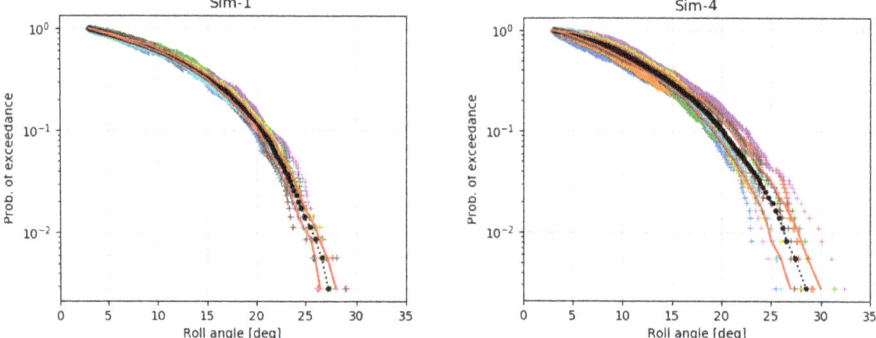

Figure 5. Extreme value distributions of the roll motions from programs Sim-1 (left) and Sim-4 (right) for 10 realisations of the wave. Simulations were run in following seas with Hs = 4 m, Tp = 13.5 s and average speedVs = 7.9 kn.

It can be observed that there is an inherent variability in the distributions. This variability seems to depend on the programs and the sea states. In general, the simulations exhibit wider spreads for the large roll angles. This can be observed especially in the trends of the confidence intervals. Overall, the two programs produce similar results for both sea states

5.3. Deterministic Analysis of Results

A deterministic comparison of the numerical and experimental results needs some special care. The experiments were carried out using a free running model, so the speed was not constant. The speed variations in the experiment and in the simulation are usually different, so after a short while the synchronisation of the wave experienced by the physical and the numerical model will be off. The method chosen to compensate for this involves the numerical model using the wave as measured on the location of the physical model in the tank at the subject time instant. This approach was introduced by van Walree and de Jong [25]. The solutions of the equations of motion, and hence the accelerations and velocities, was not adapted.

When simulations are carried out in exactly long-crested head or following seas, it is necessary to give some small disturbance in the lateral direction to initiate the roll excitation and hence the motion. Two methods were tried; the first is to give a small heel angle (0.1°) early in the simulation (at t = 10 s) and the second is to do the simulation in a heading of 179° or 179.5°, rather than 180°. The results proved that both methods work and give identical results.

6. Overview of Experimental Results

The conditions in which the experiments were carried out have been summarised in Table 4. An overview of the results of the experiments in head and following seas has been made in Figure 6. The figure shows the distributions of the extreme roll values and the number of parametric roll events divided in different amplitude classes. The first figure also features three-parameter Weibull fits that are normally used for extreme value predictions. Noted again is the fact that the loading condition of the vessel is different for the head seas and the following seas cases.

Table 4. Overview of tested wave conditions.

Wave Direction	Av. ship Speed Vs-av [kn]	Sign. Wave Height Hs [m]	Peak Period Tp [s]	Peakedness Parameter γ [-]	Test Duration [h]
Head	7.6 7.9	7.0 8.0	13.5	3.3	3
Following	7.9	4.0	13.5	3.3	3

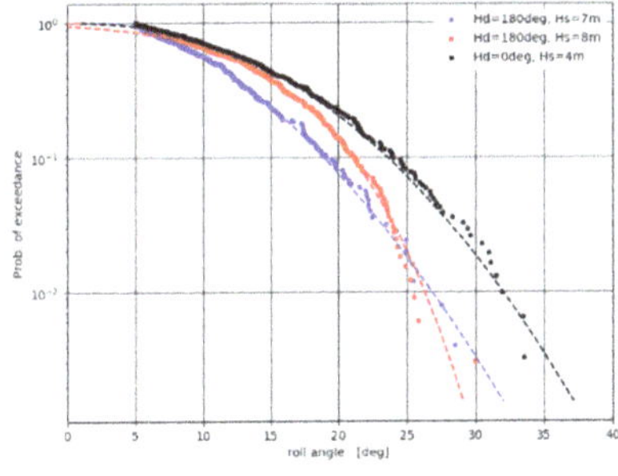

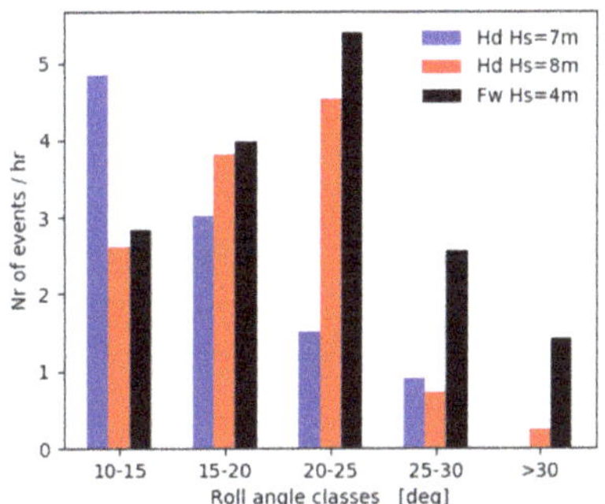

Figure 6. Results of the experiments in three sea states; two head waves (HW) and one following wave (FW), all at a speed of about 8 kn. One realisation of the wave with a duration of 3 h was run in the experiments. Left: probability of exceedance of extreme values of the roll motions. Right: number of parametric roll events per hour divided in classes of roll angles.

This figure shows some interesting results. Although the roll angles in the 8-m sea state are on average larger than in the 7-m sea state, the opposite appears to be true for the extreme values. The distributions of the parametric roll events are quite different in these two sea states. The lower sea state shows many events with a low amplitude, and the number of events is steadily decreasing for the higher amplitude classes. This is not the case in the higher sea state; there is a strong peak of events with a maximum amplitude of 20°–25°.

The trends in following seas show a similar distribution of the parametric roll events to those measured in the 8-m head sea case: there is also a strong peak for parametric roll events in the 20°–25° bin. The extreme values are particularly large for this condition, with quite a few events with a roll angle > 30°.

Note that especially for the large roll angles, the probability of exceedance might vary between different wave realizations (simulation results definitely show this trend). Since the experimental results shown in the figures come from only one realization of 3 hr duration, care should be taken while interpreting the results of particularly the tail of the distributions.

7. Comparison of Results in Head Seas

7.1. Probabilistic Analysis

The probability of exceedance of the extreme roll values, derived from the four simulation programs and from the experiment, is shown in Figure 7 for two sea states. Both the mean and the 95% confidence interval obtained from the simulations are shown. From the experiment, the result consists of one realisation of 3 h duration, as that is the only available data. The four programs produce generally similar results for both sea states, though there is some spread among them. The amount of spread is larger for the lower sea state, Hs = 7 m, than for the higher one, Hs = 8 m. For the lower sea state, three programs, Sim-2, Sim-3 and Sim-4, produce very similar mean values, while Sim-1 diverges from this group. The confidence intervals from the four programs vary, the one from Sim-4 being in general the widest. As for the higher sea state with Hs = 8 m, the four programs also produce similar mean values overall, although the mean from Sim-4 deviates from the others for the roll angles above 22°. The confidence intervals vary in their widths, with the one from Sim-4 being again the widest.

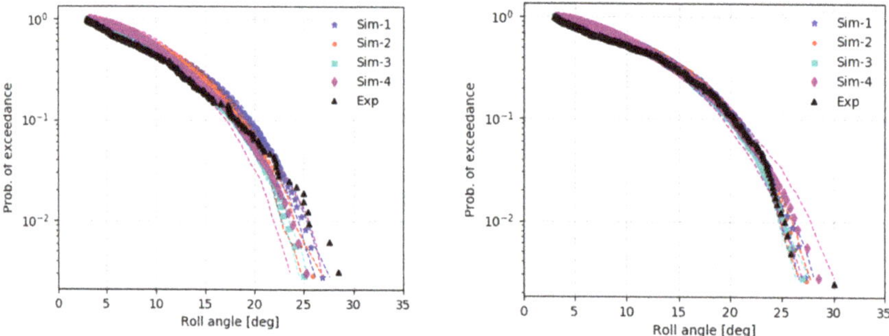

Figure 7. Extreme value distributions of the roll motions from the four programs and experiment. The mean (markers) and the 95% confidence interval (dashed lines) obtained from 10 realisations of the wave are shown for the simulations. Durations of the experiment and each realisation in the simulations are 3 h. Left: Head waves, Hs = 7 m, Tp = 13.5 s, speed vs. = 7.6 kn. Right: Head waves, Hs = 8 m, Tp = 13.5 s, speed vs. = 7.9 kn.

The variability in the extreme roll values was not determined in the experimental campaign. In that sense, the comparison of the numerical results to the experiments is incomplete. Considering the available data illustrated in Figure 7, it can be observed that the simulations produce non-conservative results for the extreme roll in the case of the lower sea state with Hs = 7 m. For the higher sea state, all the results of the simulations are conservative.

Figure 8 shows the number of parametric roll events per hour, divided into classes of roll angles from the four programs and the experiment for the two head sea cases. There is overall a reasonable agreement among the four programs, though the error bars of the 95% confidence interval are large in some cases. The simulations show a significantly larger number of events in the medium classes than the experiments. The number of events classified as parametric roll ($\varphi > 30°$) is very low, so no conclusions can be drawn on this. The typical distributions of the number of events in the experiments (a steadily decreasing number of events in the increasing classes for the 7-m sea state, and a large number of events in the 20°–25° class for the 8-m sea state) is much better reproduced by the simulations in the 8-m sea state than in the 7-m sea state. The distribution of events over the classes resulting from the simulations is quite similar for the two sea states.

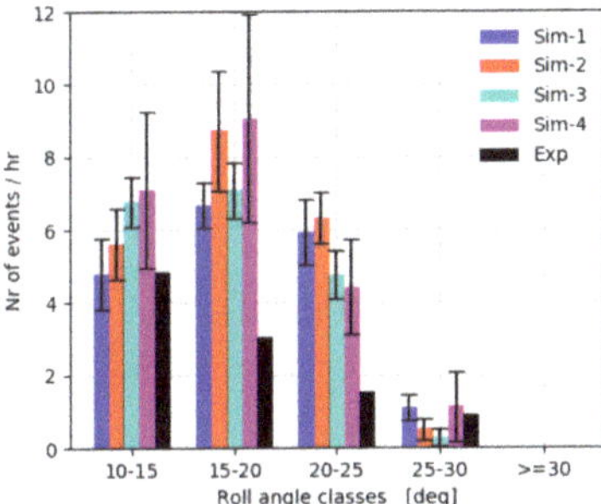

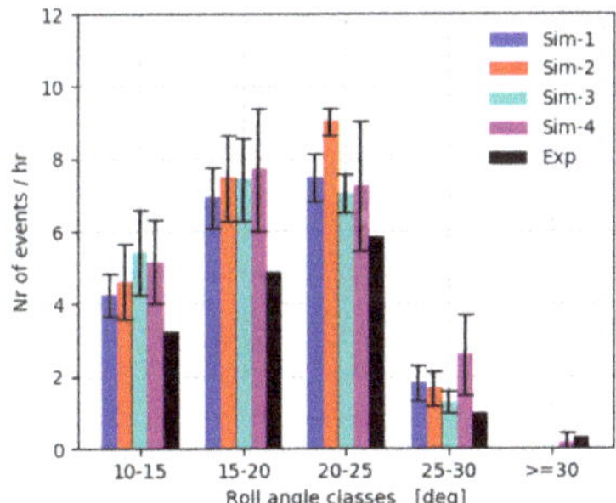

Figure 8. The number of parametric roll events per hour divided in classes of roll angles. Results of four sets of simulations, each the total of 10 simulations of 3 h duration, compared to results of experiments (3 h). The bars indicate the means and the error bars indicate the 95% confidence intervals obtained from 10 realizations for each simulation program. Left: Head sea with Hs = 7 m, Tp = 13.5 s, speed Vs = 7.6 kn, Right: Head sea with Hs = 8 m, Tp = 13.5 s, speed Vs = 7.9 kn.

7.2. Deterministic Analysis

The first case selected for the deterministic analysis was a run in the Hs = 7 m sea state. This run was chosen because the roll motion is very low in the first part and there is a parametric roll event in the second part. The resulting time traces of two simulations and of the measured signals are shown in Figure 9. The figure shows that the heave and pitch motions are well predicted. The Sim-2 model predicts a parametric roll event resulting from the high wave group at around t = 300 s, while Sim-3 and the experiment do not produce an event at that instance. The large event predicted by Sim-2 could be a consequence of the start-up. Both simulation programs predict an important event in the interval $600 < t < 800$ s; this event was not present in the experiments. The large event within $1000 < t < 1200$ s is predicted accurately by both models, although the simulated events have a longer duration than the one in the experiment.

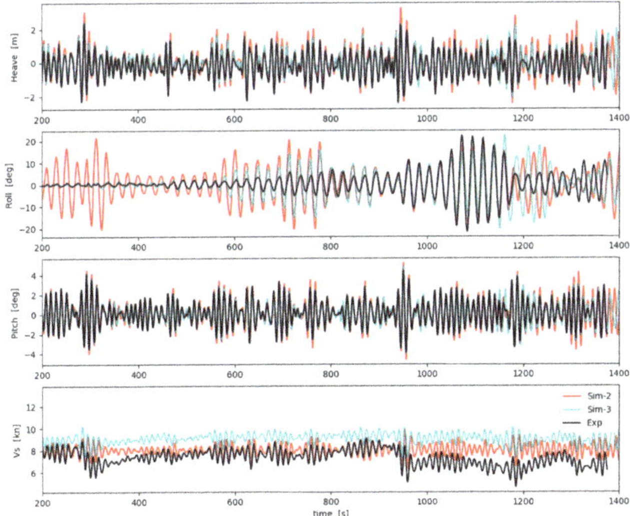

Figure 9. Results of deterministic comparison of a parametric roll event in irregular head seas (Jonswap spectrum, Hs = 7.0 m, Tp = 13.5 s, γ = 3.3), Vs ≈ 8 kn. Heave, roll, pitch and vessel speed from two programs, Sim-2 and Sim-3, are compared to the measurement.

The second case was a run in the Hs = 8 m sea state. This run was chosen because the waves were high, but parametric roll did not occur. The simulation programs were proven to have the tendency to overpredict the occurrence of parametric roll; this was concluded before from the results in regular waves [3]. Figure 8 shows that the same is true in irregular waves.

The time traces of the simulations are compared to the measured signals in Figure 10. The results of both simulations indeed show several parametric roll events, although at different parts of the run. The measurements show no indication of parametric roll. The peaks in the heave and pitch motions resulting from the Sim-2 simulations are quite large, while those from the Sim-3 simulations correspond much better to the experimental values.

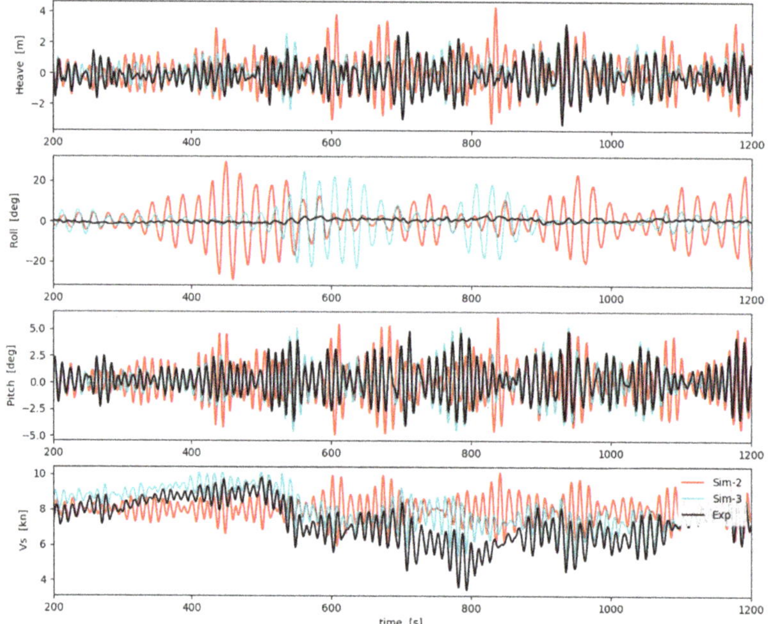

Figure 10. Results of deterministic comparison of a parametric roll event in irregular head seas (Jonswap spectrum, Hs = 8.0 m, Tp = 13.5 s, $\gamma = 3.3$), vs. $\approx$ 8 kn.

8. Comparison of Results in Following Seas

8.1. Probabilistic Analysis

The statistics of the extreme roll values from the four simulation programs and the experiment are shown in Figure 11. The mean and the 95% confidence interval from the simulations are also shown in this plot. The means of the exceedance probabilities from the four programs are similar for $\phi < 20°$; after this value they split into two groups. The means estimated by Sim-1 and Sim-4 indicate lower probabilities than those by Sim-2 and Sim-3 for the roll angles larger than 20°. All four programs estimate lower probabilities than the measurement for the roll angles above 30°, considering the available measurement data. It is concluded that the simulation programs are non-conservative in this condition. This conclusion is supported by the bar diagram shown in Figure 11 (right); the number of events in the experiments is lower than those resulting from the simulations for all classes, except for the class $\varphi > 30°$.

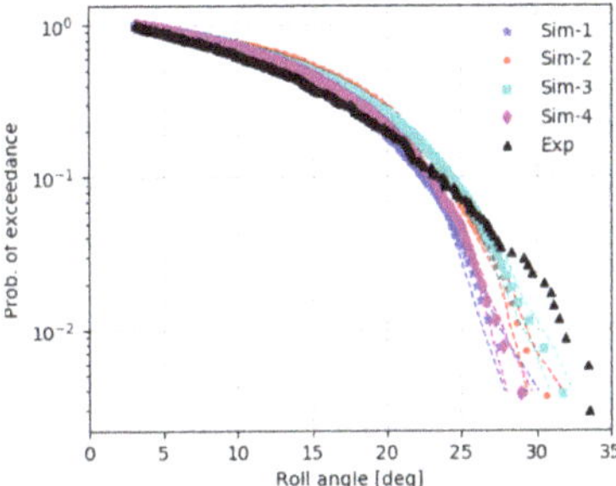

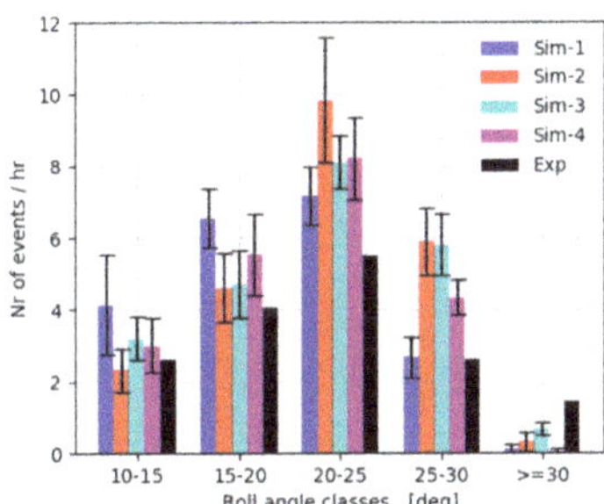

Figure 11. Statistics of extreme values of the roll motions (left) and the number of parametric roll events/hour subdivided in classes (right). Results of 4 sets of simulations, each the total of 10 simulations of 3 h duration, compared to results of experiments (duration 3 h). The mean (markers) and the 95% confidence interval (dashed lines) obtained from 10 realizations of the wave are shown for the simulations (left). The mean (bars) and the 95% confidence intervals (error bars) obtained from 10 realizations are plotted on the right. Sea state: Following seas, Hs = 4 m, Tp = 13.5 s, speed vs. = 7.9 kn.

8.2. Deterministic Analysis

Figure 12 shows the first case selected for deterministic comparison in the following seas. This run was chosen because there is an important parametric roll event in the second part of the run. The time traces of heave, roll, pitch and vessel speed from the experiment and two programs, Sim-2 and Sim-3, are illustrated in the figure. The agreement to the experiments is astonishing considering the results in the head seas. There is an important difference in the loading condition, and hence in the roll natural period, between the following and head sea cases. The natural roll period for the experiments in head seas was $T\varphi = 23.4$ s, while for the experiments in following waves it was $T\varphi = 34.1$ s (Table 1). It might be that at this lower encounter frequency of the waves at which parametric roll is initiated, the ratio of the diffraction force to the Froude–Krylov force is much lower, and that the modelling in the simulation programs (that ignore non-linear diffraction forces) is then closer to reality. This suggestion is detailed in Section 10.

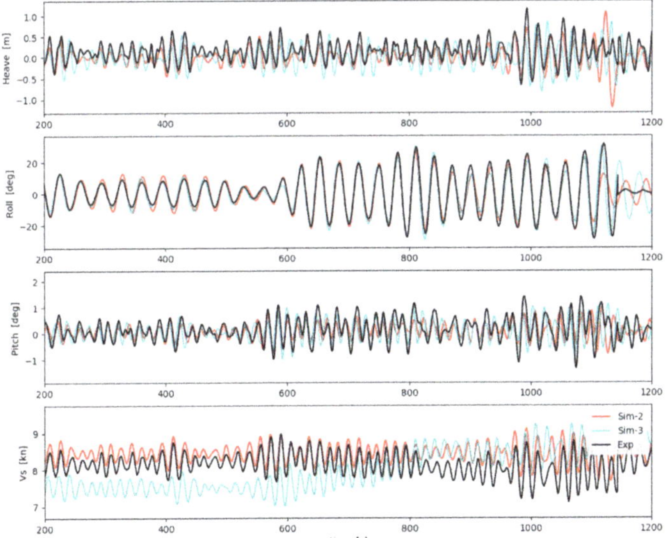

Figure 12. Results of a first case of a deterministic comparison of a parametric roll event in irregular following seas (Jonswap spectrum, Hs = 4.0 m, Tp = 13.5 s, $\gamma = 3.3$), vs. ≈ 8 kn. Heave, roll, pitch and vessel speed calculated by Sim-2 and Sim-3 are compared to the measurement.

The results of the second deterministic comparison in the following seas are shown in Figure 13. The duration is 1145 s. This run was chosen because it starts with low roll motions, and then a parametric roll event with a maximum roll motion of 33.5° develops. This event dies out, and about 500 s later a second event develops with a similar maximum roll value.

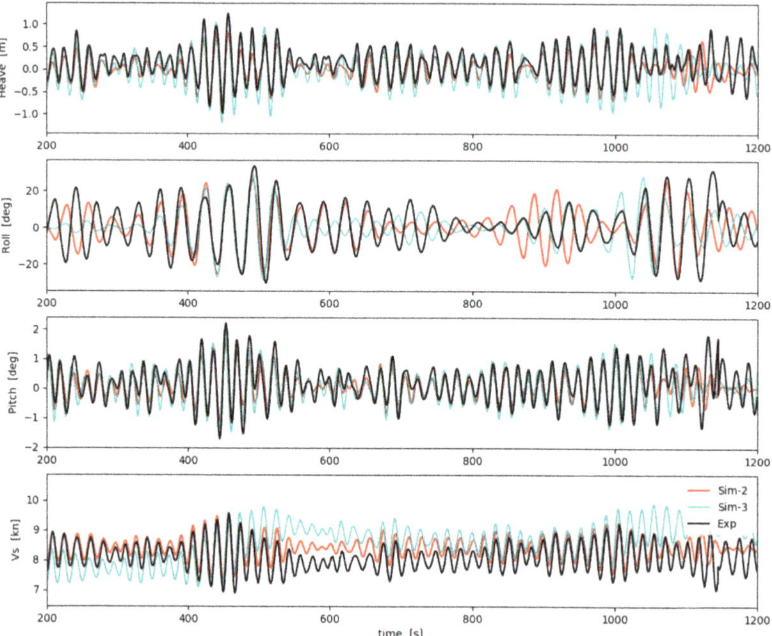

Figure 13. Results of the second case of a deterministic comparison of a parametric roll event in irregular following seas (Jonswap spectrum, Hs = 4.0 m, Tp = 13.5 s, γ = 3.3), vs. ≈ 8 kn.

Time traces of heave, roll, pitch and vessel speed, from the experiment and two simulations programs, Sim-2 and Sim-3, are shown in Figure 13. Similar to the results in Figure 12, the agreement is much better than in head seas. Both parametric roll events are captured, and the peak roll values of the events are accurately predicted.

9. Discussion

9.1. Objective

The intention of this study was to compare existing simulation programs "as is", and using settings according to the best practice of the owners of each of the programs. The background of this choice was that a ship operator could ask any of these companies to do these simulations, expecting similar results that are also close to experimental values.

Although the differences in the results of the four simulation programs are globally not large, the focus is on the probability of exceedance of a roll angle larger than 30°. Only an event with such a roll angle is classified as a parametric roll event. Looking at the results in this way, there are important differences in the four simulation programs. The order of the results of the programs, in the sense of the one predicting the largest probability for $\varphi > 30°$ to the one predicting the lowest probability, is not the same for the three conditions.

It is acknowledged that the results of this study raise questions regarding the cause of the differences. An investigation into the cause of the differences was outside the scope of the present study, but it is the logical next step to take. Essentially, this next step consists of two phases: first, to find reasons for the differences between the simulation programs as they are now and to reduce them, and secondly to improve the accuracy of all the programs. The discussion presented here suggests possible improvements for this second phase.

9.2. Modelling the Waves

It is concluded from the results of this study that, although roll damping is a crucial effect, using the same roll damping model does not guarantee identical results when different simulation programs are being used. Apparently, other aspects are important as well. One of the candidate aspects is the model for the irregular seas. All programs use a linear combination of wave frequencies, but the number and the choice of values for the frequencies also play roles. A comparison in regular waves of these same programs, as done by Kapsenberg [3], showed important differences for the threshold wave amplitude, but the roll angle of parametric roll in 2.5- and 3.0-m wave amplitudes was similar. On the other hand, the deterministic comparison was made with a defined number of frequencies, amplitudes and phase angles. In that sense, the input was identical for programs Sim-2 and Sim-3. This identical input resulted in similar heave and pitch motions for the case of the 7-m sea state (Figure 9), but there were rather different results for those parameters in the case of the 8-m sea state (Figure 10).

9.3. Variation in the Stability

The physical explanation of parametric roll is classically based on roll stability variations when a vessel sails in waves (Paulling [26]). The time-dependent roll stability leads to the Mathieu equation, Equation (5), which is the simplest mathematical model for predicting the probability of parametric roll. The Mathieu equation is a linear differential equation, in which an oscillating restoring term with amplitude $k{\cdot}m$ is added to the constant restoring coefficient k. The solution of the Mathieu equation can be illustrated in a figure, known as the Ince–Strutt diagram, Figure 14. This figure shows areas with stable (bounded) and unstable (unbounded) solutions that predict parametric roll events reasonably well.

$$\ddot{\varphi} + k[1 - m\ \cos(\omega t)]\varphi = 0 \quad (5)$$

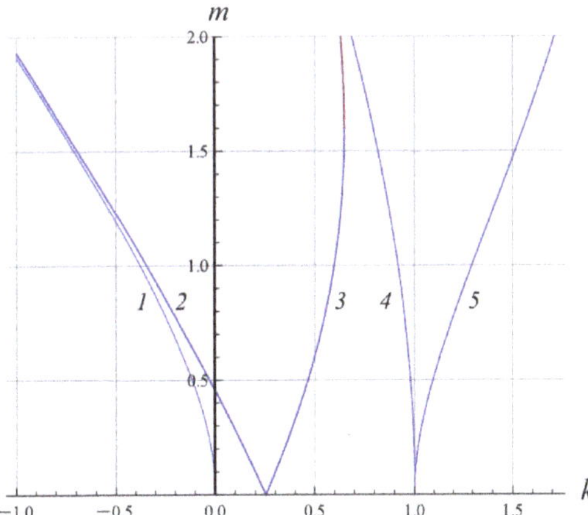

Figure 14. Results of the Mathieu equation, the Ince–Strutt diagram. The figure shows stable (white) and unstable (shaded) regions, depending on restoring term k and amplitude of the stability variation m. (Figure reproduced from Butikov [27]).

The time-dependent roll stability term is a necessary prerequisite in a mathematical model to predict parametric roll. Practical implementations of this effect in mathematical models are usually based on the calculation of the wave elevation along the hull for each time step. This calculation often uses the undisturbed wave and the motions of the ship to determine the wetted surface. The force is then determined by a pressure integration; the pressure above the calm water surface is determined by the hydrostatic component only, or by using Wheeler stretching [22] for the dynamic part. This non-linear restoring term is often accompanied by a non-linear roll damping term, but other components of the equations of motion are based on linear theories.

9.4. Roll Damping

The use of experimental values for the roll damping does not guarantee that all problems related to this parameter have been solved. An important part of the roll damping is due to the bilge keels. The potential flow damping is low, due to the long natural roll period. The roll damping contributions, as a function of the forward speed, are shown in Figure 15. The roll damping in this figure is based on Ikeda's method. The calculation is essentially based on a forced motion in calm water, and as such the results are directly comparable to the roll decay and forced moment tests done in this campaign. In waves, the situation is different; the forces on the bilge keels are partly due to the velocities of the incoming and diffracted wave. This contribution is neglected in Ikeda's method.

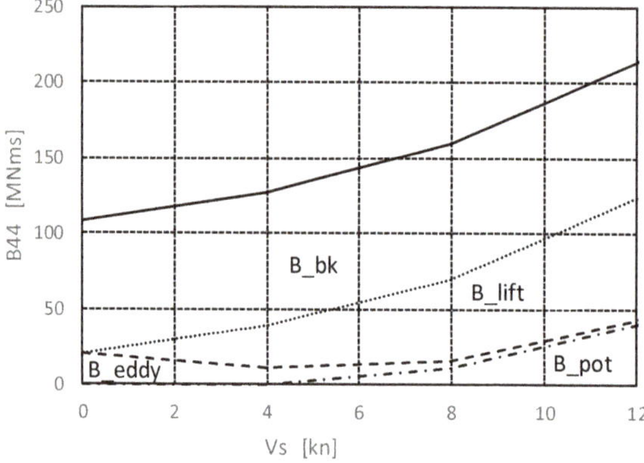

Figure 15. Roll damping contributions for the KCS, LC-1, at the natural roll period (Tφ = 23.4 s) as a function of the forward speed. The figure shows the contribution of different components: Potential flow component including the effect of forward speed, the eddy damping component, the lift component and the bilge keel component. For Vs = 8 kn, the bilge keels provide 56% of the total roll damping.

9.5. Relation between Parametric Roll Amplitude and Wave Height

The relation between the amplitude of the parametric roll angle and the wave height is very non-linear. First, there is a threshold wave height below which parametric roll does not occur. Secondly, it appears that the roll angle stabilises at some wave height, and does not further increase. These effects were also quite apparent in the results of the regular waves tests (see Kapsenberg et al.) [3].

It has been accepted [28] that the magnitude of the roll damping is largely responsible for the value of the threshold wave height. Therefore, it could be advisable to increase speed in order to increase the roll damping and hence to reduce the risk of parametric roll. The effect of forward speed was noted in the experiments in regular waves; tests at a speed of 10 kn did not show a tendency for parametric roll. If this statement is accepted, it should matter a lot if a second or a third order model is used to describe the damping, as shown in Figure 1 (the damping for $\varphi = 0$ is a factor different between these two models).

It is not clear why the roll motion stabilises at a certain wave height. There is a publication that suggest that the amplitude even decreases in very high waves; Hashimoto et al. [29] showed experimental results that indicated a decrease of the parametric roll angle for a wave steepness > 0.05.

Since the wave-length of interest is in the order of the ship length, this is a wave with a height of more than 11 m, which is the order of the draft of the KCS. This is an extreme condition by all standards. A less extreme explanation might lie in the third order behaviour of the damping curve (Figure 1). A parametric roll excitation that increases linearly with the roll angle explains both the threshold wave amplitude and the stabilisation of the roll angle. This point will be developed in the next paragraph.

9.6. Nonlinear Diffraction

The non-linear effect of the incoming wave is accounted for by calculating the wetted surface of the hull at each time step. The Froude–Krylov force is then based on a pressure integration over the wetted hull. The presented simulation programs all include just the linear diffraction forces. It has been suggested by Dallinga [30] and Bu et al. [31] that the non-linear diffraction forces play a role in the occurrence of parametric roll events. Dallinga [30] used linear calculations on a heeled vessel to estimate non-linear diffraction effects. Bu et al. [31] used a more complicated (and more CPU-intensive) hydrodynamic model, a body-exact method, to calculate the hydrodynamic forces at each time instant using a new panelisation. They compared results of stability calculations in waves using the body-exact method to results of a linear program and a heeled vessel, similar to Dallinga. They found perfect agreement for small angles of heel (up to 12°) and some differences for large angles. Limited results were shown for parametric roll calculations; for one condition it appeared that the predicted roll amplitude was similar when comparing the two methods. The effect on the onset of parametric roll was not studied.

It is proposed here to use the approach of a linear diffraction program, and to approximate non-linear diffraction by using a heeled vessel in the input. Calculations were carried out for the KCS vessel in the LC-1 condition, at a speed of 8 kn, in head seas at three angles of heel ($\varphi = 2.5°$, 5.0° and 10°). The natural roll frequency for this condition is $\omega_\varphi = 0.266$ rad/s. The condition that the wave-encounter period is twice the roll natural frequency corresponds to an earth fixed wave frequency $\omega_0 = 0.448$ rad/s. Figure 16 (left) shows the results of the calculation at this frequency. The excitation and diffraction roll moments (F4) are split into a component in-phase with the roll motion, Re{F4}, and a component in-phase with the roll velocity, Im{F4}. There appears to be a significant effect on both components, whether including the diffraction force or not. For both components, it results in a significant reduction of the magnitude. The same figure has been made for the KCS in the LC-12 loading condition in following waves [Figure 16 (right)]. The natural roll frequency for this loading condition is $\omega_\varphi = 0.184$ rad/s; this corresponds, in following waves and a speed Vs = 8 kn, to an earth fixed wave frequency $\omega_0 = 0.456$ rad/s. Figure 16 (right) shows that the effect of adding the diffraction roll moment to the excitation is much smaller than for the head seas case.

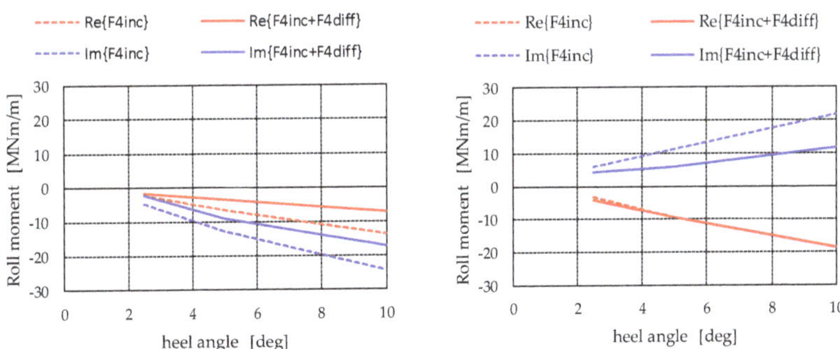

Figure 16. Roll moment due to incoming wave only, and due to incoming + diffracted wave as a function of the initial heel angle. The moment has been split into two components; one component in-phase with the roll motion, Re{F4}, and the other component in-phase with the roll velocity, Im{F4}. Left: Results are for KCS, LC-1, $\omega_0 = 0.45$ rad/s, Hd = 180° and Vs = 8 kn; Right: KCS, LC-12, $\omega_0 = 0.45$ rad/s, Hd = 0° and Vs = 8 kn.

Note that this frequency corresponds to a wave-length that is similar to the critical condition in head seas, the wave-length to ship-length ratio is 1.3.

Another observation that can be based on Figure 16 is that both moments, due to the incoming and the diffracted wave, are linear, with respect to the heel angle. This means that this moment (in head and following seas) can be modelled as:

$$F4 = \varphi F4_a \cos(\omega_e t + \epsilon_{F4}) \tag{6}$$

We assume that the roll motion φ is also oscillating as a harmonic function, with amplitude φ_a and frequency ω_φ. We define the phase angle of the roll as zero, so the phase angle in Equation (5) is relative to the roll motion. If we introduce this harmonic oscillation of the roll motion in Equation (5), we have:

$$F4 = \frac{1}{2}\varphi_a F4_a\Big[\cos\Big(\omega_e t + \epsilon_{F4} - \omega_\varphi t\Big) + \cos\Big(\omega_e t + \epsilon_{F4} + \omega_\varphi t\Big)\Big] \tag{7}$$

The classical condition for parametric roll is:

$$\omega_e = 2\omega_\varphi \tag{8}$$

When this is substituted into Equation (6), we arrive at:

$$F4 = \frac{1}{2}\varphi_a F4_a\Big[\cos\Big(\omega_\varphi t + \epsilon_{F4}\Big) + \cos\Big(3\omega_\varphi t + \epsilon_{F4}\Big)\Big] \tag{9}$$

This equation shows that a vessel sailing in waves with an encounter frequency of twice the roll natural frequency experiences an excitation in the roll natural frequency and in a frequency of three times the roll natural frequency. Both components contribute to the energy input of the vessel, which should be compensated for by the roll damping.

Similar to the derivation above, the expressions for the non-linear damping and the oscillating part of the restoring force also result in $\omega_\varphi t$ and $3\omega_\varphi t$ terms in the equation of motion. The ship will respond mainly on the ω_φ terms; a force oscillating with three times the natural frequency does not provoke a significant response.

9.7. *Effect of Water on Deck*

Another component of nonlinear excitation may come from the water on the submerged deck. This topic has been investigated for a vessel at zero speed (FPSO) by Greco et al. [32]; they found that the water on deck had a limited effect on parametric roll. In this case, it has been found that water on the deck has a significant effect on extreme roll angles. There will of course be no effect on the onset of parametric roll. Once parametric roll occurs, and as the roll motion grows, the edge of the deck surface may be submerged, and the water on deck is detrimental to the stability of the vessel, and will increase the roll response. For the irregular following seas with Hs = 4 m, the simulated roll motions and a corresponding snapshot of a submerged deck edge are shown in Figure 17. A deck-in-water model, considering hydrostatic and Froude–Krylov pressures, is applied to the submerged deck surface. The figure shows that water on the deck increases the extreme roll angles larger than 30°. This result demonstrates the importance of accurate deck modelling in the numerical simulations, as well as in the model experiments, to determine the probability of parametric roll, as it is defined here.

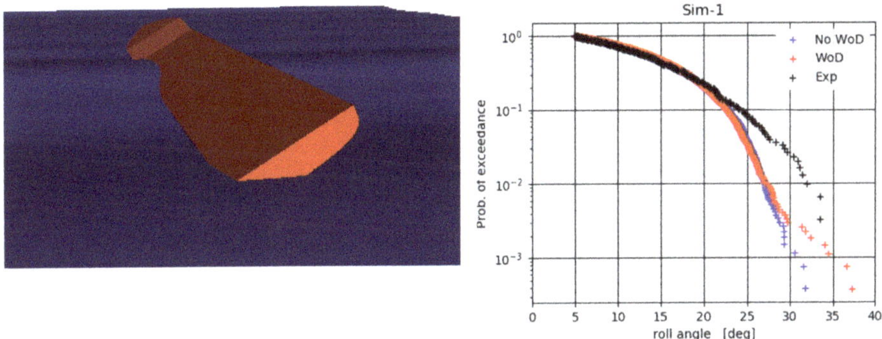

Figure 17. Snapshot from a simulated maximum roll angle with water on the deck (**left**). Probability of exceedance for the values of the positive roll peaks with water on deck in the simulation (WoD) or no water on deck (No WoD) is shown together with the experimental result. The plot is based on 10 realisations of 3 h durations each, condition: following waves, Hs = 4 m, Tp = 13.5 s, Vs ≈ 8 kn.

10. Conclusions

This article presents results of a comparison study of different numerical simulation programs, and dedicated model experiments on the topic of parametric roll. The conditions for the experiments and simulations consisted of head and following seas in high and moderate sea states, respectively. The speed of the subject vessel, the Korean Container Ship KCS, was about 8 kn, which is a realistic speed considering the sea conditions.

It appeared that parametric roll occurred frequently, both in the simulations and in the experiments. Similar to an earlier similar comparison of the same vessel in regular waves, the simulations showed more parametric roll events than were measured during the experiments. This was in particular the case for events with medium extreme values in the head seas condition.

It cannot be concluded that these simulations are always conservative or always non-conservative. Sometimes they are conservative, sometimes they are not. A complicating factor is that the extreme roll motions in a 7-m sea state appear to be larger than those in an 8-m sea state (with the same period and at the same speed).

The differences between the various simulation programs are not very large, considering the extreme value distributions of the roll motions. However, if one considers the probability of a roll angle large enough to classify the event as parametric roll (roll angle > 30°), there are important differences. This is also illustrated by the bar diagrams giving the number of events in a certain roll angle range [Figures 8 and 11 (right)].

For a validation study, one needs to compare converged statistical results. Convergence was shown for the simulations by considering 10 realisations of 3 h each. For the experiments, it could be shown that convergence was not achieved for 3-h experiments. To prove convergence, a series of experiments with different seeds is necessary, but a duration of 3 h might not be necessary.

The considered simulation programs do not model non-linear effects of the diffraction forces. It has been shown, by using a linear diffraction program for a heeled vessel, that the effect of the non-linear diffraction force is important (more so in head seas than in following seas). This suggests that a useful improvement of simulation programs can be made by adding this effect.

Author Contributions: Conceptualisation, G.K.; methodology, G.K., B.D. and C.W.; software, B.D., C.W., S.K.; validation, G.K., B.D., C.W. and S.K.; formal analysis, G.K., B.D. and C.W.; investigation, G.K.; resources, G.K., B.D., C.W. and S.K.; data curation, C.W. and S.K.; writing—Original draft preparation, G.K., B.D. and C.W.; writing—Review and editing, C.W., B.D., S.K. All authors have read and agreed to the published version of the manuscript.

Funding: This research was funded by the Cooperative Research Ships consortium [2] in the project DYNASTY. Permission to publish these results is gratefully acknowledged.

Acknowledgments: The permission to use the simulation results of Naval Group is gratefully acknowledged.

Conflicts of Interest: The authors declare no conflict of interest.

References

1. IMO. *Report of the Drafting Group on Intact Stability, Finalization of Second Generation Intact Stability Criteria*; IMO Document SDC 7/WP; IMO: London, UK, 2020.
2. IMO. *Report of the Correspondence Group on Intact Stability, Finalization of Second Generation Intact Stability Criteria*; IMO Document SDC 4/5/1/Add; IMO: London, UK, 2016.
3. Cooperative Research Ships. Available online: http://www.crships.org/web/show (accessed on 7 May 2020).
4. Kapsenberg, G.K.; Abeil, B.; Kim, S.; Wandji, C.; Ruth, E.; Pages, A. On parametric roll predictions. In Proceedings of the 17th International Ship Stability Workshop (STAB2019), Helsinki, Finland, 10–12 Jun 2019.
5. France, W.; Levadou, M.; Treakle, T.W.; Paulling, J.R.; Michel, R.K.; Moore, C. An investigation of head-sea parametric rolling and its influence on container lashing system. *Mar. Technol.* **2003**, *40*, 1–19.
6. Spanos, D.; Papanikolaou, A. Benchmark Study on Numerical Simulation Methods for the Prediction of Parametric Roll of Ships in Waves. In Proceedings of the 10th International Conference on Stab of Ships and Ocean Vehicles (STAB2009), St Petersburg, Russia, 22–26 June 2009.
7. Reed, A.M. 26th ITTC Parametric Roll Benchmark Study. In Proceedings of the 12th International Ship Stability Workshop (STAB2011), Washington, DC, USA, 12–15 June 2011.
8. Ikeda, Y.; Himeno, Y.; Tanaka, N. *On Eddy Making Component of Roll Damping Force on Naked Hull*; Report No. 00403; Un. of Osaka Pref., Dept. of Naval Arch.: Osaka, Japan, 1978.
9. Ikeda, Y.; Himeno, Y.; Tanaka, N. *Components of Roll Damping of Ship at Forward Speed*; Report No. 00404; Un. of Osaka Pref., Dept. of Naval Arch.: Osaka, Japan, 1978.
10. Ikeda, Y.; Himeno, Y.; Tanaka, N. *On Roll Damping Force of Ship—Effect of Friction of Hull and Normal Force of Bilge Keels*; Report No. 00401; Un. of Osaka Pref., Dept. of Naval Arch.: Osaka, Japan, 1978.
11. Ikeda, Y.; Himeno, Y.; Tanaka, N. *A Prediction Method for Ship Roll Damping*; Report No. 00405; Un. of Osaka Pref., Dept. of Naval Arch.: Osaka, Japan, 1978.
12. Ikeda, Y.; Komatsu, K. *On roll Damping Force of Ship—Effect of Hull Surface Pressure Created by Bilge Keels*; Report No. 00402; Un. of Osaka Pref., Dept. of Naval Arch.: Osaka, Japan, 1979.
13. Himeno, Y. *Prediction of Ship Roll Damping—State of the Art*; Report No. 239; Un. of Michigan, Dept. of Naval Arch. and Marine Eng.: Ann Arbor, MI, USA, 1978.
14. Hashimoto, H.; Omura, T.; Matsuda, A.; Yoneda, S.; Stern, F.; Tahara, Y. Some remarks on EFD and CFD for ship roll decay. In Proceedings of the 13th International Conference Stability of Ships and Ocean Vehicles (STAB2018), Kobe, Japan, 16–21 September 2018.
15. Yu, L.; Ma, N.; Hirakawa, Y.; Taguchi, K.; Akagi, K. Free running model experiment and numerical simulation on the occurrence and early detection of KCS parametric roll in head waves. In Proceedings of the 13th International Conference Stability of Ships and Ocean Vehicles (STAB2018), Kobe, Japan, 16–21 September 2018.
16. Yu, L.; Taguchi, K.; Kenta, A.; Ma, N.; Hirakawa, Y. Model experiments on the early detection and rudder stabilization of KCS parametric roll in head waves. *J. Mar. Sci. Technol.* **2018**, *23*, 141–163. [CrossRef]
17. Ruiz, M.T.; Villagomez, J.; Delefortrie, G.; Lataire, E.; Vantorre, M. Parametric Rolling in Regular Head Waves of the KRISO Container Ship: Numerical and Experimental Investigation in Shallow Water. In Proceedings of the 38th International Conference on Ocean, Offshore and Arctic Eng. (OMAE2019), Glasgow, UK, 9–14 June 2019.
18. Shin, Y.-S.; Belenky, V.L.; Paulling, J.R.; Weems, K.M.; Lin, W.M. Criteria for parametric roll of large containerships in longitudinal seas. In Proceedings of the SNAME Annual Meeting, Washington, DC, USA, 30 September 2004.
19. Neves, M.A.S.; Rodríguez, C.A. A non-linear mathematical model of higher order for strong parametric resonance of the roll motion of ships in waves. *Mar. Syst. Ocean Technol.* **2005**, *1*, 69–81. [CrossRef]
20. SIMMAN 2008 Workshop. Available online: http://www.simman2008.dk/KCS/container.html (accessed on 8 May 2020).
21. Hasselmann, K.; Barnett, T.P.; Bouws, E.; Carlson, H.; Cartwright, D.E.; Enke, K.; Ewing, J.A.; Gienapp, H.; Hasselmann, D.E.; Meerburg, A. *Measurements of Wind-Wave Growth and Swell Decay during the Joint North Sea Wave Project (JONSWAP)*; Ergänzungsheft 8–12; Deutsches Hydrographisches Institut: Hamburg, Germany, 1973.

22. Wheeler, J.D. Method for calculating forces produced by irregular waves. *J. Pet. Technol.* **1970**, *22*, 359–367. [CrossRef]
23. Lewandowski, E.M. Comparison of Some Analysis Methods for Ship Roll Decay Data. In Proceedings of the 12th International Ship Stability Workshop (STAB2011), Washington, DC, USA, 12–15 June 2011.
24. Wandji, C. Review of Probabilistic Methods for Dynamic Stability of Ships in Rough Seas. In Proceedings of the 17th International Ship Stability Workshop (STAB2019), Helsinki, Finland, 10–12 June 2019; pp. 47–56.
25. van Walree, F.; de Jong, P. Deterministic validation of a time domain panel code for parametric roll. In Proceedings of the 12th International Ship Stability Workshop (STAB2011), Washington, DC, USA, 12–15 June 2011.
26. Paulling, J.R. The transverse stability of a ship in a longitudinal seaway. *J. Ship Res.* **1961**, *4*, 37–49.
27. Butikov, E.I. Analytical expressions for stability regions in the Ince–Strutt diagram of Mathieu equation. *Am. J. Phys.* **2018**, *86*, 257–267. [CrossRef]
28. Francescutto, A.; Bulian, G. Nonlinear and stochastic aspects of parametric rolling modelling. In Proceedings of the 6th International Ship Stability Workshop, Glenn Cove, NY, USA, 13–16 October 2002.
29. Hashimoto, H.; Umeda, N.; Matsuda, A.; Nakamura, S. Experimental and Numerical Studies on Parametric Roll of a Post-Panamax Container Ship in Irregular Waves. In Proceedings of the 9th International Conference on Stability of Ships and Ocean Vehicles, Rio de Janeiro, Brazil, 25–29 September 2006.
30. Dallinga, R.P. Predicting Parametric Roll, SWZ|Maritime, 135, July–August 2014; pp. 44–47. Available online: https://www.swzmaritime.nl/pdf-archive/2014-edition-8/ (accessed on 15 May 2020).
31. Bu, S.-X.; Gu, M.; Lu, J.; Abdel-Maksoud, M. Effects of radiation and diffraction forces on the prediction of parametric roll. *Ocean Eng.* **2019**, *175*, 262–272. [CrossRef]
32. Greco, M.; Lugni, C.; Colicchio, G.; Faltinsen, O.M. Numerical Study of Bilge-Keel Effect on Parametric Roll and Water on Deck for an FPSO. In Proceedings of the 34th International Conference on Ocean, Offshore and Arctic Eng. (OMAE2015), St. John's, NL, Canada, 31 May–5 June 2015.

Journal of
Marine Science and Engineering

Article

The Smart Detection of Ship Severe Roll Motions and Decision-Making for Evasive Actions

Maria Acanfora * and Flavio Balsamo

Department of Industrial Engineering, University of Naples "Federico II", 80125 Napoli, Italy; flavio.balsamo@unina.it

* Correspondence: maria.acanfora@unina.it

Received: 30 April 2020; Accepted: 4 June 2020; Published: 6 June 2020

Abstract: This paper presents a numerical model for the smart detection of synchronous and parametric roll resonance of a ship. The model implements manoeuvring equations superimposed onto ship dynamics in waves. It also features suited autopilot and rudder actuator models, aiming at a fair depiction of the control delay. The developed method is able to identify and distinguish between synchronous and parametric roll resonance, based on the estimation of encounter wave period from ship motions. Therefore, it could be useful as a smart tool for manned vessels and, also, in the perspective of unmanned and autonomous vessels (in the paper it is assumed a hypothetical remote crew). Once the resonance threat is identified, different evasive actions are simulated and compared, based on course and speed change. Calculations are carried out on a ro-ro pax vessel vulnerable to parametric roll. We conclude that, in roll resonance situations, and in the absence of roll stabilisation systems on-board, course change could be the most effective countermeasure.

Keywords: ship dynamics; ship manoeuvring; numerical simulations; parametric and synchronous roll; smart vessels

---

1. Introduction

Recent interest in smart ships brings out new possibilities in vessel operations but, at the same time, involves new challenges [1,2]. Regarding the smart tool development for supporting crew decisions on manned ships, the smart vessel concept extends to unmanned and autonomous ships with several levels of automation. The unmanned ship is expected to be operated by remote actions since there would be no crew on-board. Therefore, the vessel has to be equipped with an extensive set of sensors concerning videos, acceleration and motion tracking, accounting for redundant tools [3,4]. The decision-making processes performed by the crew, conventionally based on physical perceptions, have to be moved to a digital perception of the reality. The autonomous ship is expected to operate without any human action, independently from the presence of crew on-board. The vessel is provided with smart tools capable of elaborating sensor signals and making decisions on ship routing and operation, autonomously. This is particularly relevant for decision-making processes, conventionally performed by humans, and the resulting choices.

One of the main concerns about smart vessels regards the assessment of ship safety [5–7], with the development and establishment of dedicated rules. Such regulations on autonomous/remote ships are currently under development by the classification societies [8]. The ship usually faces several troublesome situations during navigation. This paper focuses on excessive motions arising from parametric and synchronous roll, that are responsible for large accelerations with severe consequences for ship safety. Parametric roll resonance depends on the variation of the immersed hull in waves and can happen mainly in head/following seas, but even in oblique waves. The most severe condition for triggering parametric resonance is characterised by an encounter wave period that is half of the natural roll period. Weaker parametric resonance also can be observed for an encounter wave period equal to a natural roll period, although this might be confused with synchronous roll resonance, especially in oblique seas. Indeed, synchronous roll resonance, depending on the agreement of encounter wave period and natural roll period, can happen for oblique and beam waves, while for head and following seas no roll motion is expected. Herein, weaker parametric resonance scenarios (characterised by a natural roll period equal to the wave period) are not under investigation, nevertheless, the developed method is expected to identify them as synchronous roll threats. Specifically, ships prone to parametric roll could develop large roll motions and, thus, lead to dangerous scenarios such as cargo shifting [9,10] or even capsizing. Nevertheless, synchronous roll, which indeed can happen more frequently and for all ship types, could lead to unpleasant consequences [11,12].

Considering the perspective of a smart ship, according to the guidelines developed by DNV-GL [8], a breakdown of ship functions involves the following features: condition detection, condition analysis, action planning and action control (i.e., execution). Route planning is an important aspect to take care of since it aims at critical scenario forecasting before setting out to sea. Regarding the avoidance of excessive roll motions and accelerations, route planning mainly employs polar diagrams or similar methods [13–17]. However, it has been recognised that real-time applications of such tools for condition detection and analysis would be difficult [18], especially when a ship navigates in unexpected or accidental operational conditions not covered by the diagrams. Several methods were developed and presented by the authors of [19–22] mainly for the detection of parametric roll resonance, aiming at providing on-board decision support systems during navigation. However, they disregard the synchronous roll. Dealing with roll resonance phenomena, the measurement and identification of the wave period and heading during navigation are crucial. One of the novel aspects of this paper is the development of a reliable autonomous system, able to replace the crew perception of the sea state, aiming at the detection and analysis of the resonance threats. The research focuses on three different approaches for determining encounter wave periods in irregular waves, based on the measurements of heave and pitch motions. This follows the idea of exploiting ship motions as estimators of sea state characteristics [23,24]. These three approaches refer respectively to: FFT (Fast Fourier Transform) analysis, time-domain assessment, and Hilbert transform. Except for the Hilbert-based approach, the first two were already introduced in a previous research paper [25], where they were applied separately. Herein, all three proposed approaches are simultaneously applied and compared, aiming at the evaluation of a more reliable prediction. Therefore, the current paper concerns the enhancement of an autonomous routine for condition detection and analysis; moreover, it also aims at providing novel insights regarding action planning and control execution that, for the purpose of the paper, are assumed up to the crew. Indeed, in the reference paper [25], the analysis was limited to the identification of the excessive roll motion threats once the whole time history was achieved from the numerical simulations, with no possibility of controlling the speed and heading of the vessel during the run. The first part of the current research deals with the implementation of the method for condition detection and analysis within the time domain simulations for real-time threat identification. The investigation of action planning and control execution features the development of a numerical model that includes the main ship control systems for executing the evasive action by modifying the speed or course. The latter requires the implementation of the rudder actuator and autopilot models [26]. This implementation aims at enhancing the estimation of ship behaviour once the control is executed. The outcomes of selected case studies, involving the detection of parametric and synchronous roll resonance, followed by the executions of different evasive actions, provide novel knowledge on the best countermeasure to different critical scenarios.

The ship under study is a ro-ro pax ferry named Seatech-D. Despite the smart tool developed herein being intended for both manned and unmanned ships, it might be questioned that a passenger ferry is not a realistic candidate for becoming an unmanned or autonomous vessel for near-future industrial applications. Indeed, in this research, the hull is mainly used for testing the developed method: regarding Seatech-D, experimental campaign data regarding a self-propelled turning circle in irregular waves and straight runs in oblique irregular waves are available [27,28]. Therefore, for the scope of the current applications, it is considered appropriate.

The paper is structured as follows. First, a brief description of the numerical model for ship dynamics is given, with particular relevance to the rudder and steering actions. Then, the parametric and synchronous roll detection routine is applied to the numerical simulations. Once a threat is identified, the control of the vessel is performed and the obtained results on evasive actions are further discussed. Finally, conclusions are drawn.

2. Numerical Model for Ship Control in Irregular Waves

The numerical model is based on a combination of sea-keeping and manoeuvring motions in irregular seas, as presented by the authors of [29,30], accounting for all the pertinent non-linearities. The previous researches employed the computer program LAIDYN coded in Fortran [29], well known in the technical literature and validated through several benchmark studies [31–33]. Differently, the current numerical model is implemented in a Matlab/Simulink environment and it features novel approaches regarding autopilot and steering actions, thus aiming at a more realistic application of ship control attitude in adverse sea states.

A suitable superposition of sea-keeping and manoeuvring motions guarantees a fair agreement of the numerical outcomes with the available experimental data. This is carried out by assuming that the added mass coefficients for an infinite frequency are valid for both sea-keeping and manoeuvring models; while the damping coefficients, corresponding to the sway–yaw motions, refer only to the manoeuvring derivatives (i.e., radiated wave effects are excluded). The damping actions, in the remaining degrees of freedom, are implemented by the convolution integral technique [34]. Rudder forces develop in agreement with the instantaneous angle of attack and speed at the blade location, as explained by the authors of [29,30]. Differently from the reference papers, the effects of propeller action on the rudder flow are disregarded herein. This would induce an error on the velocity at the rudder blade, while in the current simulations no appreciable differences were found by comparing the numerical and the experimental turning circle results. Concerning the scope of the research, the small approximation regarding the propeller action on the rudder flow is considered tolerable.

The propeller thrust is achieved by a fixed revolution approach, which involves a modification of the ship speed due to the wave actions. The obtained wave resistance in the numerical simulation (i.e., all wave actions in the x-axis, body reference frame), does not account for second-order effects. The current numerical model gives the possibility to modify ship heading (through rudder deflection) and propeller revolution during the simulation. Particularly, the rudder angle changes accounting for a realistic delay in both the autopilot and the steering control laws.

The main objective of ship manoeuvring is to control the heading angle of the ship. This is conventionally carried out, in the majority of vessels, by rudder actions that ensure course keeping and course change. The control system must alter the control surfaces to a desired heading angle; the rudder deflection angle can be set manually (as usually done by the helmsman) or autonomously employing an autopilot. The current model implements a PID (proportional, integral, derivative) autopilot, capable of controlling rudder deflection to achieve the desired heading angle. The proportional, integral and derivative constants are set by the "trial and error" approach, to ensure a suitable control action in waves; the integral action is almost negligible compared to the others.

It is important to underline that the desired rudder angle, as required by the autopilot, cannot be achieved instantaneously due to the damping (provided by the surrounding fluid) and the inertia (due to the rudder shape and weight distribution). Thus, the torque for turning the rudder, provided by a hydraulic actuator, has to compensate for several effects and the rudder turn dynamics are modelled by a second-order linear system. This aims at a more realistic control approach [26].

3. Case Study

The ship under investigation is a monohull ro-ro pax ferry, named Seatech-D, whose main particulars are given in Table 1. This hull was developed and investigated at the Aalto University in terms of resistance, stability, and sea-keeping tests. Additional details can be found in Stigler and Matusiak [28,29]. Particularly, for section plans and ship drawings refer to Stigler [28].

Table 1. Principal particulars of Seatech-D.

Dimension/Hull	Seatech-D
Length between perpendiculars, L (m)	158.00
Breadth, B (m)	25.00
Depth, D (m)	15.00
Draft forward, T_F (m)	6.10
Draft aft, T_A (m)	6.10
Displacement, Δ (tons)	13,766
Center of gravity above keel, KG (m)	11.834
Long. coordinate of the center of gravity measured from aft perpendicular, LCG (m)	74.77
Transv. radius of gyration in air, k_{XX} (m)	10.06
Long. radius of gyration in air, k_{YY} (m)	39.36

Concerning this hull, experimental data regarding straight runs in oblique seas and turning circles in irregular waves are available [27,28]. The hull model is not equipped with bilge keels.

The comparisons with the straight runs in oblique waves have been already presented [11], showing a fine match of the results (assessed by superposition of the numerical and experimental data, checking the agreement of the overlapping curves). Figure 1 shows a turning circle simulation, carried out by the numerical model under development herein. Irregular sea refers to JONSWAP spectrum with a zero-crossing period T_Z = 5.5 s, and significant wave height H_S = 4.8 m, and it is discretised by N = 30 wave components; ship speed is set Vs = 8.40 m/s and rudder deflection is δ = 20° (in ship scale). Turning path results are in model scale 1:39, the remaining outcomes refer to ship scale. There is an overall fair agreement between the turning diameter and the roll motion behaviour, as observed from the reference experimental and numerical data [27,28].

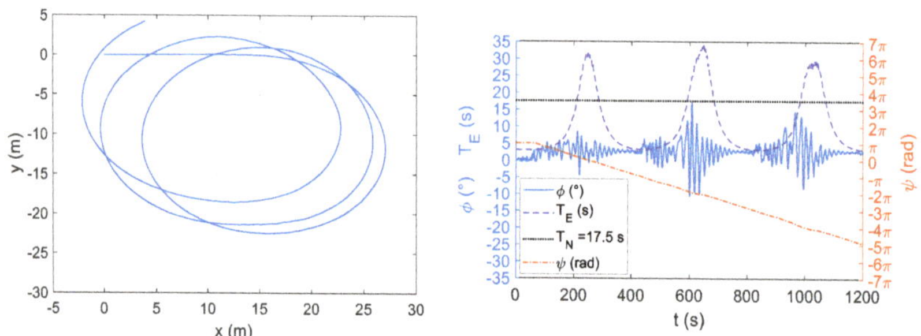

Figure 1. Turning circle in irregular waves (JONSWAP T_Z = 5.5 s, H_S = 4.8 m, N = 30 wave components, Vs = 8.40 m/s, δ = 20° in ship scale). Turning path on the left-hand side (model scale); Roll and sway angles on the right-hand side (ship scale).

All details about the turning circle conditions are given (in ship scale) in the Figure 1 caption. The ship performs three turns in less than 1200 s and, consequently, she experiences three times an amplification of the roll motions due to the resonance between the natural roll period (T_N = 17.5 s) and encounter period T_E. The latter accounts for the variation of wave heading and ship speed during the turn. This test ensures the reliability of the implemented models for rudder actions (for a constant deflection) and ship manoeuvring, superimposed onto the vessel dynamics in irregular waves.

The rudder actuator model implements inertia, damping, restoring and PID coefficients that are needed for the rudder turn execution, derived from design considerations. These coefficients are assumed to ensure that the time to deflect the rudder in still water from $\delta = -30°$ to $\delta = 35°$ remains below 28 s [35], in the absence of detailed data. A numerical application of this deflection test for Seatech-D is shown in Figure 2: rudder deflection starts from $\delta = -30°$ at 101 s and reaches approximately $\delta = 35°$ at 127 s. The rate of turn $\dot{\delta}$ does not exceed 7.5 deg/s.

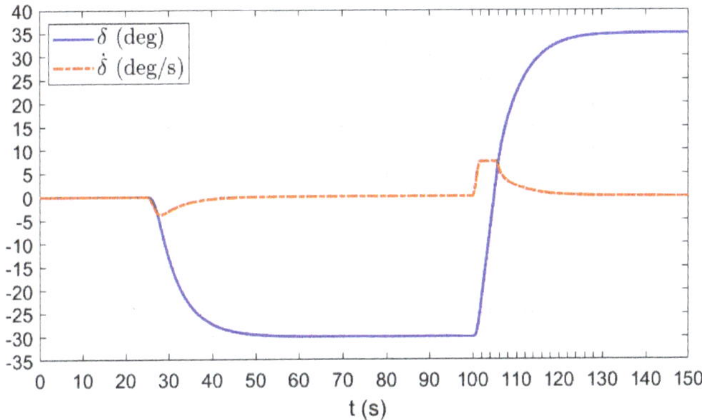

Figure 2. Example of a rudder deflection test from −30° to 35° at operational speed in still water.

Despite the simplifying assumptions adopted herein, the outcomes of the turning circle and rudder deflection tests provide a satisfactory behaviour of the numerical model in representing ship dynamics and the steering action in a realistic manner.

3.1. *Autonomous Detection of the Excessive Roll Motions*

The autonomous detection system for excessive roll threats in severe seas features a constant monitoring of the sea state, i.e., it estimates the wave encounter period through measurements and analysis of heave and pitch motions. The system measures roll motions alongside, focusing on average roll amplitude, and all data are processed every 90 s. Once the estimated wave encounter period falls within 40–60% (or 80–120%) of the known natural period (T_N) of the ship, and mean roll amplitude exceeds a certain threshold, the autonomous detection system forwards an alert signal of parametric (or synchronous) roll. This alert could be sent to an on-board crew or a remote one, depending on a manned or unmanned ship (in the remaining of the paper it is assumed a hypothetical remote crew). This resembles the first part of the approach developed by Acanfora [25], i.e., autonomous detection and alert identification.

Following the alert signal, the crew examines the ship behaviour and operational condition and, consequently, it takes an evasive action by changing the speed (controlling the propeller revolution) or by changing the heading (setting the autopilot and, thus, affecting rudder deflection). The simulation of the countermeasure actions and the corresponding effectiveness analysis represent the novel aspect of this research. Besides, in the current applications, it is assumed that the crew always takes an evasive action once a roll resonance threat is detected, although it could decide not to intervene.

3.2. *Techniques for Monitoring Encounter Wave Period and Roll Amplitude*

The need for a reliable assessment of the encounter wave period is directly correlated to a correct interpretation of the reason behind the development of large roll motions. When the encounter period of the waves and the natural roll period of the ship fall close to the ratio 1:2 (that is $T_E = T_N/2$), with large roll amplitudes even in relatively moderate wave heights, then a parametric resonance could reasonably be identified. Pitch and heave motions are adopted as indicators of sea state by processing the time history samples every 90 s with a frequency of 5 Hz. Indeed, these are two signals that exhibit a well-established linear behaviour, related to the encounter wave period. The peak periods of each sample of pitch and heave ($\overline{T}_\theta$ and $\overline{T}_\zeta$ respectively) are assumed as the current encounter wave period $\overline{T}_w$. The adoption of three techniques for each signal, listed and briefly explained below, can help in identifying the most reliable approach and can improve redundancy. FFT analysis ($\overline{T}_\theta$ and $\overline{T}_\zeta$ are the peak values of the pitch and heave spectra, obtained by the 90 s samples of pitch and heave, respectively). Mean period of the peaks from the time history (M is the number of the peaks, referring to heave-subscript ζ and pitch-subscript θ respectively, within the 90 s samples):

$$\begin{cases} \overline{T}_\zeta = \frac{1}{M_\zeta}\sum_{1}^{M_\zeta} T_{\zeta,i} \\ \overline{T}_\theta = \frac{1}{M_\theta}\sum_{1}^{M_\theta} T_{\theta,i} \end{cases}$$

Hilbert transform analysis of pitch $\theta(t)$ and heave $\zeta(t)$ samples [36]:

$$z_\theta = \theta(t) + iH[\theta(t)] = \theta(t) + i\frac{1}{\pi}P.V.\int_{-\infty}^{+\infty} \frac{\theta(\tau)}{t-\tau}d\tau = A_\theta(t)e^{ik_\theta(t)}$$

$$\overline{T}_\theta = 1/T\int_T \frac{2\pi}{dk_\theta(t)/dt}d\tau$$

$$z_\zeta = \zeta(t) + iH[\zeta(t)] = \zeta(t) + i\frac{1}{\pi}P.V.\int_{-\infty}^{+\infty} \frac{\zeta(\tau)}{t-\tau}d\tau = A_\zeta(t)e^{ik_\zeta(t)}$$

$$\overline{T}_\zeta = 1/T\int_T \frac{2\pi}{dk_\zeta(t)/dt}d\tau$$

where A is the signal instantaneous amplitude and k is the signal instantaneous phase, referring to heave-subscript ζ and pitch-subscript θ respectively, within the 90 s samples.

Along these lines, three values of the period for each motion sample are available. Redundant values of $\overline{T}_w$, estimated utilising $\overline{T}_\theta$ and $\overline{T}_\zeta$, can be achieved. A single value of the period is obtained as a mean value among the three; however, in the case of a large disagreement among them, they are considered separately.

The parametric roll (*PAR*) alert is set as follows:

$$PAR_{index} = \begin{cases} 1\ if\ \phi_A \geq \phi_{alert}\ and\ 0.4 < \overline{T}_w/T_N < 0.6 \\ 0\ if\ \phi_A < \phi_{alert} \end{cases}$$

where ϕ_A is the average value of roll amplitude, estimated from the measurement of roll motion in the same time range of heave and pitch. Thus, the Hilbert transform approach is employed:

$$z_\phi = \phi(t) + iH[\phi(t)] = \phi(t) + i\frac{1}{\pi}P.V.\int_{-\infty}^{+\infty} \frac{\phi(\tau)}{t-\tau}d\tau = A_\phi(t)e^{ik_\phi(t)}\ \phi_A^2 = 1/T\int_T A_\phi^2(t)d\tau$$

When there is a scenario involving synchronous (S) roll resonance, the corresponding alert index is set as:

$$S_{index} = \begin{cases} 1\ if\ \phi_A \geq \phi_{alert}\ and\ 0.8 < \overline{T}_w/T_N < 1.2 \\ 0\ if\ \phi_A < \phi_{alert} \end{cases}$$

Regarding an alert scenario, a threshold value between 10–15° should be expected [25]. Since this alert refers to possible developments of roll resonance in an early stage, a moderate roll amplitude is considered adequate for setting the alert. During the current simulation, it is set to ϕ_{alert} = 12.5°, as a mean value between 10–15°.

4. Results and Discussions

4.1. Parametric Roll Scenario

The autonomous detection technique operates on two redundant wave period values (one on the heave, one on the pitch, as mean values among the three techniques introduced above); therefore, two alert indexes are provided. When at least one of these two values becomes equal to one, then the system forwards the alert signal immediately to the remote crew. Once the alert is sent, the reaction time would depend on the preparedness of the remote crew, while the control action would depend on its expertise and experience.

The following applications show an example of the autonomous detection of large parametric roll motions together with some possible remote crew actions. The case study involves an irregular sea state modeled by a JONSWAP spectrum with H_S = 4.5 m and T_Z = 9.5 s (peak period T_P = 12.2 s). It is implemented by exploiting the technique proposed by Acanfora [11], adopting N = 30 wave components. This irregular sea scenario, together with a ship sailing in the head sea direction with a speed of V_S = 8.4 m/s (that is around 16 knots), can lead to the development of parametric roll. The initial ship heading in the simulation, is set at ψ = 175° to introduce a small perturbation (for a faster development of parametric resonance within the numerical model). The autonomous detection technique applies to the sample critical scenario, regarding which the remote crew might intervene with three different actions to modify the operational condition: change of heading of 15°, change of heading of 30° or speed reduction of 5 knots. Concerning a fair comparison between the outcomes, each numerical run is based on the same wave realisation. Figure 3 shows the effects on ship dynamics when changing the vessel course of 15°. Figure 3A (left-hand side) regards ship dynamics and control actions, while Figure 3B (right-hand side) focuses on the alert detection and on the effects of the control application. It is possible to observe that large roll motions, in head sea, start developing after 12 min of navigation (i.e., around 720 s). The PAR_{index} becomes equal to 1, i.e., the alert signal is sent to the remote crew, with 90 s of delay. This corresponds to the sample time that the system takes to assess the current alert state of the ship.

Regarding these applications, in the absence of precise and dedicated data, it is assumed that the remote operator reacts within 30 s after the alert signal and quickly sets the autopilot to change the course by 15°. The ship change of heading to ψ = 160° does not produce any appreciable change to roll motions i.e., the hull is still within the parametric resonance condition. An additional change of course by 15°, leading to ψ = 145°, is effective in bringing the hull out of resonance, with moderate roll motions. Under such circumstances, it takes time to judge that the first change of heading was ineffective, thus, parametric roll lasts quite a long time before moving out from the resonance range. The best approach, in terms of incisive countermeasures to parametric roll, is the immediate change of course by 30°, soon after the alert is identified (see Figure 4). Once the command is fed to the autopilot, the PAR_{index} returns to zero after 4 min, which means that the parametric roll extinguishes after 2.5 min (i.e., 90 s in advance of the PAR_{index}). However, this choice would result in a quite significant re-scheduling of the ship route, although it guarantees a quick and effective action.

Another feasible approach, to avoid route modification, is the action involving a speed reduction, thus tolerating a slowdown due to the adverse weather conditions. Seen in Figure 5, a speed reduction of 5 knots is adopted as a countermeasure against parametric roll. However, the results in Figure 5A (left-hand side) and Figure 5B (right-hand side) show that this approach is not fully effective. The roll motion mitigates, but it does not extinguish completely. This is supported by the evidence that the PAR_{index} does not remain equal to zero after the speed reduction (especially the PAR_{index} related to pitch motion). Therefore, among the proposed actions, the speed reduction appears the least reliable and the least safe. A further reduction in speed would not guarantee a feasible alternative, not only from a routing point of view, it would be detrimental for vessel directional stability and course keeping, also leading to less dampened ship motions.

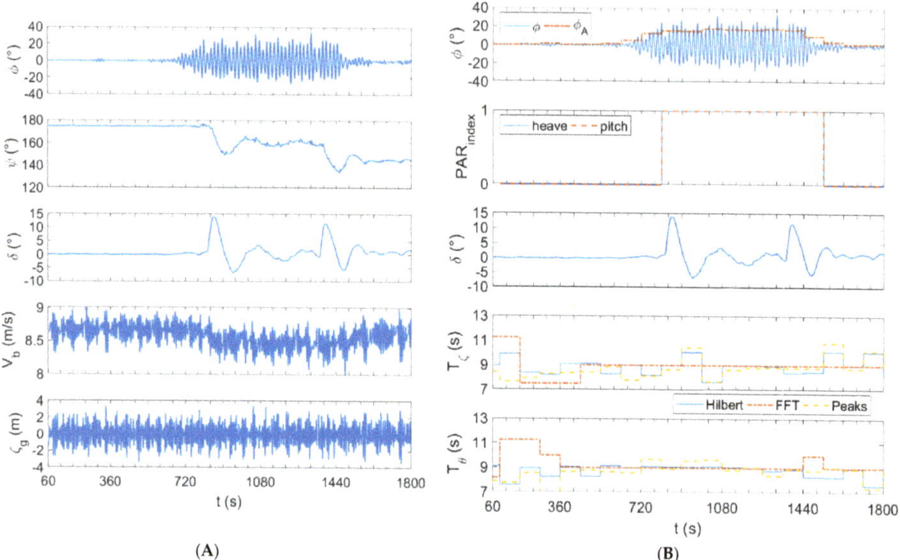

Figure 3. Parametric roll and course change of 15° plus 15° in irregular waves (JONSWAP $T_Z = 9.5$ s, $T_P = 12.2$ s, $H_S = 4.5$ m, N = 30, $V_S = 8.40$ m/s, ψ = 175°). (**A**) Ship dynamics outcomes; (**B**) Alert and control outcomes.

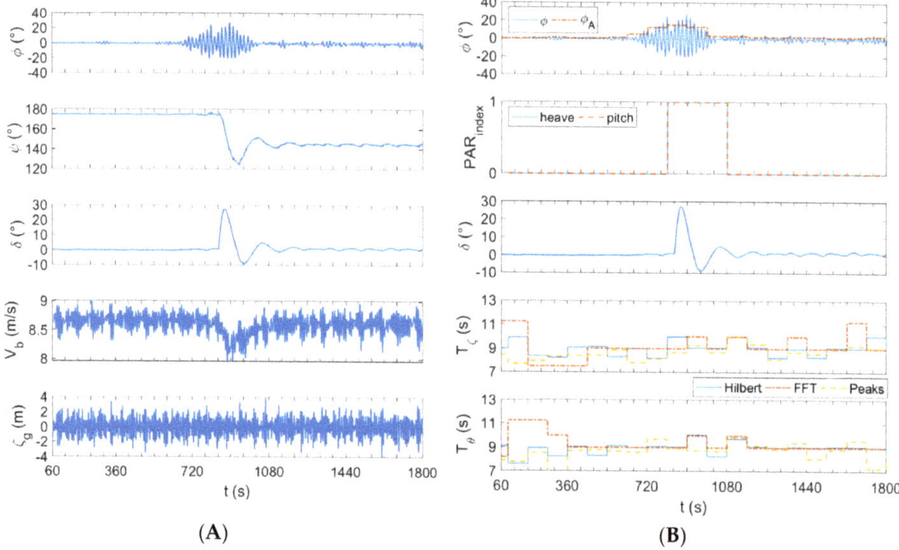

Figure 4. Parametric roll and course change of 30° in irregular waves (JONSWAP $T_Z = 9.5$ s $T_P = 12.2$ s, $H_S = 4.5$ m, N = 30, $V_S = 8.40$ m/s, ψ = 175°). (**A**) Ship dynamics outcomes; (**B**) Alert and control outcomes.

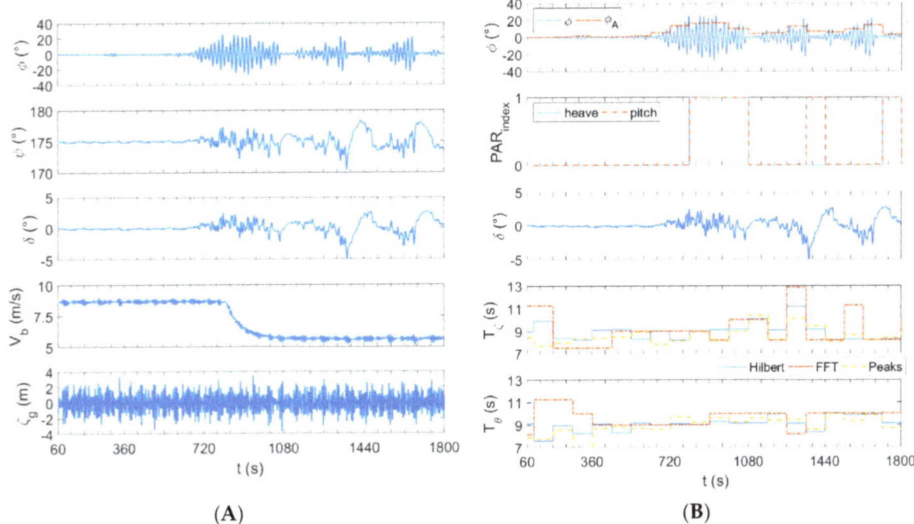

Figure 5. Parametric roll and speed reduction in irregular waves (JONSWAP T_Z = 9.5 s, T_P = 12.2 s, H_S = 4.5 m, N = 30, V_S = 8.40 m/s, ψ = 175°). (**A**) Ship dynamics outcomes; (**B**) Alert and control outcomes.

4.2. Synchronous Roll Scenario

Large roll motions can generally arise in the event of synchronous resonance. This phenomenon is critical for all ship types and it is a more frequent issue during ship navigation than parametric roll resonance. Therefore, the autonomous detection system has to provide alerts also in the case of synchronous roll threats.

Synchronous resonance is found for the operational condition in irregular sea characterised by T_Z = 5.5 s (peak period T_P = 7.1 s), H_S = 6 m, with ship speed V_S = 8.40 m/s and heading ψ = 45°. While for parametric roll case all three techniques provided comparable periods, thus improving the redundancy of the system, actually this is not confirmed by the outcomes of synchronous resonance case. Figure 6 yields an overview of the accuracy of the wave period estimation related to the alert index S_{index}. It appears that $\overline{T}_w$, obtained as the mean value among FFT, Hilbert and Peak calculated periods, leads to underestimated alert conditions. Indeed, if we evaluate the S_{index} separately for every single period, the FFT-based value is the most effective in alert detection, although it shows a larger variation in time. Conversely, the Peak-based value does not provide any alert signal in the whole time range. However, all approaches fail around 200–300 s, when ϕ_A exceeds 12.5°, and none of them identifies the alert. This issue also could be linked to the heading corresponding to the case study. The parametric resonance case was investigated in the head sea range, where heave and pitch motions are predominant, while the synchronous resonance case was investigated in the stern-quartering sea, where all motions develop. Moreover, the latter case also included larger autopilot actions for course keeping in oblique waves, characterized by low-frequency manoeuvring motions. All this might affect the fine development of heave and pitch motions in accordance with the wave period, when the ship operates in the stern-quartering sea. Nevertheless, these considerations bring out some limits in the proposed techniques that should be further analysed for assessing the most robust approach.

Alternatively, the alert could be assessed by focusing on roll amplitude only, leaving the autonomous detection of S_{index} versus PAR_{index} as discretionary indicators. Thus, the remote crew should judge between synchronous or parametric resonance by looking at the different weights of $\overline{T}_w$. Nevertheless, based on the application outcomes for the Seatech-D hull, the identification of the type of resonance has a great relevance in the type of countermeasure to adopt.

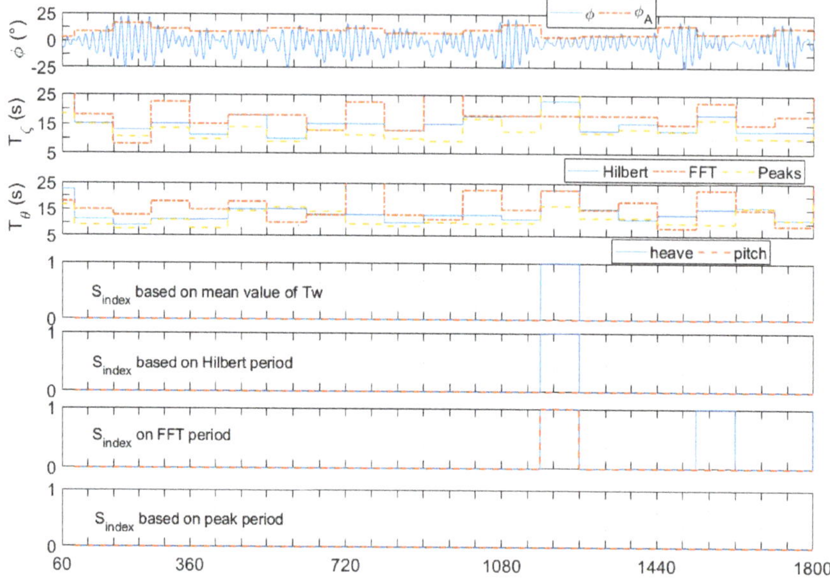

Figure 6. Synchronous roll and comparison of alert based on wave period estimation; irregular waves (JONSWAP T_Z = 5.5 s, T_P = 7.1 s, H_S = 6 m, N = 30, V_S = 8.40 m/s, ψ = 45°).

Particularly, once the alert is provided (around 1200 s) it appears that synchronous roll resonance extinguishes more easily than parametric resonance. Indeed, small changes in the heading (see Figure 7) and a modest reduction in ship speed (see Figure 8) guarantee the resolution to large roll motions. Figure 7 shows the dynamics of the ship to a course change of 15° given at 1230 s: roll amplitudes reduce significantly.

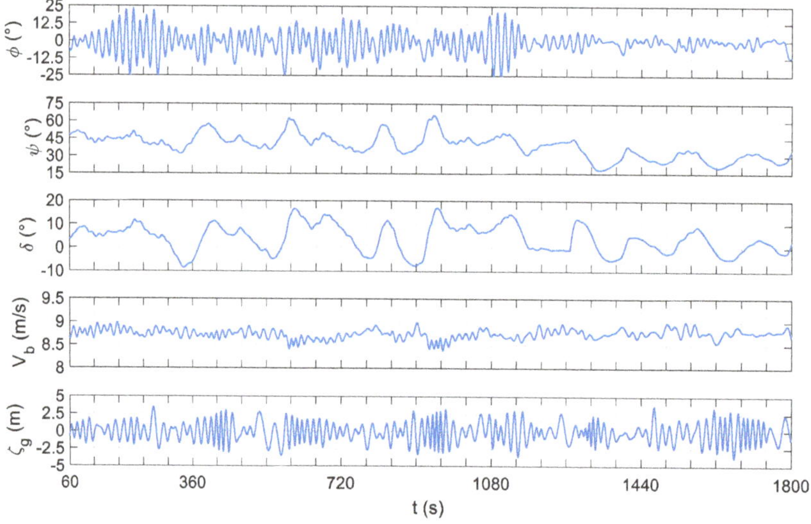

Figure 7. Synchronous roll avoidance by a course change of 15°; irregular waves (JONSWAP T_Z = 5.5 s, T_P = 7.1 s, H_S = 6 m, N = 30, V_S = 8.40 m/s, ψ = 45°).

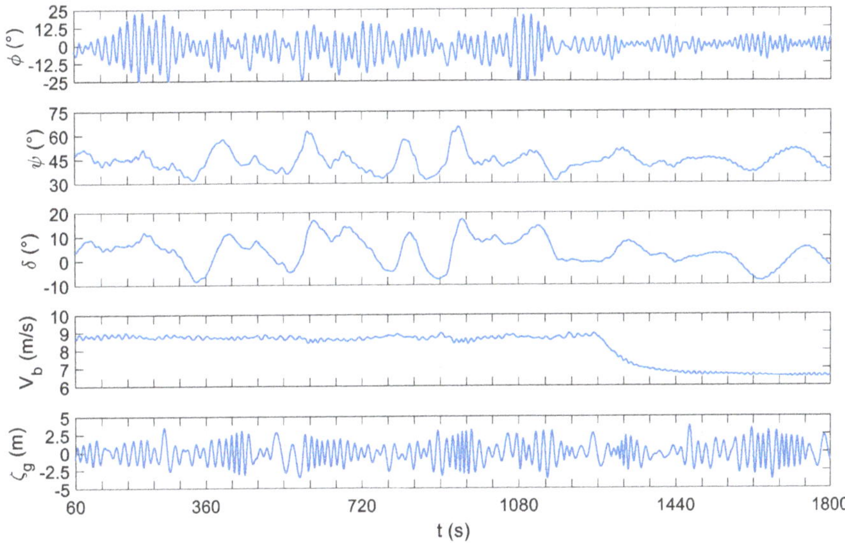

Figure 8. Synchronous roll avoidance by a speed reduction of 3.5 knots; irregular waves (JONSWAP T_Z = 5.5 s, T_P = 7.1 s, H_S = 6 m, N = 30, V_S = 8.40 m/s, ψ = 45°).

It is possible to observe that the oblique sea navigation requires a demanding rudder control for the course keeping in waves. This also is appreciable in Figure 8, where no course change is expected, but the rudder still has to operate to keep the initial heading. Moreover, Figure 8 points out that a speed reduction of around 3.5 knots is as effective as a course change in mitigating roll motions. Small deviations to initial operational conditions appear sufficient to react to synchronous resonance.

5. Conclusions

In this paper novel techniques for the smart assessment of large roll motions were implemented by a numerical simulation model trained to idealize the influence of evasive actions on ship dynamics. Parametric and synchronous roll resonance threats were investigated as applied to a ro-ro pax ship.

The estimation of the wave encounter period exploited the use of the ship as a floating buoy: short records of pitch and heave motions were processed in three different ways for monitoring their periods that, in our analysis, correspond to wave periods. The differences among the three adopted techniques appeared more pronounced for the synchronous roll, while for a parametric roll comparable results were obtained. However, the numerical simulation outcomes suggest that the most effective technique was the one based on FFT. The results of the applications pointed out that a proper discernment of the resonance mechanism between synchronous and parametric rolls is a key factor for choosing and executing an effective countermeasure. It is worth underlining that this paper did not account for the effects of active roll stabilisation devices; clearly, in case they were available on-board, their adoption could help in mitigating roll motion amplitude. During the simulations, a sufficient change of ship course that corresponded to an adequate modification of the heading angle, appeared more appropriate than the speed reduction in evading resonance phenomena. Particularly, for extinguishing large roll motions induced by a parametric resonance, the speed reduction approach was completely ineffective. Indeed, a course change of 30° was necessary to evade large roll amplitude, while for the synchronous roll case a change of course by 15°was sufficient.

The numerical model provided fairly accurate results, however, a further step of development should involve towing tank simulations on a scaled ship hull, to test the applicability of the proposed detection method and to check experimentally the more reliable approach for the estimation of the wave period.

The proposed method was developed keeping in mind future industrial applications on unmanned cargo ships prone to parametric and/or synchronous roll, to help a remote crew work against excessive roll motions and accelerations. Nevertheless, the method currently could be applied to manned ships as a smart tool for supporting crew decisions in distress situations involving roll resonance phenomena.

Author Contributions: Conceptualization, M.A. and F.B.; methodology, M.A.; software, M.A.; validation, M.A. and F.B.; formal analysis, M.A. investigation, M.A. and F.B.; data curation, M.A.; writing—original draft preparation, M.A.; writing—review and editing, M.A. and F.B; visualization, M.A and F.B. All authors have read and agreed to the published version of the manuscript.

Funding: This research received no external funding.

Conflicts of Interest: The authors declare no conflict of interest.

References

1. AAWA. *Remote and Autonomous Ships: The Next Steps*; AAWA: London, UK, 2016.
2. Felski, A.; Zwolak, K. The ocean-going autonomous ship—Challenges and threats. *J. Mar. Sci. Eng.* **2020**, *8*, 41. [CrossRef]
3. Glenn Wright, R. Intelligent autonomous ship navigation using multisensor modalities. *TransNav. Int. J. Mar. Navig. Saf. Sea Transp.* **2019**, *13*, 503–510. [CrossRef]
4. Long, M.T.; Bruhn, W.C.; Burmeister, H.C.; Long, M.T.; Moraeus, J.A. Conducting look-out on an unmanned vessel: Introduction to the advanced sensor module for Munin's autonomous dry bulk carrier. In Proceedings of the 10th International Symposium ISIS 2014 "Integrated Ship's Information Systems", Hamburg, Germany, 4–5 September 2014.
5. Wróbel, K.; Montewka, J.; Kujala, P. Towards the assessment of potential impact of unmanned vessels on maritime transportation safety. *Reliab. Eng. Syst. Saf.* **2017**, *165*, 155–169. [CrossRef]
6. Komianos, A. The autonomous shipping era. operational, regulatory, and quality challenges. *TransNav. Int. J. Mar. Navig. Saf. Sea Transp.* **2018**. [CrossRef]
7. Wróbel, K.; Montewka, J.; Kujala, P. Towards the development of a system-theoretic model for safety assessment of autonomous merchant vessels. *Reliab. Eng. Syst. Saf.* **2018**, *178*, 209–224. [CrossRef]
8. DNV-GL. *Remote-Controlled and Autonomous Ships*; DNV-GL: Oslo, Norway, 2018.
9. Acanfora, M.; Montewka, J.; Hinz, T.; Matusiak, J. On the estimation of the design loads on container stacks due to excessive acceleration in adverse weather conditions. *Mar. Struct.* **2017**, *53*. [CrossRef]
10. France, W.; Levadou, M.; Treakle, T.W.; Paulling, J.R.; Michel, R.K.; Moore, C. An investigation of head-sea parametric rolling and its influence on container lashing systems. *Mar. Technol.* **2003**, *40*, 1–19.
11. Acanfora, M.; Rizzuto, E. Time domain predictions of inertial loads on a drifting ship in irregular beam waves. *Ocean Eng.* **2019**, *174*, 135–147. [CrossRef]
12. Krata, P.; Szlapczynska, J. Ship weather routing optimization with dynamic constraints based on reliable synchronous roll prediction. *Ocean Eng.* **2018**, *150*, 124–137. [CrossRef]
13. Krueger, S. Evaluation of the cargo loss of a large container vessel due to parametric roll. In Proceedings of the 9th International Marine Design Conference, Dubrovnik, Croatia, 15–18 May 2006.
14. Levadou, M.; Gaillarde, G. Operational guidance to avoid parametric roll. In Proceedings of the Design and Operation of Containerships, London, UK, 1 January 2003; pp. 75–86.
15. Hashimoto, H.; Umeda, N.; Ogawa, Y.; Taguchi, H.; Iseki, T.; Bulian, G.; Ishida, S.; Toki, N.; Matsuda, A. Prediction methods for parametric rolling with forward velocity and their validation—Final report of scape committee. In Proceedings of the 6th Osaka Colloquium on Seakeeping and Stability of Ships, Osaka, Japan, 26–29 March 2008.
16. Shigunov, V.; Moctar, O.E.; Rathje, H. Operational guidance for prevention of cargo loss and damage on container ships. *Sh. Technol. Res.* **2010**, *57*, 8–25. [CrossRef]
17. Acanfora, M.; Montewka, J.; Hinz, T.; Matusiak, J. Towards realistic estimation of ship excessive motions in heavy weather. A case study of a containership in the Pacific Ocean. *Ocean Eng.* **2017**, *138*. [CrossRef]

18. Belenky, V.; Yu, H.-C.; Weems, K. Numerical procedures and practical experience of assessment of parametric roll of container carriers. In Proceedings of the 9th International Conference on Stability of Ships and Ocean Vehicles, Rio de Janeiro, Brazil, 25–29 September 2006.
19. Galeazzi, R.; Blanke, M.; Poulsen, N.K. Early detection of parametric roll resonance on container ships. *IEEE Trans. Control Syst. Technol.* **2013**, *21*, 489–503. [CrossRef]
20. Galeazzi, R.; Blanke, M.; Poulsen, N.K. Detection of parametric roll for ships. In *Parametric Resonance in Dynamical System*; Fossen, T.I., Nijimeijer, H., Eds.; Springer: Berlin/Heidelberg, Germany, 2012; ISBN 978-1-4614-1042-3.
21. McCue, L.S.; Bulian, G. A numerical feasibility study of a parametric roll advance warning system. *J. Offshore Mech. Arct. Eng.* **2007**, *129*, 165. [CrossRef]
22. Hashimoto, H.; Taniguchi, Y.; Fujii, M. A case study on operational limitations by means of navigation simulation. In Proceedings of the 16th International Ship Stability Workshop (ISSW2017), Belgrade, Serbia, 5–7 June 2017; pp. 5–7.
23. Nielsen, U.D.; Brodtkorb, A.H.; Sørensen, A.J. Sea state estimation using multiple ships simultaneously as sailing wave buoys. *Appl. Ocean Res.* **2019**, *83*, 65–76. [CrossRef]
24. Nielsen, U.D. Estimations of on-site directional wave spectra from measured ship responses. *Mar. Struct.* **2006**, *19*, 33–69. [CrossRef]
25. Acanfora, M.; Krata, P.; Montewka, J.; Kujala, P. Towards a method for detecting large roll motions suitable for oceangoing ships. *Appl. Ocean Res.* **2018**, *79*, 49–61. [CrossRef]
26. Fossen, T.I. *Guidance and Control of Ocean Vehicles*; Wiley: Chichester, UK, 1996; ISBN 0-471-94113-1.
27. Matusiak, J.; Stigler, C. Ship roll motion in irregular waves during a turning circle maneuver. In Proceedings of the 11th International Conference on Stability of Ships and Ocean Vehicles, Athens, Greece, 23–28 September 2012; pp. 291–298.
28. Stigler, C. *Investigation of the Behaviour of a RoPax Vessel in Stern Quartering Irregular Long-Crested Waves*; Aalto University: Espoo, Finland, 2012.
29. Matusiak, J. *Dynamics of a Rigid Ship*; Aalto University Publication Series: Science + Technology: Espoo, Finland, 2013; ISBN 9789526052045.
30. Acanfora, M.; Matusiak, J. On the estimations of ship motions during maneuvering tasks in irregular seas. In Proceedings of the 3rd International Conference on Maritime Technology and Engineering, Lisbon, Portugal, 4–6 July 2016; Volume 1.
31. ITTC The Specialist Committee on Prediction of Extreme Ship Motions and Capsizing Final Report and Recommendations to the 23rd ITTC. In Proceedings of the 23rd International Towing Tank Conference, Venice, Italy, 8–14 September 2002; Volume II, pp. 619–748.
32. Spanos, D.; Palmquist, M. Benchmark Study on Numerical Simulation Methods for the Prediction of Parametric Roll of Ships in Waves. In Proceedings of the 10th International Conference on Stability of Ships and Ocean Vehicles, Sankt Petersburg, Russia, 22–26 June 2009; pp. 1–9.
33. Shigunov, V.; el Moctar, O.; Papanikolaou, A.; Potthoff, R.; Liu, S. International benchmark study on numerical simulation methods for prediction of manoeuvrability of ships in waves. *Ocean Eng.* **2018**, *165*, 365–385. [CrossRef]
34. Cummins, W.E. *The Impulse Response Function and Ship Motions*; Institut für Schiffbau der Universität Hamburg: Hamburg, Germany, 1962; Volume 57.
35. RINA. *RINA Rules, Steering Gear, Part C, Charpter 1, Section 11*; RINA: Genoa, Italy, 2017.
36. Chen, G.; Wang, Z. A signal decomposition theorem with Hilbert transform and its application to narrowband time series with closely spaced frequency components. *Mech. Syst. Signal Process.* **2012**, *28*, 258–279. [CrossRef]

Article

Dynamic Prediction and Optimization of Energy Efficiency Operational Index (EEOI) for an Operating Ship in Varying Environments

Chao Sun [1,2], Haiyan Wang [3,*], Chao Liu [2] and Ye Zhao [4]

1 Marine Engineering College, Dalian Maritime University, Dalian 116026, China; aba11@163.com
2 Systems Engineering Research Institute, Beijing 100036, China; doming81@163.com
3 Merchant Marine College, Shanghai Maritime University, Shanghai 200136, China
4 Shanghai Rail Transportation Equipment Co., Ltd., Shanghai 200245, China; zhaoyeshanghai@163.com
* Correspondence: wanghaiyan@shmtu.edu.cn; Tel.: +86-21-3828-2974

Received: 22 September 2019; Accepted: 5 November 2019; Published: 8 November 2019

Abstract: The demands for lower Energy Efficiency Operational Index (EEOI) reflect the requirements of international conventions for green shipping. Within this context it is believed that practical solutions for the dynamic optimization of a ship's main engine and the reduction of EEOI in real conditions are useful in terms of improving sustainable shipping operations. In this paper, we introduce a model for dynamic optimization of the main engine that can improve fuel efficiency and decrease EEOI. The model considers as input environmental factors that influence overall ship dynamics (e.g., wind speed, wind direction, wave height, water flow speed) and engine revolutions. Fuel consumption rate and ship speed are taken as outputs. Consequently, a genetic algorithm is applied to optimize the initial connection weight and threshold of nodes of a neural network (NN) that is used to predict fuel consumption rate and ship speed. Navigation data from the training ship "YUMING" are applied to train the network. The genetic algorithm is used to optimize engine revolution and obtain the lowest EEOI. Results show that the optimization method proposed may assist with the prediction of lower EEOI in different environmental conditions and operational speed.

Keywords: merchant shipping; EEOI; genetic algorithms; neural networks; ship dynamics

1. Introduction

According to reports from the International Maritime Organization (IMO) [1], unless serious action is taken, annual CO_2 emissions from the international shipping industry may increase by 1.5~3.5 times by 2050. For this reason, vigorous developments regarding ship energy conservation and emission reduction technologies are essential [2,3].

The Energy Efficiency Operational Index (EEOI) was introduced by the IMO in 2009 to evaluate the CO_2 emission performance of a ship. It served as a benchmark tool in monitoring ship and fleet efficiency performance. The indicator enables operators to measure the fuel efficiency of a ship in operation and to gauge the effect of any changes in operation [4]. The lower the EEOI is, the better the performance is.

The CO_2 emission performance is directly affected by the environmental conditions and ship sailing state. However, the navigation environment varies with time and sea area. Therefore, the EEOI cannot be maintained at a lower level all the time. At the same time, fouling of the submerged part of the hull, degradation of the power plant and fluctuation of engine parameters will bring disadvantages to ship performance and enlarge the EEOI. Hence, there is a significant economic and environmental benefit to developing a general optimization method that may help to decrease the EEOI in varying environmental conditions.

Harilaos N. Psarafits et al. stated that the ship speed is a decision variable for fuel consumption and emissions [5]. In fact, most methods applied to improve ship efficiency are to adjust ship speed. Route and speed optimization are operational procedures that may be used to improve shipping efficiency and EEOI. Inge Norstad et al. used a recursive smoothing algorithm to optimize the speed for ship routing and scheduling for a tramp ship [6]. Wang Shuai'an et al. conducted an in-depth analysis on the relationship between fuel consumption and ship speed based on the historical operational data of a container liner and optimized the ship speed by using nonlinear programming [7]. Ming-Chung Fang et al. applied a heuristic method to optimize ship routing in different weather conditions [8].

Lower steaming is another kind of validated method for decreasing fuel consumption and the EEOI. However, such operational strategy is limited by many factors, including the ETA (estimated time of arrival), fuel price, charter rates, influence of speed reduction on engine efficiency, and so forth [9,10].

Joan P. Peterse developed a fuel consumption model by way of a machine learning method and ship navigation data [11]. Benjamin applied a neural network (NN) trained by noon report data to establish a propulsion power model under hydrostatic conditions [12]. Wang Kai et al. used the wavelet packet neural network to predict the sea conditions for a short-journey of a sailing ship and introduced a real-time energy efficiency model under predicted sea conditions [13].

This paper suggests a method that could be used to identify lower EEOI using real time operational data. An EEOI during a fixed time span is introduced to indicate fuel efficiency. A back propagation neural network, which is trained by navigation data, is used to predict ship speed and fuel consumption rate under real environmental conditions. The genetic algorithm is applied to obtain the lowest EEOI by optimizing the main engine revolution in a fixed time span in real conditions.

2. EEOI Prediction Model

In the guide issued by IMO in 2009 [14], it is shown that the EEOI can be expressed as:

$$EEOI = \frac{M_{CO_2}}{Q} = \frac{\sum_{j=1}^{n} F_{C_j} \cdot C_{F_j}}{M \cdot D}, \tag{1}$$

where M_{CO_2} is the carbon dioxide emissions (t), Q is the transport volume (t), F_{C_j} is the fuel consumption in the jth voyage segment (t), C_{F_j} is the carbon dioxide conversion factor in the jth voyage segment, M is the cargo capacity (t), and D is the transport distance (nm).

Transport volume is constant over a single voyage. Therefore, EEOI can be defined as the function of fuel consumption and transport distance. In practice, fuel consumption rate and ship speed vary. The fuel consumption over transport distance can be obtained by integration of the fuel consumption rate and ship speed. If a periodical time interval is applied to the integration, real time EEOI can be obtained, as shown in Equation (2):

$$EEOI = \frac{C_F \int_0^T f_C dt}{M \int_0^T v_s dt}, \tag{2}$$

where T is the cycle time (s), f_C is the fuel consumption rate (t/s), and v_s is the ship speed (m/s).

There is no voyage segment parameter, because there is only one segment in the short cycle time during a single voyage.

2.1. Factors

The navigation environment is a complex nonlinear system of systems, the interaction of which includes ship speed and fuel consumption (see Figure 1). For this reason, sufficient capture and clustering of big data analytics by neural networks is essential.

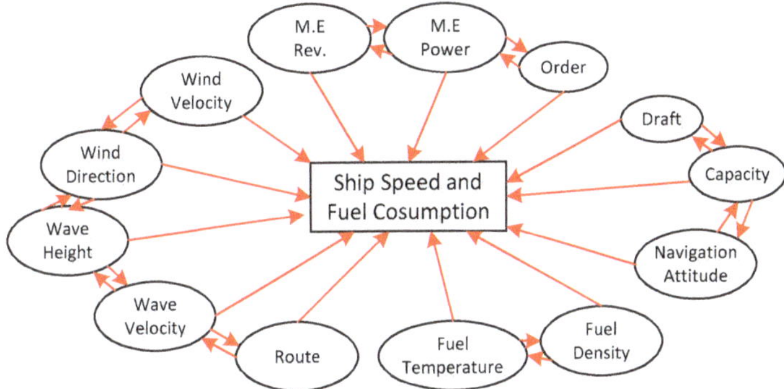

Figure 1. The main factors affecting ship speed and fuel consumption.

It is noted that ship draft, navigation attitude and route can be ignored because these parameters are constant over a voyage.

2.2. Big Data Analytics

Big data collected by the surveillance system are characterized by high Volume, low Veracity, high Variety, high Velocity, and high Value [15]. The navigation data collected during sailing have the same 5V characteristics. Irrelevant data and errors may result in excessive noise signal traces. It is therefore essential to pre-process data by following the procedures outlines in Sections 2.2.1–2.2.4 below [16].

2.2.1. Big Data Trimming

Because the vessel may be docked or berthed, some subsets of the data are not only useless but also may damage the prediction ability of the model. Therefore, these subsets of the data must be pruned. As shown in Figure 2, the subsets of the data marked by the red ellipse are data collected in the time period when the ship is in a port or berthing state. The power of the main engine is zero in these states, and subsets should be pruned. In addition, the acceleration and deceleration processes at the beginning and end of the data segment should be removed because data do not contribute much in terms of modeling the entire voyage conditions.

2.2.2. Big Data Synchronization

Un-synchronized data, i.e. data that are not consistent in term of time scales, may result in inconsistent model input dimensions. In such situations, there is large probability that the input matrix may be singular matrix and not suitable for long-term vacancies.

From a practical perspective, each data block corresponds to a single trip (see Figure 3). Generally, all dimensional data should have the same length at the time axis. That is to say, the data from every factor should entirely overlap. Therefore, the method of synchronizing is to eliminate the part where the data from factors do not overlap and to choose valuable data from the overlapping zone. Then, an equal-length data set can be obtained, as shown in Figure 3.

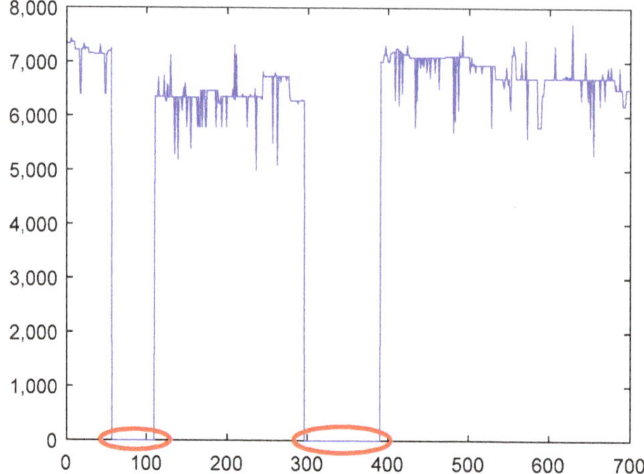

Figure 2. Data trimming diagram.

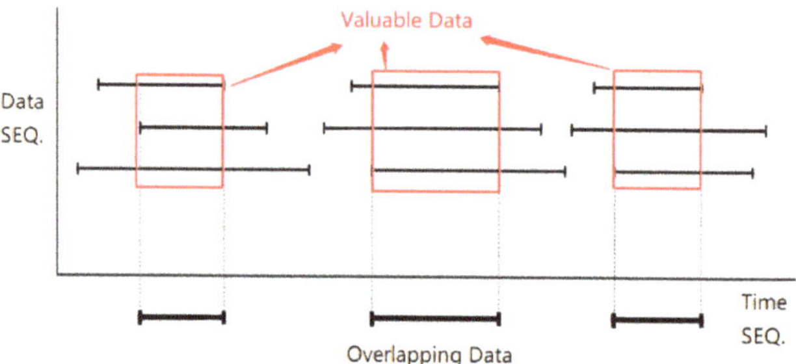

Figure 3. Data synchronization diagram.

2.2.3. Big Data Normalization

Big data are usually not within the same order of magnitude. If they are directly applied for training the neural network and the weights of the network nodes will vary greatly, bringing adverse effects to data analysis results. In order to eliminate the differences, the selected data must be normalized to achieve comparability among different scales. A common normalization process is to linearly scale the data and compress it to a closed interval [0,1], as shown in Equation 3:

$$x^* = \frac{x - \min}{\max - \min}, \tag{3}$$

where x^* is the normalized data value, x is the original data value; max is the maximum possible value in the original data, and min is the minimum possible value in the original data.

2.2.4. Dimensional Reduction Processing

Based on knowledge from ship navigation dynamics and the actual operating experience of the ship, some variables of the input set may interact and relate to each other. To mine the main variables and extract key features, it is necessary to carry out dimensional reduction.

In this paper, an intuitive correlation analysis method was used to reduce the dimensionality of data [17,18]. Accordingly, Equation 4 was used to calculate the correlation coefficient matrix R:

$$R = \begin{bmatrix} r_{11} & r_{12} & \cdots & r_{1n} \\ r_{21} & r_{22} & \cdots & r_{2n} \\ \vdots & \vdots & \vdots & \vdots \\ r_{m1} & r_{m2} & \cdots & r_{mn} \end{bmatrix}, \ r_{xy} = \frac{S_{xy}}{S_x S_y} \tag{4}$$

where r_{xy} is the sample correlation coefficient, S_{xy} is the sample covariance, S_x is sample X's standard deviation, and S_y is sample Y's standard deviation.

The data correlation has been visualized, as shown in Figure 4. The components of the vector presented are ship speed, distance, main engine revolution (M.E. Rev.), fuel consumption (Fuel Cons., which is the total quantity in a fixed time span), fuel consumption rate (F.C. Rate), M.E. power, EEOI, telegraph order command (Order), wind speed (Wind Spd.), wind direction (Wind Dir.), wave height (Wave Hgt.) and water flow speed (Flow Spd.). Analysis of data analytics has shown that five variables, including the telegraph order, engine power, fuel consumption, fuel consumption rate and engine revolution are strongly correlated with each other. Therefore, these variables were combined into one influencing factor, "M.E. Rev". According to Section 2.1, the ship speed and fuel consumption rate are the two output variables of the model. The water flow speed, wave height, wind direction and wind speed have different degrees of correlation with the outputs but have weak correlations with each other or other factors. Therefore, the number of dimensions of input variables finally reduced to 5, including the engine revolution, wind speed, wind direction, wave height, and water flow speed. These variables were taken as the inputs to the neural network.

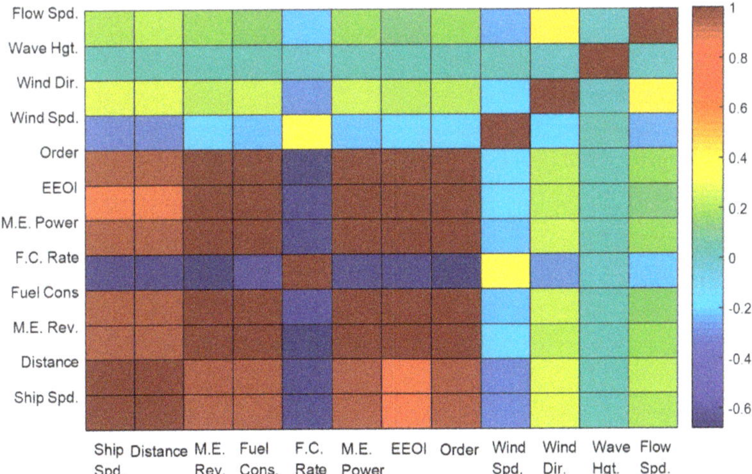

Figure 4. Data correlation diagram.

2.3. Construction of the Neural Network

2.3.1. Basic Structure

According to the previous section, the neural network has 5-dimensional inputs and 2-dimensional outputs. The classic 3-layer BP (Backward Propagation) neural network selected, and the number of hidden layers was calculated using an empirical formula, as shown in Equation (5):

$$n_h = \sqrt{n_i + n_o} + l = \sqrt{5+2} + 3 = 5, \tag{5}$$

where n_h is the number of neurons in the hidden layer, n_i is the number of nodes in the input layer, n_o is the number of nodes in the output layer, and l is the number of layers. n_h is generally a constant from 0 to 12. The network structure is shown in Figure 5. In this structure, the transfer function of the hidden layer is the Sigmoid function, and the transfer function of the output layer is a linear function.

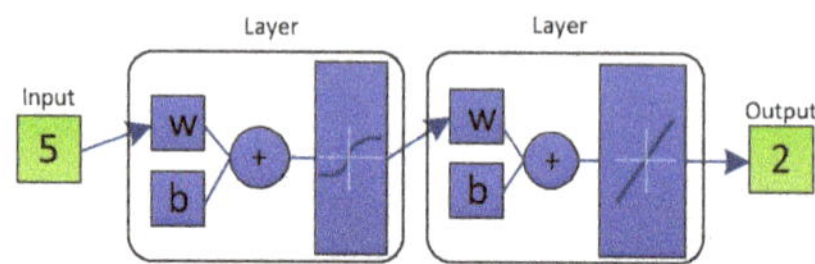

Figure 5. The network structure.

2.3.2. Initial Structure Optimization by the Genetic Algorithm

A genetic algorithm was applied to optimize the initial connection weights and thresholds which directly affect the approximation ability and adaptability of the network [19,20]. The genetic algorithm was used to globally optimize the initial connection weights and thresholds of each node in the EEOI prediction model.

The encoding method of the individual population in the genetic algorithm adopted a simple and efficient binary coding method. The coding length was set to 30. Each chromosome consisted of a piece of binary code of the connection weight and threshold, which are generally limited in the search space [21].

The parameters of the genetic algorithm were specifically set as follows—the fitness function was the mean square error between the actual output of the neural network and the expected output; the number of chromosomes was set to 50; the maximum number of iterations was set to 100; the number of variables was set to 42; the crossover rate of the chromosome was set to 0.92; and the mutation rate of the chromosome was set to 0.0238.

The genetic algorithm performance tracking is shown in Figure 6. After 40 generations, the objective value was almost constant. The variation in the fitness value was small after 40 generations. Therefore, the result after 100 generations can be seen as the optimized value.

After 100 generations of inheritance, the initial weights and bias matrices of the BP network were obtained, as shown in Appendix A.1 Section. The optimized link weights and bias were assigned to the EEOI prediction model, and the optimized neural network has been trained using historical data to update the weights and bias.

2.4. *Training the Network*

The pretreated 9890 navigation data were used to train the neural network. The mean square errors of the expected and actual output values were taken as the total error signal. The error back propagation algorithm was used. The number of iterations was set to 10000. The training was stopped at the 223th iteration when the mean square error reached 0.911×10^{-7}. The convergence speed was greatly improved. The training parameter curve is shown in Figure 7.

The BP neural network parameters after training were shown in Appendix A.2 Section.

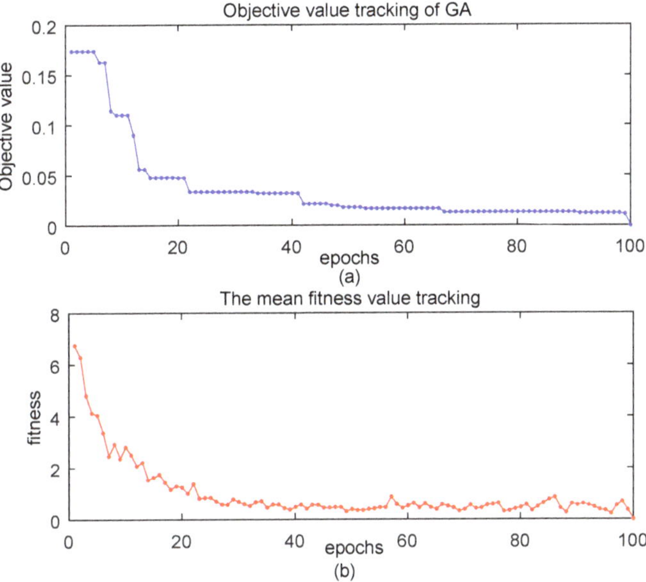

Figure 6. Genetic algorithm performance tracking. (**a**) Objective value tracking (**b**) The mean fitness value tracking.

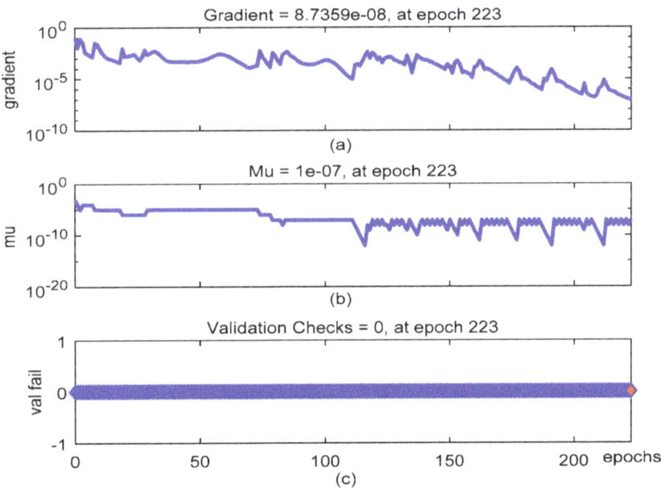

Figure 7. The training parameter curve. (**a**) Gradient (**b**) Mu (**c**) validation fail.

3. Ship Propulsion Model

The trained neural network can predict fuel consumption and ship speed on board. It can also be applied to the optimization method of EEOI. However, the optimized engine revolution cannot be directly used on board before it has been sufficiently validated. Therefore, a ship propulsion model comprising of a fixed pitch propeller driven by a low-speed direct reversion marine diesel engine was built in Simulink to rapidly verify the algorithm [22]. The propulsion model (see Figure 8) could provide some extra data for the neural network. In this model, the ME module represents the main engine, which is a large-scale, low-speed, two-stroke diesel. The prop module represents the propeller. The ship module models the ship motion. The resistance module is the model of ship's resistance when sailing. The wind/wave module simulates the wind and wave force. The Setting Unit is used to set orders to the model. The monitoring module is used for parameter observation.

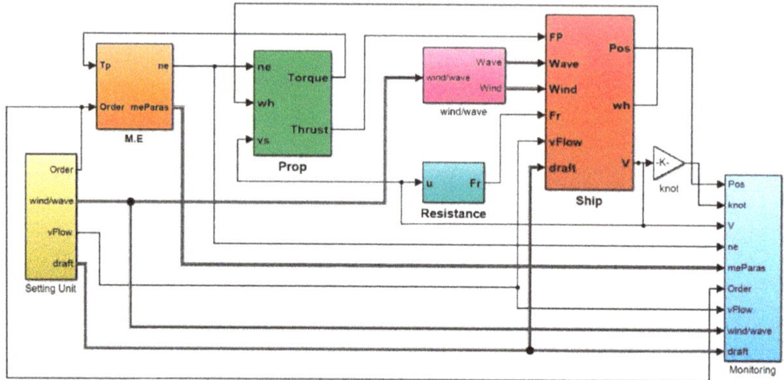

Figure 8. Ship propulsion model in Simulink.

3.1. Main Engine

The mean value diesel engine model used is a combination of the quasi-steady and volumetric models and it has been comprehensively used for non-linear control and state observations. Because of its faster computing speed, the model is widely used. In order to simulate maneuvering conditions, such as starting, braking and reversing, the model has to be modified to adapt to all-regime running conditions.

Under reversing working conditions, the engine revolution has to enter a zero crossing situation. The static friction torque, which is usually ignored under normal conditions, should be considered. Therefore, the modified friction model was introduced, as shown in Equation (6):

$$T_f = \begin{cases} T_{fn}(|n_e|) * \mathrm{sgn}(n_e) & |n_e| \geq n_s \\ T_e & |n_e| < n_s \& |T_e| < T_s, \\ T_s * \mathrm{sgn}(T_e) & others \end{cases} \tag{6}$$

where T_f is the friction torque (Nm), T_e is the driven torque (Nm), which is the sum of the indicated torque, starting torque, and propeller resistance torque, T_s is the maximum static friction torque (Nm), T_{fn} is the viscous friction torque (Nm), n_e is the engine revolution (rpm), and n_s is the speed dead zone for changing friction torque direction (rpm). For doing so, friction torque flutter is avoided.

Starting and braking is usually fulfilled by supplying compressed air into cylinders. According to a force analysis of the crank connecting rod mechanism [22], the mean torque value of compressed air for a multi-cylinder diesel can be defined as

$$T_{sa} = \mathrm{sgn}(cam)\frac{D^2(p_{sa} - p_s)S_t N_{cyl}}{16}\left[\frac{3}{2} + \frac{1}{\lambda}\left(1 - \sqrt{1 - \frac{3}{4}\lambda^2}\right)\right], \tag{7}$$

where cam is the direction of the air distributor cam, p_{sa} is the starting air pressure (Pa), p_s is the intake manifold pressure (Pa), S_t is the stroke (m), N_{cyl} is the number of cylinders, λ is the ratio of the crank to the connecting rod, and D is the diameter of the cylinder (m).

3.2. Propeller

A propeller has four quadrant working conditions that match ship motions. Those are known as—forward ship velocity/positive propeller rotation, forward velocity/negative rotation, backward velocity/negative rotation, and backward velocity/positive rotation. Figure 9 shows the thrust coefficient k_p and torque coefficient k_m as functions of the bounded advanced ratio.

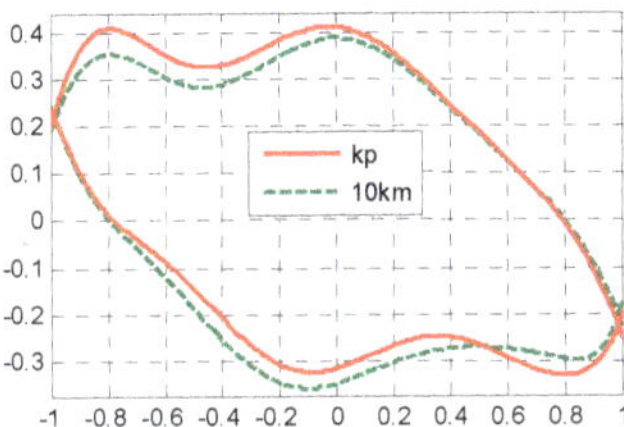

Figure 9. Propeller characteristics across four quadrants.

The bounded advanced ratio was defined as:

$$\lambda_p = (1-w)v_s / \sqrt{(1-w)^2 v_s^2 + D_p^2 n_p^2}, \tag{8}$$

where λ_p is the bounded advanced ratio, w is the wake factor, v_s is the ship velocity (m/s), D_p is the diameter of the propeller (m), and n_p is the rotational speed of the propeller (rev/s), which is equal to that of the engine, owing to the propeller directly connecting to the engine.

The torque (T_p) and thrust (P_p) of propeller were defined as:

$$T_p = k_m(1-t)\rho D_p^3\left[(1-w)^2 v_s^2 + D_p^2 n_p^2\right]. \tag{9}$$

$$P_p = k_p(1-t)\rho D_p^2\left[(1-w)^2 v_s^2 + D_p^2 n_p^2\right], \tag{10}$$

where t is the thrust-deduction fraction, and ρ is the density of sea water (kg/m^3).

3.3. Ship Longitudinal Motion

In this study, only longitudinal motion has been considered. Hence ship velocity was calculated by

$$\dot{v}_s = (P_p - R_f)/(m + m_\lambda), \tag{11}$$

where m is the mass of the ship (kg), and m_λ is the mass of the entrained water (kg). The value of m_λ depends on the mass and velocity of the ship. Usually, it may be 0.2 times the mass of the ship when the ship is sailing at a normal speed. R_f is defined as the resistance of the ship (N) and it is defined as:

$$R_f = -C_t \rho S v_s^2, \tag{12}$$

where C_t is the total friction coefficient, S is the wet area of the ship (m^2) calculated on the basis of the ship's principal dimensions.

3.4. Wind and Wave Forces

Wind and wave forces may affect ship speed. However, it is difficult and time consuming to build an exact wind and wave force model. Moreover, instantaneous force waveforms of wind and wave have no significant influences on the ship's longitudinal motion. The wind speed can be seen as superposition of the mean wind speed and disturbance wind speed. Furthermore, the disturbance wind speed can be seen as white noise. So, the wind model can be simplified. On this basis a simplified model, which is widely used in ship maneuvering simulator, was implemented here. The wind force has been related to the wind speed, angle, and area of the ship exposed to wind, as shown in Equation (13).

$$R_w = 0.5\rho_a A_w U_w C_w(\alpha_w), \tag{13}$$

where R_w is the wind resistance, (N), ρ_a is the air density, (kg/m^3), A_w is the ship's orthographic projection area above the water line, (m^2), U_w is the relative wind speed (m/s), $C_w(\alpha_w)$ is the wind force coefficient, and α_w is the relative wind direction angle (deg).

When modeling the wave force, only the first-order wave force was considered and accordingly the resistance of the wave was defined as:

$$R_v = 2aB\frac{\sin b \cdot \sin c}{c}s(t), \tag{14}$$

where $a = \rho g\left(1 - e^{-kd}\right)/k^2$, $b = kL/2\cos\chi$, $c = kB/2\sin\chi$, $s(t) = (kh/2)\sin(\omega_e t)$, $k = 2\pi/\lambda_w$. λ_w is the wavelength (m), χ is the relative wave direction, and ω_e is the encountered frequency of the ship and waves (Hz).

4. Optimization of M.E. Revolutions

The trained network was used to construct the objective function. The main engine (ME) revolution was taken as the optimizing variable. The fitness function is the difference between the actual and predicted EEOI values after optimization. The fitness function $FitV$ is defined as shown in Equation (15):

$$\begin{aligned} FitV &= \max[EEOI - EEOI'] = \max[EEOI - GABP(n_e)] \\ &= \max\left[EEOI - \frac{C_f * F_{c,pre} * 1853.2}{M_{cargo} * D_{pre}}\right] \end{aligned} \tag{15}$$

where $EEOI'$ is the last value of $EEOI$, which is predicted by $GABP(n_e)$, which represents the BP NN built in Section 2, n_e is the M.E. RPM, C_f is the carbon dioxide conversion factor, M_{cargo} is the ship-loading quantity (t), $F_{c,pre}$ is the predicted fuel consumption (f), and D_{pre} is the predicted sailing distance (nm).

Three steps were taken to build fitness function. The first step was to predict the engine revolution and fuel consumption rate using the neural network model built in Section 2. The second step was to integrate them to determine the sailing distance and the fuel consumption during sailing. The last step was to calculate the difference between the actual EEOI and the predicted EEOI after adjusting the engine revolution. After the optimization was finished, the optimized main engine revolution was set to the real engine. The ship propulsion model was applied to represent the actual ship and the actual engine. The network was trained online at the same time of optimization. The entire EEOI index dynamic optimization algorithm flow is shown in Figure 10.

After the objective function was established, the genetic algorithm was applied to perform global optimization. The boundary condition was that the engine revolution was controlled between 80 and 130 rpm, and the power was controlled between 75% and 100% MCR (Maximum Continuous Rating point). The navigation environment was set as shown in Table 1.

Table 1. Initial value of the navigation environment.

Parameter	Value
Wind speed (m/s)	9
Wind angle (deg.)	62
Wave height (m)	5
Water flow speed	−6
Main engine revolution (rpm)	10.3

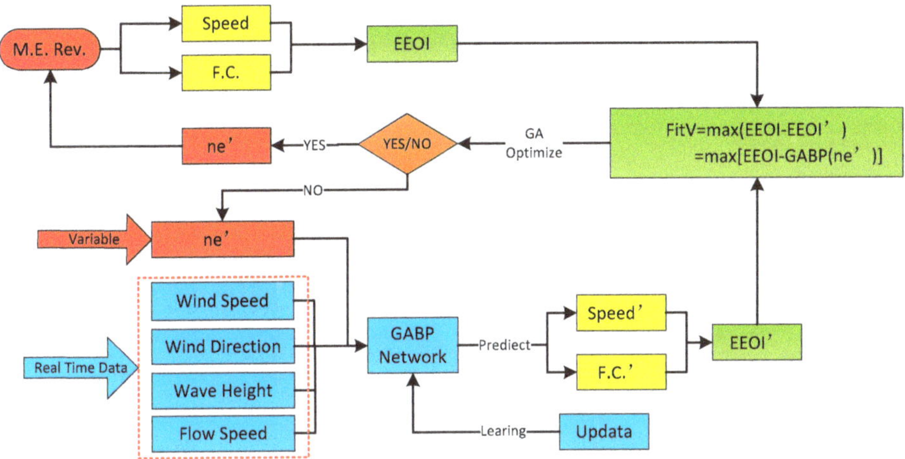

Figure 10. The dynamic optimization algorithm flow.

The EEOI dynamic optimization algorithm was used to optimize the engine revolution. The algorithm performance tracking is shown in Figure 11. The predicted EEOI was lowest when the engine revolution was kept at 122 rpm. Therefore, the optimized engine revolution was used to set the real engine as the new order.

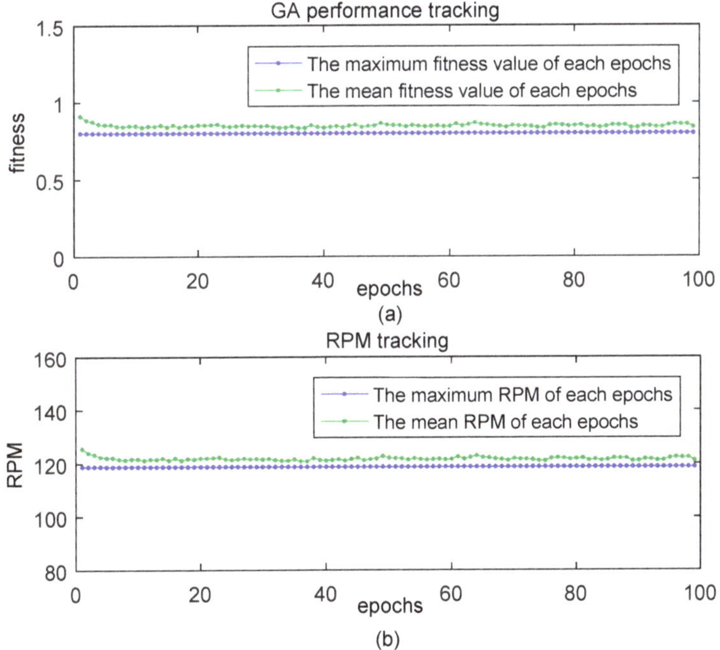

Figure 11. Performance tracking of the engine revolution optimization algorithm. (**a**) GA performance tracking (**b**) RPM tracking.

5. Results

The results presented in this paper are based on the training ship "YUMING," which belongs to Shanghai Maritime University. "YUMING" is a bulk carrier and can accommodate 160 personnel. Its parameters are shown in Table 2.

Table 2. Parameters of the YUMING vessel.

Parameter	Value	Parameter	Value
Length (m)	189.9	Breadth (m)	32.3
Depth (m)	15.7	Block Coefficient	0.85
Design draft (m)	10.3	Design Displacement (t)	58123.9
Type of M.E	6S50ME	Rated Revolution (rev/min)	127
Stroke (m)	2	Rated Power (kW)	7948
Diameter of propeller (mm)	5850	Number of blades	4

5.1. Simulation of the Ship Propulsion Model

To verify the performance of the model, simulations were conducted under the same navigation environments experienced by YUMING during sea trial. Accordingly, the ship forward draft and aft draft were set to 7.10 m and 7.95 m, respectively. The revolution of the main engine was set to 126.5, 122, 114.8, and 100.2rev/min, respectively. A comparison demonstrating good agreement of data that resulted from the actual ship sea trials and simulation is shown in Table 3. Accordingly, the ship propulsion model may be considered credible.

Table 3. Comparison between the simulation model and real ship data.

Navigation Data						Actual	Simu.	Actual	Simu.
Load	Engine Rev (RPM)	Wind Speed (m/s)	Wind Direction (°)	Wave Height (m)	Flow Speed (m/s)	Power (kW)	Power (kW)	Ship Speed (kn)	Ship Speed (kn)
100%	126.5	7.0	300	0.2	0.4	6986.5	7073	14.165	14.12
90%	122	6.0	30	1	−0.6	6390.5	6337	13.748	13.61
75%	114.8	7.7	80	0.4	0.2	5335.0	5270	12.700	12.53
50%	100.2	8.1	90	0.3	−0.4	3508.0	3506	10.945	10.58

5.2. Results of Optimization

The EEOI prediction model, ship propulsion model and genetic optimization algorithm were used to simulate the whole closed-loop optimization algorithm of the main engine revolution under different environmental conditions. The ship was assumed to operate in ballast condition, with forward and aft drafts were set according to the sea trial report. The cycle time, shown in Equation 2, was set to 3600 s. Simulations were conducted under different environmental conditions. Three kinds of engine revolution were used under the same navigation environment, including un-optimized engine revolution, optimized engine revolution, and engine revolutions lower than the optimized level (see Table 4).

Table 4. Energy Efficiency Operational Indicator (EEOI) optimization results.

No.	Navigation Environment				Normal		Optimized		Decreased	
	Wind Speed (m/s)	Wind Angle (°)	Wave Height (m)	Flow Speed (m/s)	Rev. (rpm)	EEOI	Rev. (rpm)	EEOI	Rev. (rpm)	EEOI
1	9	62	5	−6	127	9.66	122	6.79	119	9.01
2	3	−12	2.4	3.5	127	7.23	121	6.44	120	6.50
3	0	0	0.2	−2.1	127	7.92	121	6.53	119	7.01
4	2.1	120	1.5	1.5	127	7.35	123	6.50	120	6.54
5	5	−55	2.0	−0.9	127	7.86	120	6.56	117	6.68
6	1.7	−175	1.3	1.0	127	7.43	122	6.51	120	6.58
7	4.5	97	3.2	2.3	127	7.32	123	6.44	120	6.50
8	6.1	135	2.3	−1.5	127	7.75	122	6.43	119	6.83
9	3.8	21	4.2	2.9	127	7.35	121	6.54	120	6.54
10	7.1	61	2.2	−1.5	127	8.10	122	6.86	118	7.02

(The "Normal" column means that the ship sails on an un-optimized engine revolution. The "Optimized" column means that the ship sails on an optimized engine revolution. The "Decreased" column means that the ship sails on an engine revolution lower than the optimized.)

6. Discussion

Under the same environmental conditions and the same main engine revolution, the simulation data of the ship propulsion model showed satisfactory agreement with the actual sea trial report of the ship (see Table 3). Therefore, the simulation model for ship longitudinal motion is reliable and can be used to verify the EEOI optimization algorithm.

As shown in Table 4, the EEOI value with optimized engine speed has been at its lowest under different navigation conditions, even though the "Decreased" column had corresponds to lower engine revolutions. The optimized revolution is lower than the normal revolution but higher than decreased revolution. This shows the optimized engine revolutions are the best in a local revolution range. Moreover, the optimized engine revolutions are different in different navigation environments. That shows the optimization method is valuable in different environment conditions. Therefore, there are reasons to believe that the optimization concept and method used in this paper can help to improve the EEOI within a certain revolution range. In the future, different cycle times will be tested to find the best cycle time. Additionally, a more accurate wind and wave model will be introduced into the simulation.

Author Contributions: Conceptualization, C.S., C.L. and H.W.; methodology, H.W.; software, Y.Z. and H.W.; validation, C.S., H.W. and Y.Z.; formal analysis, Y.Z.; data curation, Y.Z.; writing—original draft preparation, H.W. and Y.Z.; writing—review and editing, C.S. and C.L.; visualization, Y.Z.; project administration, C.S.

Funding: This research received no external funding.

Conflicts of Interest: The authors declare no conflict of interest.

Appendix A.

Appendix A.1. Initial Parameters Optimized by the Genetic Algorithm

After 100 generations of inheritance (see Figure 6), the input layer weight of the BP network was:

$$W_1 = \begin{bmatrix} -1.248439973 & -2.803119181 & -2.315609078 & -2.734189672 & -1.009229688 \\ -2.153366518 & -1.223545805 & 1.233045222 & -0.006755213 & -2.637285875 \\ -0.706587596 & -1.552185432 & -1.70930286 & -1.426524951 & -1.544026107 \\ -2.536474959 & 0.3447923 & -1.428494005 & -0.33143305 & 1.660091796 \\ -0.298490393 & 0.033244195 & -0.231048075 & -2.173095702 & 1.011001131 \end{bmatrix}$$

The bias of the input layer of the BP network was:

$$B_1 = \begin{bmatrix} -2.356582358 \\ -0.377748027 \\ -2.597593093 \\ -0.119641422 \\ 0.556043876 \end{bmatrix}$$

The weight of the hidden layer of the BP network was:

$$W_2 = \begin{bmatrix} -1.295896113 & -0.313681643 & 2.461507838 & -1.6802041 & 0.184608839 \\ 1.097034432 & 0.456472493 & 0.432887192 & -2.590474197 & -0.072519056 \end{bmatrix}$$

The bias of the hidden layer of the BP network was:

$$B_2 = \begin{bmatrix} 0.691785557 \\ 0.93183824 \end{bmatrix}$$

Appendix A.2. Training Results of the Neural Network

The BP neural network parameters after training (see Figure 7) were as follows: The updated connection weight between the input and hidden layer nodes of the BP neural network was:

$$W_1 = \begin{bmatrix} 18.48441067 & -166.2131336 & -18.31794663 & -1.945587221 & -180.3705467 \\ 1.670072096 & -5.701199595 & 1.046077173 & -0.113205799 & -0.387741131 \\ 1.817608606 & -0.411062349 & -1.800417414 & 0.014945118 & 0.010554454 \\ -1.72338033 & 0.342933773 & -0.045117079 & 0.005803816 & 0.033218607 \\ 1.252456707 & -0.5590336 & 2.126794179 & -0.033308813 & -0.158627126 \end{bmatrix}$$

The updated bias of the input layer of the BP network was:

$$B_1 = \begin{bmatrix} 89.86764057 \\ -0.535052367 \\ -1.096480206 \\ 2.872470465 \\ -2.594345004 \end{bmatrix}$$

The updated connection weight between hidden and output layer nodes of the BP network was:

$$W_2 = \begin{bmatrix} -0.073153078 & -0.147220138 & 2.212682572 & 3.174568662 & 2.074976114 \\ -0.006980749 & -0.247812464 & 0.153286346 & -5.545652166 & 0.129901491 \end{bmatrix}$$

The updated bias of the hidden layer of the BP network was:

$$B_2 = \begin{bmatrix} -3.461900104 \\ 5.294246266 \end{bmatrix}$$

References

1. International Maritime Organization. *Reduction of GHG Emissions from Ships Third IMO GHG Study 2014 Final Report*; MEPC 67/INF.3: London, UK, 2014.
2. Corbett, J.J.; Wang, H.; Winebrake, J.J. The effectiveness and costs of speed reductions on emissions from international shipping. *Transp. Res. Part D Transp. Environ.* **2009**, *14*, 593–598. [CrossRef]

3. Lindstad, H.; Asbjørnslett, B.E.; Strømman, A.H. Reductions in greenhouse gas emissions and cost by shipping at lower speeds. *Energy Policy* **2011**, *39*, 3456–3464. [CrossRef]
4. Hirdaris, S.; Cheng, F. The Role of Technology in Green Ship Design. In Proceedings of the 11th International Marine Design Conference, Glasgow, UK, 11–14 June 2012; Volume 1. [CrossRef]
5. Psaraftis, H.N.; Kontovas, C.A. Speed models for energy-efficient maritime transportation: A taxonomy and survey. *Transp. Res. Part C Emerg. Technol.* **2013**, *26*, 331–351. [CrossRef]
6. Norstad, I.; Fagerholt, K.; Laporte, G. Tramp ship routing and scheduling with speed optimization. *Transp. Res. Part C Emerg. Technol.* **2011**, *19*, 853–865. [CrossRef]
7. Wang, S.; Meng, Q. Sailing speed optimization for container ships in a liner shipping network. *Transp. Res. Part E Logist. Trans. Rev.* **2012**, *48*, 701–714. [CrossRef]
8. Fang, M.; Lin, Y. The optimization of ship weather-routing algorithm based on the composite influence of multi-dynamic elements (II): Optimized routings. *Appl. Ocean Res.* **2015**, *50*, 130–140. [CrossRef]
9. Chang, C.; Wang, C. Evaluating the effects of speed reduce for shipping costs and CO_2 emission. *Transp. Res. Part D Transp. Environ.* **2014**, *31*, 110–115. [CrossRef]
10. Bialystocki, N.; Konovessis, D. On the estimation of ship's fuel consumption and speed curve: A statistical approach. *JOES* **2016**, *1*, 157–166. [CrossRef]
11. Petersen, J.P.; Winther, O.; Jacobsen, D.J. A Machine-Learning Approach to Predict Main Energy Consumption under Realistic Operational Conditions. *Ship Technol. Res.* **2012**, *59*, 64–72. [CrossRef]
12. Pedersen, B.P.; Larsen, J. Prediction of Full-Scale Propulsion Power using Artificial Neural Networks. In Proceedings of the 8th International Conference on Computer and IT Applications in the Maritime Industries, Budapest, Hungary, 15–17 May 2009.
13. Kai, W.; Yan, X.; Yuan, Y.; Feng, L. Real-time optimization of ship energy efficiency based on the prediction technology of working condition. *Transp. Res. Part D Transp. Environ.* **2016**, *46*, 81–93.
14. International Maritime Organization. *Guidelines for Voluntary Use of the Ship Energy Efficiency Operational Indicator*; MEPC 59/Circ. 684: London, UK, 2009.
15. Hariri, R.H.; Fredericks, E.; Bowers, K. Uncertainty in big data analytics: Survey, opportunities, and challenges. *J. Big Data* **2019**, *6*, 44. [CrossRef]
16. Veiga, J.; Enes, J.; Expósito, R.R.; Touriño, J. BDEv 3.0: Energy efficiency and microarchitectural characterization of Big Data processing frameworks. *Future Gener. Comput. Syst.* **2018**, *86*, 565–581. [CrossRef]
17. Desai, N.; Seghouane, A.; Palaniswami, M. Algorithms for two dimensional multi set canonical correlation analysis. *Pattern Recogn. Lett.* **2018**, *111*, 101–108. [CrossRef]
18. Yi, Z.; Teng, L.; Li, K.; Zhang, J. Improved Visual Correlation Analysis for Multidimensional Data. *J. Vis. Lang. Comput.* **2017**, *41*, 121–132.
19. Yu, S.; Zhu, K.; Diao, F. A dynamic all parameters adaptive BP neural networks model and its application on oil reservoir prediction. *Appl. Math. Comput.* **2008**, *195*, 66–75. [CrossRef]
20. Han, B.; Bian, X. A hybrid PSO-SVM-based model for determination of oil recovery factor in the low-permeability reservoir. *Petroleum* **2018**, *4*, 43–49. [CrossRef]
21. Taheri, M.; Mohebbi, A. Design of artificial neural networks using a genetic algorithm to predict collection efficiency in venturi scrubbers. *J. Hazard. Mater.* **2008**, *157*, 122–129. [CrossRef] [PubMed]
22. Wang, H.; Tang, H. Modeling and Simulation of diesel propulsion system in maneuvering navigation condition. *Appl. Mech. Mater.* **2012**, *128–129*, 1168–1172. [CrossRef]

MDPI
St. Alban-Anlage 66
4052 Basel
Switzerland
Tel. +41 61 683 77 34
Fax +41 61 302 89 18
www.mdpi.com

Journal of Marine Science and Engineering Editorial Office
E-mail: jmse@mdpi.com
www.mdpi.com/journal/jmse

www.ingramcontent.com/pod-product-compliance
Lightning Source LLC
LaVergne TN
LVHW070137170726
843515LV00007B/1816

9783036506166